MW00760721

UNIVERSALITY AND EMERGENT COMPUTATION IN CELLULAR NEURAL NETWORKS

WORLD SCIENTIFIC SERIES ON NONLINEAR SCIENCE

Editor: Leon O. Chua
University of California, Berkeley

Series A. MONOGRAPHS AND TREATISES

Volume 27: The Thermomechanics of Nonlinear Irreversible Behaviors — An Introduction
G. A. Maugin

Volume 28: Applied Nonlinear Dynamics & Chaos of Mechanical Systems with Discontinuities
Edited by M. Wiercigroch & B. de Kraker

Volume 29: Nonlinear & Parametric Phenomena*
V. Damgov

Volume 30: Quasi-Conservative Systems: Cycles, Resonances and Chaos
A. D. Morozov

Volume 31: CNN: A Paradigm for Complexity
L. O. Chua

Volume 32: From Order to Chaos II
L. P. Kadanoff

Volume 33: Lectures in Synergetics
V. I. Sugakov

Volume 34: Introduction to Nonlinear Dynamics*
L. Kocarev & M. P. Kennedy

Volume 35: Introduction to Control of Oscillations and Chaos
A. L. Fradkov & A. Yu. Pogromsky

Volume 36: Chaotic Mechanics in Systems with Impacts & Friction
B. Blazejczyk-Okolewska, K. Czolczynski, T. Kapitaniak & J. Wojewoda

Volume 37: Invariant Sets for Windows — Resonance Structures, Attractors, Fractals and Patterns
A. D. Morozov, T. N. Dragunov, S. A. Boykova & O. V. Malysheva

Volume 38: Nonlinear Noninteger Order Circuits & Systems — An Introduction
P. Arena, R. Caponetto, L. Fortuna & D. Porto

Volume 39: The Chaos Avant-Garde: Memories of the Early Days of Chaos Theory
Edited by Ralph Abraham & Yoshisuke Ueda

Volume 40: Advanced Topics in Nonlinear Control Systems
Edited by T. P. Leung & H. S. Qin

Volume 41: Synchronization in Coupled Chaotic Circuits and Systems
C. W. Wu

Volume 42: Chaotic Synchronization: Applications to Living Systems
E. Mosekilde, Y. Maistrenko & D. Postnov

*Forthcoming

WORLD SCIENTIFIC SERIES ON NONLINEAR SCIENCE Series A Vol. 43
Series Editor: Leon O. Chua

UNIVERSALITY AND EMERGENT COMPUTATION IN CELLULAR NEURAL NETWORKS

Radu Dogaru
Polytechnic University of Bucharest, Romania

Published by

World Scientific Publishing Co. Pte. Ltd.
5 Toh Tuck Link, Singapore 596224
USA office: Suite 202, 1060 Main Street, River Edge, NJ 07661
UK office: 57 Shelton Street, Covent Garden, London WC2H 9HE

British Library Cataloguing-in-Publication Data
A catalogue record for this book is available from the British Library.

UNIVERSALITY AND EMERGENT COMPUTATION IN CELLULAR NEURAL NETWORKS

ISBN 981-238-102-3

Printed by FuIsland Offset Printing (S) Pte Ltd, Singapore

Preface

The Cellular Neural Network (CNN), introduced by Chua and Yang from University of California at Berkeley in the late 1980's is an attractive computational structure, particularly from the perspective of implementation in various micro and nanoelectronic technologies. The CNN paradigm includes cellular automata (CAs) as a particular case and in addition it borrows many ideas and techniques from the field of Neural Computation.

Computation in CNNs is brain-like rather than "classic" in the sense of the widespread computing architectures based on microprocessors. Emergent computation, viewed as the class of both dynamic and static patterns of activity emerging in an array of interconnected cells which are meaningful for various information processing tasks, is the equivalent of programming classic computers.

It is thus of a paramount importance to find the equivalent of the "programming rules" for such cellular devices.

The following image can be suggestive to understand the difference between cellular and classic computation: Assume that we have an array of memory cells (i.e. a Random Access Memory). In a classic computer the cells are *sequentially* updated and located by the central processing unit via some external buses while in a cellular computer each memory cell exchanges information locally only with the neighboring cells. A "gene" associated with each cell controls the exchange of information with the cells in a neighborhood. The cells are updated in *parallel* and there is no central processing unit to control the cells.

Like in a classical computer, the array of cells starts from an initial state, which contains the problem, and the solution will be found in the same array of cells after a period of time during which computation emerges. While the designer of a classic computer focuses on the central processing unit, on data coding, address buses and instruction sets, the designer of a cellular computer has to focus mostly on the cell. The "program" is now coded in what Leon Chua called "cells' gene" (i.e. the entire set of parameters defining the cell). Quite often all cells have identical structure and parameters and various tasks can be "programmed" on the same CNN chip by simply changing the genes. The following problems are raised to the cellular computer designer:

- To what degree is a cell capable to perform arbitrary local computations, i.e. the *universality* of a cell?
- What is the choice of the cell parameters such that emergent computation will occur? More sharply, one would like to find the exact values of the parameters for a given information processing task.

This book provides original answers to the above questions. It introduces novel techniques to understand and control better universality and emergence in cellular neural networks. After an introductory chapter and a chapter providing the basics ideas and concepts necessary to understand the remainder of the book, the problem of universal local computation is extensively described in Chapter 3. Our solution, based on the theory of piecewise-linear function approximations, is compact and efficient for both binary and continuous cells. A systematic approach to the second problem, grounded by the very recent theory of local activity [Chua, 1998] is provided in Chapter 4. A set of analytic tools is developed to identify a specific sub-domain called an "edge of chaos" domain in the cell parameter space such that the probability of emergent computation is maximized. Several examples of applying this method are provided. A measure for emergence in discrete time cellular systems is then introduced in Chapter 5 and then exemplified to identify several interesting behaviors in a cellular system, which is a *mutation* of the widely known "Game of Life". The importance of mutations and evolutionary approaches for designing cellular systems with emergent computation is then emphasized in the same chapter for a discrete time cellular system with continuous states. A potential application of emergent dynamic patterns for biometric authentication is presented in Chapter 6.

Why emergent computation in cellular computers when the technology of programming and designing classic computers is so well established and prolific? Here are some possible answers:

(i) The type of computation taking place in a cellular computer is the one used by living entities. Life itself is an emergent phenomenon and several examples in this book will show that simple living-like entities may emerge as a pattern of cellular activities. Brain-like computation could be better mimicked by a compact CNN rather than by a classic computer. Particularly when such computations are required in micro-robotics, or in any circumstance requiring compact yet intelligent sensor systems, the CNN could be a better choice;

(ii) The cellular systems are highly parallel and consequently they perform several orders of magnitudes faster than classic (serial) computing ones. Tera-ops processing speed (10^{12} elementary operation per second) is common for the actual generation of CNNs;

(iii) Recent developments in the area of nanotechnology indicate that cellular structures made of lattices of interconnected active devices (for example, resonant tunneling diodes, quantum dots or single electron transistors) could be easily developed. Characterized by a very high density of cells, they can fully exploit the benefits of emergent computation for various tasks in information processing.

Radu Dogaru,
November 2002

Acknowledgements

Most of the results presented in this book are the outcome of research conducted by the author since 1996 when he joined the Nonlinear Electronics Laboratory at the University of California at Berkeley being sponsored by a Fulbright scholarship.

Part of this research was sponsored by several scholarships offered by the ISIA (The Graduate School in Intelligent Systems) at the Technical University of Darmstadt – Germany and by several Office of Naval Research (U.S.A.) grants. I express my gratitude to the Volkswagen Stiftung foundation who partially supported some of the research included in this book through its program "Cooperation with Natural and Engineering Scientists in Central and Eastern Europe".

This book is in a significant part the result of the support and encouragement from the exceptional professor and scientist Leon O. Chua, director of the Nonlinear Electronics Laboratory at the University of California at Berkeley. Many thanks to my Berkeley colleagues and friends, Dr. Kenneth Crounse, Dr. Martin Haenggi, Dr. Pedro Julian, Tao Yang and many others who contributed a great deal to shape and improve some of the ideas exposed in this book.

I am also deeply grateful to Professor Manfred Glesner, Chair of the Microelectronics Institute at Technical University of Darmstadt (Germany) for the opportunity he offered me to join a highly competitive research team. I am thankful for the valuable support of Tudor Murgan, Mihail Petrov, and Octavian Mitrea in the last stages of preparation of the manuscript.

My family played an important role during the various stages of writing this book. I am grateful to my wife Ioana, for her love, patience and valuable support in the last stages of writing this book. I dedicate this book to the memory of my father Tudor, who enthusiastically encouraged me in most of my professional endeavors, this book being unfortunately the last where his direct encouragement reached me.

Last but not least, I wish to express my gratitude to Professor Felicia Ionescu and many other colleagues within the Department of Applied Electronics and Information Engineering and within the "Politehnica" University of Bucharest who expressed interest in my work.

Contents

Chapter 1
Introduction

1.1. Emergent computation as a universal phenomena

Information exchange is all around us. According to Varela [Varela *et al.*, 1974] life itself is a process of perpetual cognition (or *information exchange* with the environment). Computer scientists often call information processing and exchange with the word *computation,* which essentially express the idea of manipulating and transforming information in such a way that is meaningful for a certain purpose or *objective*. A pragmatic attribute is attached to the meaning of the word "computation" in that we often associate the word not only with the process itself but also with the effective means of achieving it. Therefore, the derived word *computer* would usually define a medium (electronic, mechanical, bio-chemical, or of other physical nature) used to achieve *computation* as an *emergent* property. Starting with the simple *abacus,* passing trough mechanical computing systems such as Babbage's first computer, until the nowadays digital computers, humans are always in a search for improved and sophisticated computing methods and mediums to satisfy their needs.

This process of developing computing machines is a *universal* and a natural one, which obeys certain life laws according to which life itself is sustained by a continuous information exchange with the external world. It appears that while life itself is the result of a continuous flow of information exchange (or *computation* in a wider sense), hierarchies of other (superior) "living" entities develop as a result of information exchange between similar entities. So, simple cells *collaborate* to produce and maintain *functional organs,* which then are linked in a network forming *individuals.* Such individuals then form *families, villages, countries, federations*, etc.

Note that in this picture, at a given level of the hierarchy, the actors (let us call them *cells*) are quite similar and they do usually *exchange information* only within a very small fraction of the entire population. This particular fraction constitutes *a neighborhood,* which is essentially of informational nature although some topological constraints may have an influence on which members belong to a neighborhood (for example, a group of scientists usually collaborate through e-mail although they are located geographically at very distances). The same structure is reflected in our brains. Our brain operates as an *emergent* computer where a network of concepts and relationships are stored. Following the above model, one concept is usually inter-related with only a few other concepts in its neighborhood. Sequentially firing the neurons associated with a mental concept or with a sensorial stimulus will allow the recall of a related concept in several "iterations" corresponding to the same number of *neighborhoods* activated in the internal network of concepts.

A similar *hierarchical* and *cellular* organization is revealed in *language,* which could be considered an "image of our brains". Basic cells (phonemes or characters in the written language) *collaborate* in an emergent manner to form *words,* which then are again combined in *phrases* and so on providing a structured reflection of the outer world. Similar *cellular* models could be provided for networks of similar electronic circuits, networks of computers and so on.

Recapitulating, *universal* characteristics of our world can be captured in the *emergent* dynamics of a *cellular* model where a population of similar cells *exchanges information* locally with the cells in their *neighborhood.* Emergent computation depends on how information is exchanged and processed in a cell neighborhood.

1.2. Emergence

The dictionary[1] definition of *emergence* is based on the verb *to emerge,* which is defined as "to come out from inside or from being hidden". Yet, from a scientific point of view is rather difficult to come up with a definition for *emergence.* In a 1990 posting over the newsgroup comp.ai.philosophy David Chalmers proposes the following definition: "Emergence is the phenomenon wherein complex, interesting high-level function is produced as a result of combining simple low-level mechanisms in simple ways" while suggesting that "emergence is a psychological property". Indeed, according to Chalmers "Properties are classed as emergent based at least in part on (1) the *interestingness* to a given observer of the high-level property at hand; and (2) the difficulty of an observer's deducing the high-level property from low-level properties". Finally Chalmers proposes the following definition: "Emergence is the phenomenon wherein a system is designed according to certain principles, but *interesting properties* arise that are not included in the goals of the designer.

Following the above line, Ronald [Ronald *et al.* 1999] proposes an *emergence test* based on a *surprise effect* and somehow similar to the Turing test for detecting *intelligence.* According to them, if using a (macro) language L2 to describe the *global* behavior of a system, designed according to a (micro) language L1 (usually providing information about the cell structure and its *local* interconnectivity) the behaviors observed in L2 are *non-obvious* to the observer – who therefore experiences surprise – one may conclude that the global behavior is an *emergent* one.

Standish [Standish, 2001] develops the ideas of Ronald observing that *the surprise effect* in the above test disappears after the first observation of an emergent behavior. Then he proposes the following definition of *emergence*:

"An emergent phenomenon is simply one that is described by atomic concepts available in the macro language, but cannot be so described in the micro language".

1.3. Cellular computing systems

The above observations are linked to the idea of developing a computing paradigm called *cellular computing.* The structure of such a computer is composed of a grid (often two-dimensional) of cells locally interconnected with their neighbors. Each cell can be in a number of states (ranging from 2 to infinity) and the state of a cell depends by itself and by the states of its neighbors through a *nonlinear* functional, which can be defined in different ways. This functional is associated with a practical implementation of the cell and it includes a set of *tunable* parameters. By similarity to biology, where the overall functionality of a living being is determined by a finite string of proteins, in [Chua, 1998] it was proposed that the set of tunable parameters will be called a *gene.* By tuning the gene parameters one can achieve *programmability* i.e. different behaviors within the same basic cellular architecture. The cell assumes an *initial state* and may have one or more external *inputs.* In a cellular computer, computation can be considered any form of *meaningful emergent* global phenomenon resulted from a proper *design* of the cell functional (gene). Usually the initial state and the inputs code the problem to be solved while the answer to this problem is coded in an equilibrium steady state but it can be also coded in the emergent dynamics of the cellular system. By *meaningful* I mean a subjective definition according to

---

[1] Longman dictionary of contemporary English, special edition, 1992.

which one can *effectively* use the results of the emergent phenomena for a purpose. For example, loading an input image in a cellular computer is corresponds to setting the initial state of each cell proportional to the brightness of its corresponding a pixel in the image. Then using a properly designed cell, the cellular computer will dynamically evolve towards a global state representing a transformation of the initial image (e.g. an image containing only edges). Or, a set of different waves could emerge in a cellular computer, each being formally assigned top a computational symbol. The dynamics of the cellular system will lead to collisions between waves, which, when meaningful can be interpreted as computing.

The first cellular computers were theoretical constructs introduced by Stanislas Ulam in the forties. He suggested John von Neumann to use what he called "cellular spaces" to build his self-reproductive machine. Later cellular models, often called Cellular Automata (CA) were developed to explain various natural phenomena. Choosing the proper *genes* in the form of *local rules* defining the behavior of each cell was the equivalent of programming in serial computers. The well known "Game of Life" rule introduced by Conway in the 70s gained popularity due to the complex and diverse patterns emerging in such a simply defined system. We should note that designing a proper set of local rules was then a matter of intuition and educated guess rather than the outcome of a well-defined procedure. It was proved that such a simple machine (this is a 2 state per cell cellular automata with a very simple local rule) is capable of universal computation (i.e. it is a universal Turing machine). Following the line of Von Neumann, a lot of research has been devoted to the study of cellular automata and local rules leading to *emergent* properties such as self-reproduction and artificial life. An overview of these non-conventional computers can be found in [Toffoli, 1998]

Recently, starting with the work of Chua and Yang [Chua & Yang, 1988] a novel cellular computing paradigm called Cellular Neural Network was developed. It inherits the basic ideas of cellular computing and in addition bore some interesting ideas from the field of *neural computation*. While most of the previously described cellular computing paradigms were conceptual, the CNN was from the beginning circuit oriented i.e. intended for practical applications as an integrated circuit. Moreover, in 1993 Roska and Chua [Roska & Chua, 1993] proposed a revolutionary framework called a CNN Universal Machine, in fact a specialized programmable cellular computer which is capable to execute complex image processing tasks and which found numerous applications in vision, robotics and remote sensing [Chua & Roska, 2001]. Nowadays this paradigm is successfully exploited in various applications dealing mainly with extremely fast nonlinear signal processing and intelligent sensors. In [Chua, 1998] it is demonstrated that the CNN paradigm include Cellular Automata as a special case. Therefore many of the research in the area of cellular automata can be easily mapped into the CNN formalism with the advantage of exploiting a range of powerful chip implementations that have been developed over the years [Roska & Rodriguez-Vazquez, 2000].

To date several types of emergent computation were identified as meaningful and useful either for computing applications (e.g. in the area of vision and image processing) or for modeling purposed (e.g. models of the cell membrane) by what I could call generically evolutionary strategies. An interesting example is the development of a relatively large library of CNN *genes (templates)* over the last decade [AnalogicLAB, 2002]. Many of these genes were discovered by chance, studying the dynamic evolution of the CNN and identifying certain dynamic behaviors with meaningful computational primitives such as *edge* or *corner detection*, *hole filling*, *motion detection*, and so on. Although some theoretical approaches, mainly inspired from the techniques of filter design were successfully employed to design new *chromosomes* design there is still much to do for a *systematic* design of the *cells* and *genes*. This book offers several solutions to this problem starting from the recent theory of *local activity* [Chua, 1998].

1.4. Universality

In a cellular system, it is of interest to find a universal nonlinear description of its cells. More precisely, any particular local function (which describes the output of a cell given the outputs of all cells in its neighborhood) can be specified using the same general functional description and a particular set of *gene* parameters which individualize the cell.

In Chapter 3 as well as in [Dogaru & Chua, 1999a], [Dogaru *et al.*, 2002a] the issue is discussed in extenso. It is shown that weighted summation, binary memory cells, comparators and absolute value function operators are the only basic ingredients to define universal cells for either the Boolean (binary inputs and outputs) and the continuous case. A universal cell is a prerequisite for a more systematic analysis of the relationship between genes and emergent phenomena in an array of coupled cells, as shown in Chapter 5.

Our approach to universality allows one to map CA local rules that were found responsible for very interesting emergent phenomena (such as Conway's "game of life") into a systematic framework where a unique *gene parameter* space has to be explored and put in connection with such emergent behaviors. Within the framework of *universal* cells the cell parameter space can be structured in sub-domains which are specific for emergent on non-emergent behaviors much like the theory of local activity is able to predict for the particular case of *Reaction-Difussion* cellular systems discussed in Chapter 4. Therefore, we believe that employing universal cells in discrete-time systems is the prerequisite for a consistent theory which allows one to predict how cellular coupling influences the structure of a *genes' parameter space*. A universal piecewise-linear description of a nonlinear system (in our case, a cell), is a simple mathematical construct which I believe opens interesting perspectives towards a consistent analytic theory of the relationship between emergence and genes.

1.5. Designing for emergence, the essence of this book

As mentioned before, an emergent phenomenon is one that cannot be described in terms of a micro language (e.g. by specifying the cell as a nonlinear dynamic system or its *gene*). However, since the choice space of *genes* is usually huge, it is of practical interest to develop *methods* to locate in this space of parameters the regions where *emergent* phenomena are likely to occur. This question is in fact difficult to answer since "the surprise effect" associated with emergence does not allow us to specify in advance *what we are we looking for* i.e. the desired emergent behavior. Novel approaches to this problem of interest are the subject of this book. The main question is: under what conditions (described in L1) is the system emergent? A possible answer to this question is the subject of this book, where *cellular neural / nonlinear networks* (CNN) are the subject of study.

I believe that emergent computation is an *evolutionary* process. For example, one may start with a cell definition leading to a non-emergent behavior and then, applying successively *genetic* operators (e.g. simple mutations) followed by observations of the global behavior one can detect *emergent phenomena* that might be of a computational nature. Often a desired behavior of the emergent phenomena is very difficult to formalize therefore one should let the system evolve and select the genes leading to useful emergent behaviors following the natural law of evolution.

Therefore it seems counterproductive to attack the cell design problem from the perspective of trying to define exactly the desired behavior of the cellular system. Instead, a better approach would be to construct cellular systems with cells properly tuned (i.e. the gene is chosen in those regions of the cell parameter space such that emergent phenomena are likely to occur) and then evolutionary explore their gene space around the initial guess solution. During this exploration some genes will be selected, the ones that generate surprisingly behaviors which in addition may match certain computational objectives.

This is the essence of our approach as described in this book. Although two different types of cellular systems (discrete-time and continuous) are investigated, the results give hints for a further unified treatment. The following steps are suggested for the *design for emergence*:

- First of all, in both discrete and continuous time cases we introduce the idea of a cell parameter space, which can be effectively defined after a cell mathematical model is specified. For the case of discrete-time CNNs, a universal cell framework is proposed in Chapter 3. According to this model, any possible cell is now specified only trough a *gene* vector $\mathbf{G} = [g_1, g_2, \ldots, g_m]$ of real parameters. For the continuous time case of Reaction-Diffusion type cellular systems the cell is a continuous-state dynamical system (a 2'nd order system was considered for simplicity). Again, certain parameters can be identified forming the cell parameter space. Although the examples in Chapter 4 employ specific equations for each of the three types of cells investigated, since the simplicial cell described in Chapter 3 (Section 3.3.) can be regarded as a universal functional, these specific cells can be also mapped into a universal CNN cell formalism.
- The second step, and the most essential, is to find a way (theoretical or algorithmic) to identify those regions in the cells parameter space such that emergent phenomena are likely to occur in the array of coupled cells. This is the main focus of this book. For the case of Reaction Diffusion type of continuous-time cellular systems, an analytical method was developed based on the powerful *local activity* concept introduced by Chua. Using this method one can effectively draw boundaries in the cell parameter space and locate those regions of interest. While no analytic method is yet available for the case of discrete time generalized cellular automata[2] an algorithmic method is proposed in Chapter 5 to identify regions in the cell parameter space of universal CNN cells such that emergent behaviors are likely to occur. It is important to note that using an appropriate measure the same behavioral classes (passive, locally active and unstable, and "edge of chaos" or locally active and stable) can be identified as for the case of continuous time systems were the theory of *local activity* was initially applied. These results suggest that indeed the *local activity* concept is a powerful one and further research may lead to a unified theory of emergence in cellular system employing this concept to identify the boundaries between behavioral domains in a universal cells parameter space.
- The third step would be to refine the outcome of the previous step in locating much finer sub-domains where application specific genes will be found using evolutionary techniques. Some ideas of applying this step are sketched in Chapter 5 (Section 5.3.3) and in Chapter 6.

A qualitative picture of the *design for emergence* problem is given in Fig.1. A proper set of classification tools allows classifying each cell parameter point in one of the three major categories (depicted in different colors). A *blue* color corresponds to cell parameters for which *local passivity* is sure. Such regions are of no interest since they are associated with non-emergent behaviors in the system formed of coupled cells. The *red* and *green* colors correspond both to *locally active* cells differentiated by their stability. Green regions correspond to *unstable* cells while *red* regions to *stable* cells. The theory used in Chapter 4

---

[2] For discrete-time Reaction Diffusion systems a analytical method for locating "edge of chaos" regions in the cell parameter space was recently reported [Sbitnev et al., 2001].

is capable to predict such boundaries like the *magenta* boundary in the figure but this theory does not guaranty that all cells within the boundary may lead to emergent behaviors. The region where emergent phenomena effectively occur (drawn with color *orange* in Fig.1) may be a much smaller region, included in the one defined by the *magenta* boundary. For example, using an algorithmic approach such boundaries can be derived as we show in Chapter 5.

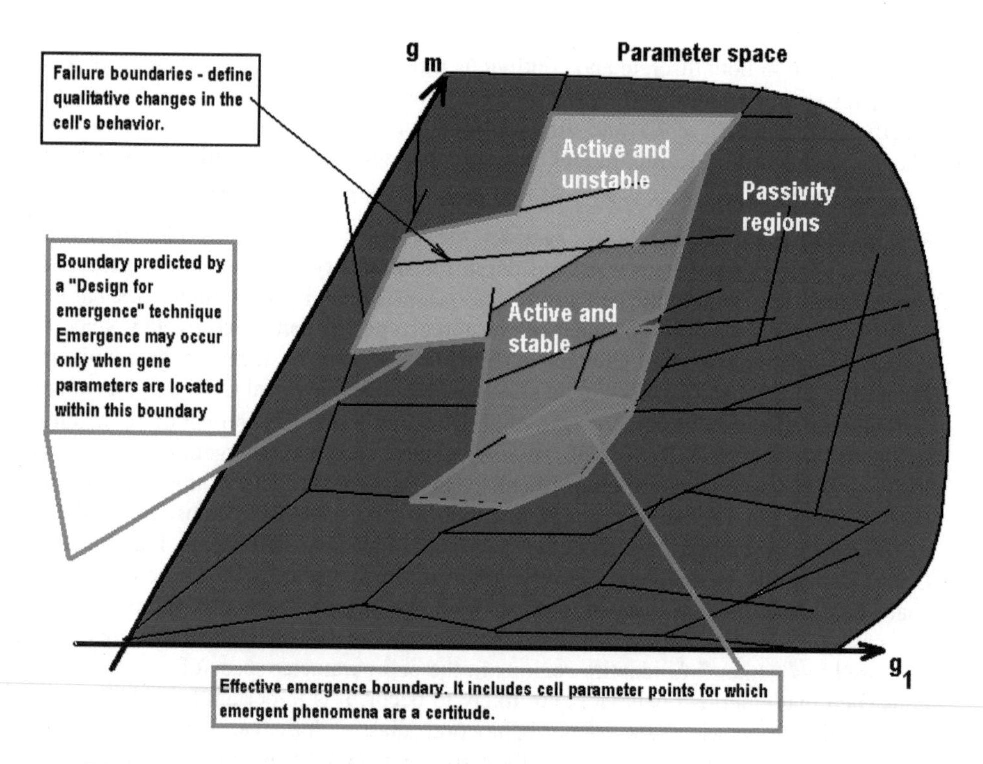

Fig.1. A good "design for emergence" technique is capable to derive boundaries in the cell parameter space $g_1,\ldots,g_m$ such that emergence is likely to occur for cell parameter points located within these boundaries.

1.6. Detecting the potential for emergence: the local activity theory

A cell could be *passive* or *active*. For example think of a seminar room. The lecturer provides the group with some ideas, consequently sending a flow of information. Some of the students will receive the information but will simply absorb it without further processing and feedback. Such students are *passive* since they do not *resonate* with the flow of information. If the students would *all* be *passive* after several time there will be nothing to be discussed in that group and there will be *no information exchange* at both *local* and *global* (i.e. entire group) level. However there may be some students in the room interested in the lecturer's speech. They will *react* to the input flow of information by *locally processing it* and sending back information to their neighbor cells (fellow students and their teacher). These are *locally active* cells in our cellular systems framework and they may contribute significantly to the *emergence* of new ideas.

Note that *emergence* will produce *surprisingly new* global ideas starting from the interaction of active cells. Still a *locally active* cell (read "student" in our context) could be a noisy one, which produces a lot of information, some of which is not necessarily related to the content of the talk. Such cells may be called *active* but *unstable* (green regions in Fig.1) to differentiate from *active* and *stable* cells (red regions in Fig.1).

The above metaphoric example intuitively introduces concepts such as *local activity* and *passivity* [Chua, 1998]. A *locally passive* cell is one that is *not locally active*. Such concepts were developed starting from theorems in the field of non-linear circuits and systems, but they are rather general and in my opinion an information-theoretic approach may open interesting perspectives for a generalized theory of the local activity. The main result of the local activity theory can be stated: "In a system made of *passive* cells emergence cannot occur". Refinements of this theory allow *locating the emergence sub-domains in the gene parameter space* therefore giving an efficient procedure to locate initial gene solutions in *designing for emergence* problems.

In the previous example one can identify three major types of cell behaviors and how they influence the global system. The first, entitled *passive,* brings nothing new in the global system. Such cells *ignore* the information around them producing a limited informational output such that on a long term the whole system made of such cells will enter a dull state without any pattern (spatial, temporal or spatio-temporal) of information. Rigorous mathematical conditions for passivity were derived in [Chua, 1998] to find the mathematical expressions defining boundaries in the cell parameter space which isolate passivity domains. This theory requires that a cell is described as a nonlinear dynamical system (usually by ordinary differential equations – ODEs) however algorithmic approaches (see Chapter 5) may be also used to design cellular systems for emergence in more general cases including discrete-time systems.

Note than in the actual theory of *local activity* the word *local* stands not for the local versus global computation in a cellular system but for a *locality in the state space of the cell* (typically an equilibrium point), around which the property of the cell (active or passive) is evaluated.

Coming back to the *passive* cells, such cells cannot *amplify* small perturbations, while *active* cells do. A network of *resistors* and *capacitors* is of no practical use since all the above devices are *passive electronic devices.* However, when operational amplifiers, transistors or other active electronic devices are added to the network – useful computational information processing emerge (e.g. amplification, filtering, demodulation, etc.). Actual computers are in fact networks of coupled active devices (logical gates), each cell operating in a binary mode.

Design for emergence includes in fact the whole spectrum of design rules in the field of electronics and its evolutionary feature is obvious. Starting from simple tube schematics in the pioneering years of the 20'th century generations of engineers evolved more and more complicated schematics allowing various types of information processing to emerge.

With the advent of emerging technologies (e.g. nanotechnologies or molecular computing), the interest for understanding and "programming" cellular computers is continuously growing. Among the most impressive results expected are computing systems capable of self-reproduction and self-repair while performing brain-like operations at a very high speed. Such kind of systems may easily find applications in intelligent sensors and autonomous micro robots but also in other various tasks requiring fast and multi-dimensional signal processing. In such technologies, networks of similar coupled cells are easy to generate. Therefore a consistent and universal theory of "designing for emergence" in cellular systems is likely to provide the effective tools for designing applications for these new classes of devices. Moreover, such theories are likely to bring more light in understanding how complex life forms can *emerge* from simple egg cells.

This book introduces the authors' recent results in cellular systems design for emergence and provides several examples of applying such techniques to certain particular systems. Several hints for applications are also sketched. These are only the first steps towards the development of a consistent theory of *designing for emergence.*

Chapter 2
Cellular Paradigms: Theory and Simulation

2.1. Cellular systems

The main features of cellular systems are *regularity* and *homogeneity*. In fact a cellular system can be defined as a structured collection of identical elements called ***cells.*** The structure is given by the choice of a ***lattice***. Such lattices are 1-dimensional, 2-dimensional and, less used, 3 or more dimensional. The following are examples of common used 2-dimensional lattices:

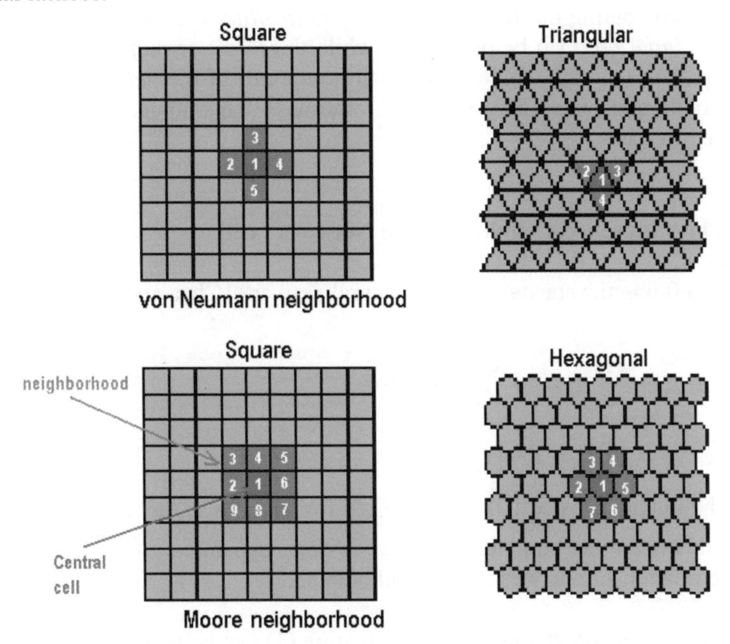

Fig.1. Cells, lattices, neighborhoods and indexes.

In the above pictures, the central cell is depicted in red while *cells in the* ***neighborhood*** are depicted in magenta. The *neighborhood* is another important concept which defines a cellular system and it represents the set of cells that are directly interacting with the central cell.

The basic computational unit in a cellular automata is called a *cell* and such cells are in fact ***nonlinear dynamic systems***. The cell dynamics can be ***continuous in time*** and in this case they are mathematically described by Ordinary Differential Equations ODEs or ***discrete in time*** and in such cases their dynamics is described by a *difference* equation.

Example 1: $y_1(t) = \sum_{k \in N} a_k \cdot y_k(t-1)$ defines the dynamics of the output $y_1(t)$ of a *discrete time* ***autonomous*** cell defined by a weighted summation of all neighboring cell outputs at the previous time clock. The neighborhood is represented by the set N and a **unique index** k is chosen to identify the neighboring cell in a particular neighborhood. The assignment of different integer values to k is conventional and some examples are provided in Fig.1. where such indexes are depicted in white for each type of neighborhood. The time variable t is here an *integer*.

The above cell is one of the simplest possible. It is autonomous since there is no external input to drive the dynamics of the cell. In the most general case a cell is described by the following variables:

Inputs – usually denoted by variable u (scalar) or $\mathbf{u}$ (vector of inputs);
States – usually denoted by variable x (scalar) or $\mathbf{x}$ (vector of states);
Initial states- a particular state variable at the initial moment, $t=0$.
Outputs – usually denoted by variable y (scalar) or $\mathbf{y}$ (vector of outputs).

In terms of a computing machine, *inputs* and *initial states* are used to supply the cellular system with the *input data* (to be processed) while the result or *output data* is available in the form of a pattern (spatial, temporal or spatio-temporal) of the *outputs*. For simpler cells sometimes outputs and states may coincide, otherwise the outputs represent a nonlinear function applied to states.

Example 2 : **The "Standard Cellular Neural Network" cell**

The ODE defining the standard cellular neural network [Chua and Yang, 1988] is:

$$\dot{x}_1 = -x_1 + \sum_{k\in N} a_k y_k + \sum_{k\in N} b_k u_k + z\,, \qquad (1)$$

where $y_k = \frac{1}{2}\left(|x_k + 1| - |x_k - 1|\right)$,

and the initial states $x_k(0)$ are specified for all cells in the neighborhood N.

Observe that this is a more complex, *continuous time*, dynamic system with a nonlinearity induced by the relationship between states and outputs.

In general (non-standard CNN cell) equation (1) can be replaced with:

$$\dot{x}_1 = -x_1 + f(G, y_k, u_k) \qquad (1')$$

where f is a nonlinear function and G is a gene as defined below.

Genes

Observe that in all previous examples the dynamics of the cell for the same input excitation and the same initial state is significantly influenced by the values given to certain parameters (denoted a_k, b_k, and z in the above example). In [Chua, 98] it was proposed to pack all these parameters in a unique vector $\mathbf{G} = [a_1, a_2, ..a_{n1}, .., b_1, b_2, ..b_{n2}, ...]$ called a ***gene*** since it determines the overall function of the cellular system much like the DNA – based gene determines the functionality of the biological systems made of cells containing that DNA.

Discrete and Continuous states / outputs

In defining a cellular system one has to define the variation domain of the state and output variables. For example, one can use ***continuous state*** cells where the outputs are defined within a bounded interval or one can also use ***discrete state*** cells where the states or/and the outputs belong to a finite set of possible values.

A **binary output** cell implements Boolean functions, i.e. it provides a logical "TRUE" or "FALSE" output for each of the 2^n possible combinations of inputs. Each input also represents a truth value. The cell can be again specified as a **discrete-time dynamical system** but it can be also specified using a **transition table** or a set of **local rules**. The last two modes of specifying a cell are specific to the Cellular Automata formalism [Toffoli & Margolus, 1987]. As we will show in the next chapter, a compact piecewise linear description (i.e. a discrete dynamical system) can be found for any Boolean or other type of input function:

Example 3: **A Boolean cell**

$$y_1(t) = sign\left(z_0 + \sum_{k \in N} b_k \cdot u_k(t) \right) \quad (2)$$

Assuming that the set $\{-1,1\}$ is used to code the truth set $\{FALSE, TRUE\}$, the cell equation (2) above defines a family of Boolean functions. For a particular Boolean function one should specify the gene parameters. For example, the AND function with 3 inputs is associated with the following gene: $\mathbf{G} = [z_0, b_1, b_2, b_3] = [-2,1,1,1]$. If the gene is replaced with $\mathbf{G} = [z_0, b_1, b_2, b_3] = [2,1,1,1]$, the OR function with 3 inputs is implemented. The Boolean cells are studied extensively in Chapter 3, while in Chapter 5 it is shown how the emergent dynamics in a CNN made of such cells can be related to the gene parameters of the cell.

Boundary conditions

While most of the cells in a cellular automata are connected with all their neighbors on a given lattice, the cells located on the boundary of a lattice have a special regime. One should define the way in which these cells interact with their neighbors. This is called **a boundary condition**. The choice of the boundary condition may have a great influence on the cellular system dynamics, therefore this is an issue that has to be clearly specified when dealing with cellular systems. Two solutions of this problem are common:

Periodic boundary conditions: Opposite borders of the lattice are "sticked together". A one dimensional "line" becomes following that way a circle, a two dimensional lattice becomes a torus.

Reflective boundary conditions: The border cells are mirrored: the consequence are symmetric border properties.

In the remainder of the book exclusively square lattices and periodic boundary conditions are used.

2.2. Major cellular systems paradigms

The cellular automata (CA)

Stanislas Ulam proposed in the forties the first cellular system. This model, called a cellular automaton (CA) was much related with the works of Von Neumann dealing with self-reproduction and artificial life. Von Neumann asked what kind of logical organisation of an automaton is sufficient to produce self-organisation. The result of his theoretical deduction is that an universal Turing machine embedded in a cellular array using cells with 29 states and a neighborhood with 5 cells is existing [Neumann von, John, 1966]. This machine (in fact a configuration of states) is called a universal constructor, and is capable to construct any machine described on the input tape and reproduce the input tape and construction machinery. Arthur W.Burks [Burks, 1968] and E.F.Codd [Codd, 1968] demonstrated the possibility to reduce the complexity of von Neumanns automaton. They introduced a machine requiring only 8-states per cell capable of self-reproduction. Later on, the idea of using Cellular Automata to study self-reproduction was developed by Langton [Langton, 1984] focusing on rather simpler machines called loops which are capable to self-reproduce in less complicated cellular systems. Much simpler models of self-reproducing loops were proposed recently. Several cellular models producing evolvable self-reproducing loops were also proposed recently (e.g. [Sayama, 2000]).

Another well known model of CA is the Conway's "Game of Life". The game became well known after an article published in 1970 [Gardner, 1970]. It is a binary, discrete time cellular automata where the cell is defined by some simple, common sense, **local rules.** It is assumed that each cell has 8 neighbors (Moore neighborhood) and each cell is either DEAD (coded as state 0) or ALIVE (coded as state 1). There are two possibilities:

1. The cell is DEAD. Then it become ALIVE in the next cycle only if 3 of its neighbors are ALIVE. Otherwise it stays DEAD (This rule somehow suggests that a new being is generated if enough people are there around. Though, too many people may compete for resources and there is no place for a "new life").
2. The cell is ALIVE. Then except if it has 2 or 3 ALIVE neighbors it will become DEAD in the next cycle. Otherwise it will stay ALIVE. This rule suggests that a cell could die either by loneliness (only 1 living cell around) or by overpopulation (more than 3 cells around competing for resources makes life impossible).

The dynamics of such CAs is surprisingly complex and intriguing and it was extensively studied [Gardner, 1983].

At http://www.bitstorm.org/gameoflife/ the reader may find an easy-to-use, platform independent example of running the Game of Life. Many other computer simulations of this popular cellular automaton are available on the Internet.

Game of Life is a serious game. It was proved [Berlekamp *et al.*, 1982] that for certain configurations of initial states the "game of life" CA embeds a universal Turing machine and therefore, in principle, a CA with "game of life" cells is capable of universal computation. We will show later in Chapter 3 that a simple nonlinear model of the "game of life" cell exists and in Chapter 5 we show that an even wider range of emergent behaviors can be traced in cellular automata or generalized cellular automata using mutations of the "game of life" cell.

The Cellular Automata are widely studied today as a convenient paradigm for modeling physical processes and for investigating emergence and complexity. Several good on-line tutorials are available, for example [Schatten A., 1999], [Rennard, 2000], [Weimar, 1996] to name just a few of the most recent.

The Cellular Neural Network (CNN) model

The Cellular Neural Network (CNN) model was proposed by Chua [Chua & Yang, 1988] as a practical circuit alternative to Hopfield and other type of recurrent networks. The CNN cell is a **continuous time and continuous state** dynamical system with some saturated nonlinearity (see equation (1)) which is well suited for implementation using analog circuits. Unlike CAs which are mostly used to prove various theories or to model physical processes the CNN was intended from the beginning to be also a useful signal processing paradigm.

An important step towards making this paradigm an application oriented one was the introduction in 1993 of the concept of CNN Universal Machine (CNN-UM) [Roska & Chua, 1993]. Within the framework of the CNN-UM a CNN kernel is employed to perform sequentially various information processing tasks. Each task is associated with a gene from a continuously growing library of more than 200 different primitive genes (and tasks). Therefore one may combine various such primitives which are stored in an analog memory much like the instruction-code memory of digital microprocessor and combined in various ways to provide complex and sophisticated applications at TerraOps computing speed. Recent electronic implementations of the CNN-UM are in fact *sensor computers* [Roska, 2000], having the capability to sense and to process an image on the same chip.

Several generations of microelectronic chips were reported so far [Roska & Rodriguez-Vazquez, 2000a], as well as development tools which allow an user to program the CNN as visual microprocessor. There is a wide range of applications, mostly in the area of image processing. Such application include image segmentation, image compression, fast halftoning, contour tracking, image fusion, pattern recognition, to name just a few.

Although initially the equilibrium dynamics of CNNs was mostly exploited for applications, recently the non-equilibrium dynamics is employed for certain interesting applications in what is currently called "computing with waves" [Roska, 2001],[Roska, 2002].

There is a lot of research around the world in the field cellular neural network most of which is reported in the Proceedings of the IEEE CNNA workshops (Cellular Neural Networks and their Applications), ISCAS or IJCNN conferences as well as in numerous journal papers or books. Some recent tutorials about CNNs are [Chua & Roska, 2001], [Chua, 1998], [Roska & A. Rodríguez-Vásquez, 2000b], [Hänggi & Moschytz, 2000].

Several CNN simulators are freely available and the reader may check [Hänggi, 1998], and [Hänggi et al., 1999] for an easy to use, computing platform independent CNN simulator. A wide range of simulators as well as news about the progress in the CNN research can be accessed from [AnalogicLAB, 2002]. The SCNN simulator from the Applied Physics Department can be used on Unix platforms [SCNN Simulator, 2000].

The Generalized Cellular Automata

In [Chua, 1998], the idea of a Generalized Cellular Automata (GCA) was introduced as an extension of the CNN so that a GCA includes CAs as a special case. The main idea of the GCA is to use a CNN in a discrete-time loop.

In other words, for each period of the clock the CNN system evolves until it eventually reaches a steady output (for some CNN genes it is possible to have complex oscillatory dynamics) starting from a given initial condition and from inputs variables that are copies of the GCA neighboring cells outputs at the and of the previous time cycle.

The additional CNN loop is thus given by the discrete time equation $u_k(t) = y_k(t-1)$. There are two cases of interest:

1. When the CNN cell is *uncoupled* (i.e. all coefficients $a_k = 0$ for $k \geq 2$ in (1)) it was proved that the nonlinear dynamic system (1') converges towards a steady state output solution and therefore aft enough period of time T the cell can be described as a simple nonlinear equation of the following form: $y_1(t) = F(\mathbf{G}, u_k(t))$ where F is a nonlinear function and G is a gene (i.e. tunable parameters). In this case, as detailed in Chapter 5, the GCA can emulate any CA, provided that there exist a method to map the local rules or transition table into the nonlinear function F. As it can be easily observed such GCA can also implement discrete-time but continuous state cellular automata.

2. When the CNN cell is *coupled*, one should first determine a set of proper genes such that the same steady state behavior occurs during the clock time T. This is often not a trivial task. Then, the behavior of the resulting GCA is more complex than that of a normal CA. The reason is that an emergent computation already takes place during the period of time T in the CNN, therefore at the discrete time moment when the output of a cell is sampled, it does not represent only the contribution of the neighboring cells as in the case of a "classic" cellular automata but rather the contribution of all CNN cells. In some sense one can say that employing a continuous time CNN during the clock time is equivalent to artificially increasing the neighborhood to the whole cellular space. This behavior may imply interesting computational consequences. At the end of Chapter 5 we suggest an application in image restoration based on this particular mode of operation of a GCA.

Reaction-Diffusion Cellular Nonlinear Networks

Reaction-Diffusion CNNs (RD-CNNs) were proposed in [Chua *et al.*, 1995] as a particular case of continuous-time autonomous[1] CNN which are space-discretized models of Partial differential Equations describing the Reaction-Diffusion physical processes.

From a circuit perspective a RD-CNN can be modeled as a collection of multiport nonlinear cells. These cells are coupled with their neighboring cells via linear resistive networks (Fig.2).

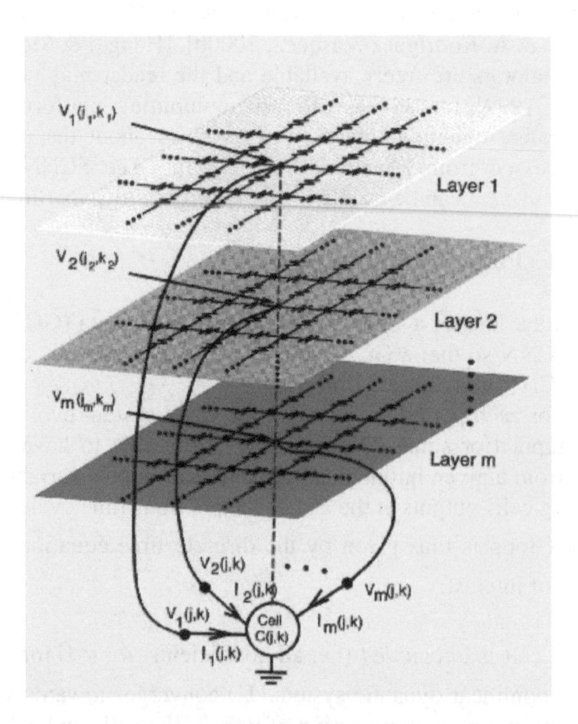

Fig.2. The topology of a Reaction-Diffusion Cellular Neural Network. A cell is a m-port described by a nonlinear ODE which models a physical ***reaction***. The coupling with neighboring cells is done via resistive grids modeling the physical process of ***diffusion***.

[1] Autonomous means without external inputs. Information processing in such systems is based on prescribing the initial state of all cells with some pattern followed by a dynamic transformation of this pattern until a stopping criterion is met.

Example 4: **The case of a Reaction Diffusion cell with von Neumann neighborhood.**

The cell equation is of the following form:

$$
\begin{aligned}
\dot{x}_1^1 &= f_1(x_1^1, x_1^2, \ldots, x_1^m, \mathbf{G}_1) + D_1\left(x_2^1 + x_3^1 + x_4^1 + x_5^1 - 4x_1^1\right), \\
&\;.. \\
\dot{x}_1^j &= f_1(x_1^1, x_1^2, \ldots, x_1^m, \mathbf{G}_j) + D_j\left(x_2^j + x_3^j + x_4^j + x_5^j - 4x_1^j\right), \\
&\;.. \\
\dot{x}_1^m &= f_1(x_1^1, x_1^2, \ldots, x_1^m, \mathbf{G}_m) + D_m\left(x_2^m + x_3^m + x_4^m + x_5^m - 4x_1^m\right)
\end{aligned}
\tag{3}
$$

where x_k^j is the state variable of cell k (using the neighborhood index as in Fig.1 with the central cell indexed as k=1) in layer j. The gene is also distributed among layers as $\mathbf{G}_1,\ldots\mathbf{G}_m$ and in addition, each layer is characterized by the *scalar diffusion coefficient* D_j which corresponds physically to the conductance of the resistors in the resistive grid in Fig.2.

RD-CNNs will be studied extensively in Chapter 4 since they have the interesting property that emergent behaviors can be predicted by carrying out a *local activity* analysis of the isolated cell (i.e. ignoring the coupling). Their simulation is discussed in Section 2.4. in this chapter.

2.3. Matlab simulation of generalized cellular automata

It is convenient to simulate various types of cellular system using the programming environment Matlab (http://www.mathworks.com/products/matlab/) produced by Mathworks. We will assume here the reader has some basic knowledge of Matlab. For readers unfamiliar with this language, a good starting point might be the primers [Sigmon and Davis, 2001] or [Sigmon, 1993].

Uncoupled GCAs

In order to simulate an uncoupled GCA one should first write the function GCA_U_CELL.M which implements the GCA cell. It is in fact a simple description of the nonlinear function $y_1(t) = F(\mathbf{G}, u_k(t))$:

```
function y=gca_u_cell(u1,u2,u3,u4,u5,u6,u7,u8,u9)
Z=[-1,-2,-4,-8,-7];
B=[1 1 1 1 1 1 1 1 1];
s=-1;

sigm=B(1)*u1+B(2)*u2+B(3)*u3+B(4)*u4+B(5)*u5+B(6)*u6+B(7)*u7+B(8)*u8+B(9)*u9;
w=Z(1)+abs(Z(2)+abs(Z(3)+abs(Z(4)+abs(Z(5)+sigm))));

y=s*sign(w);
%y=0.5*(abs(w+1)-abs(w-1));
%y=w;
```

The nine inputs of this function can be either a scalar, a vector or an array. The program above implements the Parity9 local logic function. However, as seen in Chapter 3 by simply changing the gene parameters (in our case the values of the Z,B,s) many other local Boolean functions can be implemented. Moreover, one can simply change the cell structure and use any other nonlinear function. For example, as seen in the last lines, one can employ a sign function and the result is a GCA which emulates a binary cellular automaton but also continuous valued functions with or without saturation, resulting in a GCA with continuous states.

Let now have a look at the code GCA_U.M to implement the GCA itself. It is implemented in the form of a Matlab function with two input parameters. The first is a number and represents the number of iteration steps until stop (a faster stop can be achieved by pressing the keys CTRL and C). The second can be missing and if not missing it represents the name (a string) of a file containing the initial state. The initial state gives also information about the size (number of cells) of the GCA. A periodic boundary condition is implemented, the lattice is square with a Moore neighborhood indexed as explained in the source code:

```
function gca_u(steps,init_cond)
% e.g. steps=100 (runs the GCA for 100 iterations)
% e.g. init_cond='cross199'
% The inital state should be previously saved as matrix x0. For example:
% x0=ones(199,199); x0(90:110,90:110)=1
% save one199 x0
% The neighborhood index is chosen as follows:
%     9 8 7
%     6 5 4
%     3 2 1
if nargin<2
                    % by default the "cross199" initial state is loaded if no input file is specified
  init_cond='cross199';
  x0=-ones(199,199); x0(90:110,90:110)=1;
else
  eval(['load ',init_cond]);              % load the initial state
end
[m n]=size(x0);
i=1:m; % row index
j=1:n; % column index
left_j=[n,1:n-1]; right_j=[2:n,1];        % indexes of cells on the left and right
up_i=[m,1:m-1]; low_i=[2:m,1];            % indexes of cells upper and lower

y=x0;                                     % current output is loaded with x0

for s=1:steps
                                          % Computes the inputs in the next step
u9=y(up_i,left_j);    u8=y(up_i,j);        u7=y(up_i,right_j);
u6=y(i,left_j);       u5=y;                u4=y(i,right_j);
u3=y(low_i,left_j);   u2=y(low_i,j);      u1=y(low_i,right_j);
                                          % Compute the new output using the cell function
y=gca_u_cell(u1,u2,u3,u4,u5,u6,u7,u8,u9);
                                          % display the new output
image(20*y+32); axis image; colormap jet
title(['Step: ',num2str(s),' Initial state: ',init_cond,' PRESS ANY KEY TO CONTINUE']);
                                          % wait for the next step
waitforbuttonpress
end
```

With these two simple programs you may begin to observe emergent patterns in generalized cellular automata. First, running the program as they are (in fact you should run GCA_U(100) for a 100 steps run) a sequence of growing binary patterns is observed. The color code used here associates color red with +1 and blue with –1.

One can easily change the behavior by re-editing the GCA_U_CELL.M file. For example after replacing Z=[-1,-2,-4,-8,-7] with Z=[-1,-2,-4,-6,-7] the result will be a sequence of "snow-flake" like binary patterns. Moreover, if you use the initial Z but replace the initial B with B=[1 1 1 1 -8.5 1 1 1 1] and use the output function y=0.5*(abs(w+1)-abs(w-1)) the result is a sequence of colored patterns corresponding to a continuous-state cellular automaton. Observe that much of the flexibility in programming different behaviors using the same code is due to the use of a universal piecewise-linear parametric functional (the function with nested absolute values). More issues regarding cell universality are discussed in detail in Chapter 3.

Coupled GCAs

In the implementation of the cell for the uncoupled GCA there is no temporal dynamics involved and the output is just a nonlinear function of the 9 inputs. In fact, when called from the main function GCA_U the function GCA_U_CELL updates *all* cells simultaneously since it performs matrix computation (all inputs, for example u1, are matrices of the same size as the CNN). We can view GCA_U_CELL as a layer of CNN cells, but since the model is simplified there is no temporal dynamics included. In order to implement a coupled GCA we should change the cell function such that it will implement the dynamic evolution of a continuous time CNN. The resulting code of the program GCA_C_CELL.M is listed next:

```
function y=gca_c_cell(u1,u2,u3,u4,u5,u6,u7,u8,u9)
Delta_T=0.1;                    % integration step (the smallest the best precision but takes longer)
T=4;                            % the duration of the Euler integration;
B=[0 -1 0 0 1 0 0 -1 0];        % cell parameters (gene)
A=[1 1 1 1 -3 1 1 1 1];
z=0;
%------------------------------------------------------------------------------------------
[m n]=size(u1);
i=1:m;                          % row index
j=1:n;                          % column index
left_j=[n,1:n-1]; right_j=[2:n,1]; % indexes of cells on the left and right
up_i=[m,1:m-1]; low_i=[2:m,1]; % indexes of cells upper and lower
x=0.0001*(rand(m,n)-0.5);       % initialization of the initial state with small random values
% LINE A                        % the feed-forward contribution
ffwd=z+B(1)*u1+B(2)*u2+B(3)*u3+B(4)*u4+B(5)*u5+B(6)*u6+B(7)*u7+B(8)*u8+B(9)*u9;
for s=1:round(T/Delta_T)
  y=0.5*(abs(x+1)-abs(x-1));                    % compute the outputs for all neighbors
  y9=y(up_i,left_j);  y8=y(up_i,j);        y7=y(up_i,right_j);
  y6=y(i,left_j);     y5=y;                y4=y(i,right_j);
  y3=y(low_i,left_j); y2=y(low_i,j);       y1=y(low_i,right_j);
  % LINE B                                      % the recurrent contribution
  recur=A(1)*y1+A(2)*y2+A(3)*y3+A(4)*y4+A(5)*y5+A(6)*y6+A(7)*y7+A(8)*y8+A(9)*y9;
  x=(1-Delta_T)*x+Delta_T*(ffwd+recur);         % update the state dynamics following equation (1)
                                                %display the new output
image(20*y+32); axis image; colormap jet
title(['Time: ',num2str(s*Delta_T),' PRESS ANY KEY TO CONTINUE']);
% wait for next step
waitforbuttonpress
end
```

This new cell is in fact a implementation of the standard (linear coupling) CNN model (of Chua and Yang [Chua & Yang, 1988] but in fact can be easily developed into a nonstandard model by simply changing LINE A and LINE B accordingly. The continuous time dynamics associated to Equation (1) is emulated on the discrete-time computer using a simple integration method; namely, the Euler's method. Therefore, in addition to gene parameters this new cell has to specify two dynamic parameters. The first, **Delta_T** indicates the step size and the smallest the better will be the approximation of the continuous time dynamics. The second, **T**, represents the integration time period. Usually the user may choose such a value that corresponds to a steady state dynamics in the output.

In order to simulate a coupled GCA one should use the above file in conjunction with a file called GCA_C.M which is simply obtained from a copy of the GCA_U.M file after changing the name of cell function from GCA_U_CELL to GCA_C_CELL.

Using the values in the file above, the dynamics of the CNN will be first observed, and from time to time a figure will show the GCA output. The visualization of the CNN steps is useful for first experiments and it allows one to choose the two dynamic parameters. However, if one wants to see only the GCA output, the visualization lines in the above file should be ignored (by inserting % in front of them).

As in the previous cases, a wide range of dynamic phenomena can be simulated by re-editing the above file after proper changes of the gene or dynamics parameters.

Simulation of standard cellular neural networks

As we already discussed, simulating the CNN reduces to the particular case of simulating a coupled GCA for only one discrete time step. The GCA_C_CELL is activated and it will simulate the CNN. Its parameters and even novel CNN models can be easily simulated by properly re-editing the GCA_C_CELL.M file.

2.4. Simulation of Reaction-Diffusion Cellular Neural Networks

We will exemplify the simulation for the particular case of FitzHugh Nagumo cells (See Chapter 4 for details). However, any Reaction-Diffusion system can be easily simulated by after several minor re-editing operations on the files. As in the previous cases we should first define the cell equation. In the case of the Reaction-Diffusion systems this is done by the following function (RD_CELL.M):

```
function y=rd_cell(t,x)
% Cell gene parameters - they can be further edited (here they are specific to the FitzHughNagumo cell)
a=0.1; b=1.4; e=-0.1; alfa=1; d1=.1; d2=.5;
%-------------------------------------------------------------
[sz dm]=size(x); N=sqrt(sz/2);                    % detects the CNN size
i=1:N; j=1:N;                                     % row and column index
left_j=[N,1:N-1]; right_j=[2:N,1];                % indexes of cells on the left and right
up_i=left_j; low_i=right_j;                       % indexes of the upper and lower cells

u=reshape(x(1:N^2),N,N)'; v=reshape(x(N^2+1:2*N^2),N,N)'; %repack the two state variables
u_up=u(up_i,j); u_dwn=u(low_i,j); u_lft=u(i,left_j); u_rt=u(i,right_j);
v_up=v(up_i,j); v_dwn=v(low_i,j); v_lft=v(i,left_j); v_rt=v(i,right_j);

% Implement the cell defining equations. The functions f1 and f2 are implemented separately for clarity
u_out=f1(u,v,alfa)+d1*(u_up+u_dwn+u_lft+u_rt-4*u);
v_out=f2(u,v,a,b,e)+d2*(v_up+v_dwn+v_lft+v_rt-4*v);;
y=[reshape(u_out',1,N^2), reshape(v_out',1,N^2)]'; %repack the two state variables
```

Here a two-layer Reaction-Diffusion CNN is implemented. The Reaction-Diffusion systems in Chapter 4 are all 2-layers systems. The function RD_CELL.M is prepared to be used in an Matlab specific integration routine based on the Runge-Kutta algorithm. Since the integration routine ODE23 requires a unique vector variable **x** , before preparing to write down the cell equations as in (3) we have first to unpack the vector variable into the two variables **u** (first layer) and **v** (second layer). The two variables are square matrices of dimension N each while an element u(i,j) of each matrix corresponds to the CNN cell on the (i,j) position.

The first line allows one to specify the gene parameters of the cell. Different dynamic behaviors can be simulated by changing these values and saving the file. The next lines deal with the implementation of (3). Note that the nonlinear functions f_1 and f_2 in (3) are implemented separately being specific for the FitzHugh-Nagumo cell model. They are as follows:

```
function out=f1(x,y,alfa)
out=-y+alfa*x-(1/3)*(x.^3);
```

```
function out=fhn2(x,y,a,b,e)
out=-e*(x-b*y+a);
```

Whenever a different type of cell (of order two) has to be implemented one needs only to change the above function and to provide the adequate list of parameters in RD_CELL.M. No other change is required in the file. Next let us examine the source code of the main simulator function called RD_CNN.M

```
function rd_cnn(mode,init_state)
% mode=0 -> displays 5 snapshots only, mode=1 -> displays 42 snapshots
ts=0:1:40; % Dynamic evolution (here time from 0 to 40 with snapshots from 1 to 1 time units)
%-------------------------------------------------------------
if nargin<2 u=(rand(20,20)-.5)*0.1; v=(rand(20,20)-.5)*0.1;
else eval(['load ',init_state]); end
% If there is only one parameter random initial states are generated, otherwise loaded from a file
[N N]=size(u);                                      %Evaluate the CNN dimension (square CNN only)
if mode==1 n_snp=42; elseif mode==0 n_snp=5; end; %The number of snapshots for display
% LINE A - THE ODE INTEGRATION
y0=[reshape(u',1,N^2), reshape(v',1,N^2)];          % pack the two state variables
opt=odeset('OutputSel',ts);                         % initialize the intgration routine
[t out]=ode23('rd_cell',ts,y0,opt);                 % integrate the rd_cell using ode3

u=out(:,1:N^2); v=out(:,N^2+1:2*N^2);                %unpack the state variables
mi=min(min(u)); ma=max(max(u)); mi_v=min(min(v)); ma_v=max(max(v)); [per dum]=size(t);
disp(['u_max: ',num2str(ma),' u_min: ',num2str(mi)]);

% DISPLAY RESULTS
if mode==0 set(figure(1),'Position', [263 530 770 175]);
elseif mode==1 set(figure(1),'Position',[172 95 862 641]); end
colormap jet; whitebg(1, [1 1 .5]); ti=[]; col=6; linii=ceil(n_snp/col);
for i=1:n_snp
        tsh=round(1+(per-1)*(i-1)/n_snp);
        ti=[ti t(tsh)];
        im=reshape(u(tsh,:),N,N)';
        eval(['img',num2str(i),'=64*(im-mi)/(ma-mi);']);
        subplot(linii,col,i); image(64*(im-mi)/(ma-mi)); axis image; axis off; title(['t=',num2str(t(tsh))]);
end
```

This program allows the integration of the ODE description of the cell in file RD_CELL.M and it allows also to display the dynamic evolution as a sequence of 5 or 42 snapshots depending on the selection of the input variable MODE. The method of integration is the one specific of the function ODE23, but one can simply replace it with other ODE integration routines available in Matlab such as ODE45, ODE113, etc.

The only parameter which has to be specified is the vector **ts** which specifies the moments of time for display, e.g. ts=1:5:67 means that there will be displayed snapshots taken at the moments 1,6,11,..67 i.e. from 5 to 5 time periods.

Each snapshot is a colored image and the color of each pixel codes the amplitude of the state variable *u* using the colormap code assigned to the display. The colormap code used by default is JET but it can be easily changed with one of the other available colormaps . During the simulation, the maximum and minimum values of the state variables are recorded and typed to the screen. When the snapshots are displayed these extreme values are assigned to the colors representing the minimum and respectively the maximum amplitude. For JET, the color assigned to minimum is dark blue and for maximum dark red.

Examples of simulation: Using the default parameters and typing RD_CNN(1) a dynamic evolution towards a non-homogeneous stable state is observed. If the parameters are changed as follows: a=0.0; b=1.3; e=-0.1; alfa=1; d1=.1; d2=0; in RD_CELL.M and ts=0:5:200 in RD_CNN.M the result of running RD_CNN(1) is an emergence of colliding waves. As shown in Chapter 4, a precise location of the sets of parameters producing interesting and potentially useful emergent behaviors can be done employing the local activity theory.

2.5.Concluding remarks

Cellular systems are described by simple mathematical equations. As shown above, using a high level platform such as Matlab, all major cellular systems paradigms can be simulated using very compact code. Although their description is compact, a wide palette of emergent behaviors can be observed and the main problem is how to choose such parameters to obtain computationally useful behaviors.

The next chapters deal with this problem. First, in Chapter 3 it is shown that universal CNN cells for both binary and continuous state computation can be compactly defined using piecewise-linear models. Then, in Chapter 4 we will investigate how the theory of local activity can be employed to locate sets of parameters for which emergent behaviors occur in Reaction-Diffusion type cellular system. Emergence in generalized cellular automata is the topic of Chapter 5 while in Chapter 6 a promising application in biometric authentication of the emergent patterns in generalized cellular automata is presented.

Chapter 3
Universal Cells

3.1. Universality and cellular computation, basic ideas

As shown previously, a cellular medium has several characteristics which makes it well suited to study emergent computation and complexity. Particularly attractive is the possibility to establish a direct relation within the *cell* as a basic unit of the cellular space and the *underlying dynamics* in this space. In Cellular Automata literature [Toffoli & Margolus, 1987] one deals with finite number of states and discrete time while the *cells* are specified via *local rules* or *transition tables*. Several attempts to map dynamic behaviors to cells were initiated by Packard [Packard, 1988] and Langton [Langton, 1990]. In his work, Langton proposes a parameter λ, which can be computed based on the transition table and which is supposed to give an indication on the type of dynamics occurring in the resulting cellular medium. However, as pointed out in [Mitchell *et al.,* 1993] there is no evidence of a certain relationship between the λ parameter and the emergent dynamic behavior. In fact it is obvious that such a parameter represents only a *simplified model of the entire transition table* and *does not* contain all information relevant for the cell and its couplings with adjacent cells. Therefore any connection that can be made between the value of such coefficient and the dynamic behavior is only *ambiguously specified.*

The idea of using a reduced number of continuous parameters to describe the cell model stands at the basis of our development in this chapter where the goal is to find a *universal* cell model capable to approximate any nonlinear function. A particular cell realization is then specified by a *gene* (a terminology introduced by Chua [Chua, 1998]) encoding *the whole* information about the cell and its coupling.

Our approach to emergent computation is based on an extension of the Cellular Automata model; namely, the Cellular Neural Network (CNN) and its extension the Generalized Cellular Automata (GCA) [Chua, 1998]. This model includes the Cellular Automata as special cases and has the main advantage of viewing the cell as a *nonlinear function* rather than a *transition table.* Therefore, an extension to *continuous state* cells is straightforward. In addition the cellular neural network model includes the Reaction-Diffusion CNN for which an analytical technique based on Chua's local activity theory was developed (see Chapter4) to identify *precisely* the sub-domains in the cell parameter space for which emergence will occur.

As mentioned in the introduction, the local activity theory provides tools for *precisely* identifying the *boundaries* between various types of dynamic behaviors. In general, for any type of cellular medium where the cell and its couplings are defined via a nonlinear parametric model, a theory and appropriate methods should exist such that a specific dynamic behavior can be mapped into a precisely defined sub-domain of the cell parameter space. Therefore defining a *universal cell* capable to approximate arbitrary nonlinear functions is a issue of a great interest.

In addition, the idea of representing a cell trough a parametric functional model is also practical, since it can easily lead to a compact circuit implementation with certain tunable parameters allowing the user to "program" various emergent computation regimes. As pointed out previously one of the main reason to study emergence in cellular system is to provide a foundation for the design of a new generation of computers based on cellular mediums and operating much like biological entities. It is clear that such a computer is more efficient in hardware realizations than a program running on a classic, serial computer.

In this chapter, the theory of piecewise-linear function approximation is employed first to define Boolean universal cells and then to define an improved cell model of a *universal continuous state* cell, called a *simplicial CNN cell* [Dogaru *et al.,* 2002a], [Julian, *et al,* 2002a].

Boolean universal cells

For an appropriate number of "nests" (the absolute value functions in next formula), the following cell model is capable to represent any arbitrary Boolean function:

$$y = s \cdot sign\left(z_n + \left| z_{n-1} + \left| ...z_2 + \left| z_1 + \sum_{i=1}^{n} b_i u_i \right| \right| \right| \right) \quad (1)$$

where y is the (binary) output, u_i are the inputs (binary, or continuous within [-1,1]).

As seen from (1), at most $2n$ analog parameters $\{z_i, b_i\}_{i=1,...,n}$ and one additional one-bit parameter $s \in \{-1,1\}$ suffice to specify a given Boolean function with n inputs, a result which clearly puts the *multi-nested neural cell* in the top of the most compact neural architectures. In [Hassoun, 1995] the lower bounds on the size of various "classic" neural network for arbitrary Boolean function representations are considered. Among them, the lowest bound, found for the case of a MLP (multilayer perceptron) formed of k LTG (linear threshold gates) require that $k \geq 2^n/n^2$. Since each of the LTG has at least n parameters, the number of parameters is of the order $O(2^n/n)$. We should note however, that for practical realizations of the multi-nested cell, the highly compact structure with $O(n)$ complexity comes at a price. This is the representation precision of each parameter, which is of the $O(2^n/n)$ bits in order to ensure the conservation of all *functional information* (a Boolean function is specified by 2^n bits of information). Therefore, in practice, the *multi-nested universal cell* can be implemented for up to $n = 4 \div 5$ inputs, corresponding to a representation precision of $4 \div 7$ bits per parameter, attainable by any cheap analog technology. Networks of multi-nested neural cells can be employed to solve the problems with more inputs, with a higher implementation efficiency than that of the multi layer perceptrons employing simple linear threshold gates.

The simplicial cell – universality expanded to continuous states

The nesting principle is again applied by embedding the above *Boolean universal cell* in a circuit structure resulting from the theory of simplicial subdivisions [Chien and Kuh, 1977]. The model of the simplicial neural cell is can be written as:

$$y = \sum_{l=1}^{n+1} c_{i_l} \mu_{i_l}(\mathbf{u}) \quad (2)$$

where, $\mathbf{u} = [u_1,..,u_i,..u_n]$ is the vector of n inputs, i_l is a sequence of vertices indexes ranging between 0 and $2^n - 1$ which can be easily generated by a circuit proposed in [Dogaru *et al,* 2001] so that will satisfy the requirements of the simplicial subdivision theory as shown in Section 3.3. The c_{i_l} coefficients are subject to an LMS (Widrow-Hoff) or other similar learning process so that (2) can approximate with an accepted error the *continuous state* mapping $y = F(\mathbf{u})$. It was also shown that the circuit first proposed in [Dogaru et al, 2001], can perform conveniently (using only n comparators and a ramp generator) not only the generation of the sequence of vertices indexes, but also the computation of the simplicial basis functions. The sum of products (2) is also conveniently computed using a simple integrator circuit. Although there are 2^n coefficients to learn, since at each presentation of an input stimuli only $n+1$ of the basis functions are non-zero, the learning process has the same complexity as that of a simple linear perceptron with n inputs. It is shown that under certain conditions the coefficients c_{i_l} can be represented using a one-bit resolution. The mapping $c_{i_l} = C(\mathbf{v}_{i_l})$ is then the expression of a Boolean function and it can be conveniently represented using the Boolean universal CNN cell (multi-nested cell) in (1).

The next two sections provide an in-depth analysis of the two cell models presented above and give methods to train the cells to learn a specified mapping.

3.2. Binary cells

Throughout the following chapters we will restrict ourselves to the case of an uncoupled CNN cell [Chua, 1998] described by:

Universal (Piecewise-linear) CNN cell:
Steady State Input-Output Representation

$$\sigma = \sum_{k.l \in \{i-1,i,i+1\} \times \{j-1,j,j+1\}} b_{kl} u_{kl} = \sum_{i=1}^{9} b_i u_i \tag{3}$$

$$y = \mathrm{sgn}[s \cdot w(\sigma, z_0, z_1.., z_m)] \tag{4}$$

The scalar variable σ is called an *excitation*, and in the case of the standard CNN cell it is computed as a linear correlation between the feed-forward (controlling) template vector $\mathbf{b} = [b_1,...,b_n]$, which is a repacked version of the B template [Chua, 1998], and its associated input vector $\mathbf{u} = [u_1, u_2, .., u_n]$, as defined in (1). The second notation, with the index "i" replacing the pair of indices {k,l} defining a two-dimensional neighborhood is more general and can be applied to arbitrary *choices* of CNN architectures. From this perspective, n represents the number of cell inputs and R^n is the cell input space.
Since the excitation σ in the *standard* CNN cell is obtained as a linear correlation between the inputs and the **b** template vector, the *standard* CNN cell has the capability to implement only a limited number of local Boolean functions, namely the *linearly separable* ones. Any linearly *not* separable Boolean function was previously realized using a chain of many standard cells, each implementing a linearly separable Boolean function. By combining a properly chosen set of such linearly separable functions one can implement any arbitrary Boolean function. There are mainly two methods proposed in the literature to solve this problem [Crounse *et al.*, 1997] [Nemes et al, 1998] via *a CNN universal chip.* In what follows we will show that by replacing σ with a *non-linear* piecewise-linear (PWL) discriminant function $w(\sigma, z_0, .. z_m, s)$ of σ, *any* Boolean function can be realized with the cell structure defined by the above equations. The most important result is that our

universal cell model requires no additional *correlation unit* (or weighted summation) except the one already used by the standard CNN cell defined by (3).

Observe that in our approach (3) performs a dimensionality reduction via a *projection* from the n dimensional input space to a scalar, one-dimensional axis (the *projection* axis) corresponding to the *excitation* variable σ. As a consequence, the determination of the non-linear discriminant function $w(\sigma)$ can be dramatically simplified, as shown next. By additional optimization of the *orientation vector* $\mathbf{b} = [b_1,...,b_n]$ associated with the *projection axis*, an optimal, or near-optimal, template $\mathbf{b}^* = [b^*_1,...,b^*_n]$ can be found to minimize the number m of additional parameters.

3.2.1. What would be an "ideal" binary CNN cell ?

An ideal CNN cell is an abstract concept, providing a reference for comparing various CNN cell models. As shown next, some of the features of an ideal cell may conflict with each other in practice so that certain tradeoffs between them should be accepted. The following are the most important features of an "ideal" cell:

Universality

Universality is concerned with the possibility to use the same physical structure of the CNN cell for implementing *arbitrary* Boolean functions, by simply changing (programming) the cell parameter values. It is also important that the ideal cell contains as few cell parameters as possible. As a general rule, a universal cell is less compact (i.e. it has more cell parameters and a higher complexity in physical implementation) than a cell dedicated to implement a specific function, or a restricted class of functions. For example, if one would like to use a CNN cell only for implementing the *parity function with 9 inputs*[1] (Parity9), the best choice (leading to the simplest realization) for $w(\sigma)$ would be:

$$w_{PAR9}(\sigma) = 1 - \left|-2 + \left|-4 + \left|-8 + \left|-7 + \sigma\right|\right|\right|\right|,$$

$$\text{with } \mathbf{b} = \begin{bmatrix}1 & 1 & 1 & 1 & 1 & 1 & 1 & 1 & 1\end{bmatrix} \qquad (5)$$

By choosing other **b** templates (orientation vectors), and/or by changing the numerical parameters within $w_{PAR9}(\sigma)$ we can realize many other Boolean functions. However, the cell defined by $w_{PAR9}(\sigma)$ above is *not universal*, even if the parameters are changed, because it cannot realize certain Boolean functions with 9 inputs regardless of the parameters. However, compared to the situation of the standard CNN cell defined by $w(\sigma) = \sigma + z$, the cell model (5) is much closer to universality since it can realize many linearly *non-separable* Boolean functions.

Compactness

This feature is concerned with the number of elementary physical units required to implement a CNN cell. There are two aspects influencing compactness: the number of free parameters and the complexity of the non-linear discriminant function. Although the

---

[1] The Parity 9 function returns 1 if and only if there is an odd number of active ($u_i = 1$) inputs, and –1 otherwise.

standard linear cell is not universal, it is the most compact: It has a realization requiring only n synapses associated with the linear correlation $\sigma = \mathbf{bu}^T$ and an additional threshold parameter. Compactness can be expressed as a function of n (the number of cell inputs). From this perspective, we are interested to determine universal cells having a polynomial rather than an exponential dependence on n in implementation complexity. As shown next, the compactness of a universal CNN cell can be pushed to its limits in the case of the universal multi-nested PWL CNN cell where universality is achieved with only $2n+1$ parameters.

Compactness is a feature conflicting with *robustness*. The more compact a cell is, the less robust it will be. The reason is that any local Boolean function with "n" inputs requires $N = 2^n$ bits to be unambiguously specified. A compact universal cell has to preserve this information in its definition and therefore it must distribute it among the cell parameters. For example, an arbitrary (local) Boolean function with 9 inputs (e.g. the standard local logic for two-dimensional CNNs) requires 512 bits to be specified. For the case of the *multi-nested PWL cell* where universality is achieved with $2n+1$ parameters, each parameter requires, on average, at least $\frac{512}{2 \cdot 10} \approx 26$ bits. At the opposite extreme, the pyramidal universal cell in [Dogaru *et al,* 1998e] requires 2^n parameters: In this case, since each parameter is associated with only 1 bit of information, the realization is maximally robust but clearly not compact at all.

Robustness

A CNN cell is defined by a certain set of real parameters, as shown in (3)-(4). When both the input and the output spaces are discrete (in our case, binary), the parameter space associated with a given cell structure is partitioned into compact domains separated by *failure boundaries* [Chua, 1998]. Each domain corresponds to the realization of a particular Boolean function and therefore there is a continuum of cell parameters all of them realizing the same Boolean function. A visualization of the failure boundaries for the case of a piecewise-linear cell is given in Chapter 5. Within each region, the most robust *set of parameters* can be determined as the collection of points in the cell parameter space maximizing the distance from any failure boundary. The degree of robustness of this cell parameter point depends on the volume of its associated region. Since the profile of the failure boundaries is determined by the nonlinear function $w(\sigma)$ in (4), it follows that the robustness of a CNN cell is closely related to $w(\sigma)$. Observe that universality and robustness are conflicting features. The more functions we can represent with the same structure, the more domains have to "compete" for the same parameter space and thus the less robust each function will be. Fortunately, robustness can be improved by increasing the dimension of the parameter space. However, in this case we have to accept a reduction in the cell compactness as illustrated above.

It is important to note that in general, the shape of the regions separated by the failure boundaries in a high dimensional parameter space is difficult to determine analytically. Therefore, a robust solution is relatively difficult to define in analytical terms. In the case of the *canonical PWL universal cell* (Section 3.2.3), however, an analytical robust solution can be easily identified in the $\{s, z_0, z_1, .., z_{m-1}\}$ cell parameter subspace since the failure boundaries defined by $w(\sigma)$ are simply *points* and the separating regions are simply *segments* in the one-dimensional space of the *projection tape*.

Capability of evolution

This feature is concerned with the capability of a cell to adjust its parameters so that it can "learn" new functions by *evolutionary* interactions with the external world. Basically this feature assumes that the design (or learning) algorithm is simple enough to admit an on-chip implementation so that the cell can rapidly adapt to novel tasks. Here, by evolution we mean *mutations* that may take place in the cell parameter space so that the cell can realize novel Boolean functions. There are two major components; namely, evolution by *design* and evolution *by interactions.* The first case corresponds to the genetic inheritance in the living systems, and is achieved by loading the CNN cell with a particular gene, chosen from a genome (previously determined by a human expert) so that the associated CNN will perform a prescribed task (e.g. contour detection). While any useful CNN cell model should provide a cell parameter identification (design) procedure, in many cases the interactive aspect of evolution is usually neglected. Evolution by interactions assumes that cells are allowed to mutate (change) their parameters as a result of their interaction with other cells, or with certain input stimuli provided by a specific problem. The goal is to optimize a cost function that cannot be described analytically in terms of a local Boolean function. For example, in defining the function of a "corner detector" [Chua, 1998] there are certain subjective issues which may lead human experts to generate different CNN genes to implement such subjective task. In such cases, one may consider a set of images with various types of corners and let the CNN cell *evolve* towards an optimum gene which minimizes the error between the desired output images and the actual ones. We will focus here on more versatile CNN genes that can support *evolution by interactions.* Such features may be also of interest from the perspective of building *evolvable systems* [Mange & Tomassini, 1998] capable of complex tasks such as self-repair and self-reproduction of their components.

3.2.2. Orientations and Projection Tapes

Local binary computation

In what follow we will consider the general case of implementing arbitrary Boolean functions of n input variables. According to the convention used in the CNN literature, a "0" (or false) logic level is coded with –1, while a "true" (or "1") logic level is coded with +1.

Let us now consider several methods for representing a Boolean function:

(a) ***Using a truth table***

A truth table has $N = 2^n$ rows, corresponding to the same number of possible configurations of the input vector $\mathbf{u} = [u_1, u_2, .. u_n]$. For each possible input configuration, the binary output (-1 or 1) associated with the Boolean function is presented in an additional column **Y**. In principle, as long as both inputs and the output are presented in the table, there is no need for a special ordering of rows and columns. However, it is a common practice to accept a certain *ordering convention.* We will adopt the system in [Chua, 1998], where the leftmost column contains the most significant input bit and the rightmost column the least significant one. Therefore, in the first row all inputs are -1, while they are all +1 in the last row of the table. It is convenient to associate an index *j*, where $j = 0, ... 2^n - 1$ to each row, with the convention that $j = 0$ corresponds to the last row in the table. For the above *ordering convention* the following equations define entirely the input entries in the truth table:

$$u_{i,j} = \begin{cases} +1 & \text{if } \mathbf{mod}\left(j, 2^{n-i+1}\right) < 2^{n-i} \\ -1 & \text{else} \end{cases} \quad (6)$$

where $\mathbf{mod}(m,n)$ is the remainder of m/n.

The binary output corresponding to each row *j* is denoted as $\gamma_j \in \{-1,1\}$.

(b) As a *decoding tape*

Assuming the *ordering* scheme described above, one should note that there is no need to write down the input entries. Therefore, a much more compact representation of a Boolean function was proposed in [Chua, 1998] in the form of an *N*-dimensional vector $Y(ID) = [\gamma_{N-1}, \gamma_{N-2}, ..., \gamma_0]$, where $N = 2^n$. Using a color code ("dark gray" in a gray scale representation or "red" in a colored figure for $\gamma_j = +1$, and "white" or "blue" for $\gamma_j = -1$) each Boolean function can be represented as a colored strip called a *decoding tape,* which is reminiscent to that of a *gene* in biological systems.

The decoding tape is a useful concept, giving us a one-dimensional view of the Boolean function rather than a complicated description in the form of a truth table or a spatial representation. In fact, as we will see in the next chapter, the *decoding tape* can be treated as a special case of a more general construct called a *projection tape.* The integer *ID* is a function identification number, which is the decimal equivalent of the binary vector Y.

Let us consider the Boolean function "Parity3"[2]. Its associated decoding tape is $Y(105) = [-1,1,1,-1,1,-1,-1,1]$. The equivalent binary number is obtained by substituting -1 with 0. The resulting equivalent binary string is given by: $01101001_2 = 1 + 2^3 + 2^5 + 2^6 = 105_{10}$.

(c) As a hyper-dimensional hypercube

In the input space, each vector $\mathbf{u}_j = [u_{1j}, u_{2j}, ..., u_{nj}]$ corresponds to a vertex V_j of an *n*-dimensional hypercube. By assigning to each vertex a binary value $Y(V_j) = \gamma_j$, the result is a spatial representation of the Boolean function *Y* as a geometrical object. To specify a Boolean function we use the color red (and the symbols "*") to code all vertices corresponding to $\gamma_j = 1$ and the color blue (and the symbols "o") to code all vertices corresponding to $\gamma_j = -1$. For example, in Fig. 1, the Boolean "Parity2" function (ID=6) with 2 inputs is represented by the colors of the $2^2 = 4$ vertices of a square.

In Figs. 2, and 3, two other Boolean functions with 3 inputs (ID=105, and ID=142) are represented via the $2^3 = 8$ vertices of a cube. The entire set of 256 Boolean functions with 3 inputs is represented graphically by a cube in R^3 in [Chua, 1998]. The spatial representation of a Boolean function can be extended to any type of data, including ill-defined problems normally arising from various pattern classification tasks. In such cases, each vertex V_j corresponds to a (stimulus - desired output) pair, where the stimulus is the input vector $\mathbf{u}_j = [u_{1j}, u_{2j}, ..., u_{nj}]$ and the desired output is $Y(V_j) = \gamma_j$.

---

[2] The Parity3 function returns +1 if and only if one or three inputs are +1 while the remaining one(s) are −1.

Projections

The main advantage of the spatial representation of a Boolean function as a hypercube with color-labeled vertices is that it can provide hints on how to separate the "blue" vertices form the "red" ones, i.e how to define a valid discriminant function $w(\mathbf{u})$. It is clear that for any Boolean function, one can define an infinite number of non-linear discriminant functions to separate the red vertices from the blue vertices. The *universality* requirement imposes that a canonical formula should describe it so that any change in its associated Boolean realization must correspond only to a change of the parameters. It is also desirable that these parameters be determined via a simple and fast algorithm.

The general equation of the discriminant hyper-surface $w(\mathbf{u})=0$ for an input space dimension larger than $n=2$ leads to some complicated topologies, making it rather difficult to map into a canonical piecewise-linear formula [Kahlert & Chua, 1992] with an arbitrary number of inputs. In the case of linearly separable Boolean functions, various convergent solutions exist, such as the classical "perceptron learning" algorithm, the LMS algorithm [Widrow & Hoff, 1960] [Haykin, 1999] or the use of linear programming techniques such as the Simplex algorithm. In this case, $w(\mathbf{u})=0$ corresponds to the equation of a hyper-plane: $b_1u_1+b_2u_2+...+b_nu_n+z=0$.

In the case of linearly *non-separable* functions, the main difficulty to derive a discriminant function analytically can be traced to the high dimensionality of the input space, which gives rise to complicated geometrical shapes, difficult to understand. Multi-layer perceptrons [Hassoun, 1995] combine linear threshold gates (also called neurons) so that the discriminant functions are obtained by intersecting a certain number of (non-parallel) hyper-planes, each of which is associated with a particular neuron unit in a layer. However, the multi-layer perceptron is not compact since it requires $O\left(2^n/n^2\right)$ correlation units, each having n synapses.

In what follows we will take a different approach in finding the discriminant function. According to our approach, (3) provides a unique projection from the n-dimensional input space into a one-dimensional real axis. This projection corresponds to only one linear neuron unit, exactly as in the case of the linear perceptron. What changes with respect to the (linear) perceptron is the discriminant function. By choosing a *canonical piecewise-linear* (PWL) function [Chua & Kang, 1977], a simple design algorithm can provide a guaranteed solution for realizing *any* Boolean function. It is just a matter of additional optimization to find a better orientation vector so that the number of *absolute value* function terms in the PWL discriminant is minimized. In any case, a guaranteed solution always exists and by additional optimization, an extremely compact realization can be achieved.

By using the canonical PWL cell model (3)-(4), the complex separating hyper-surface $w(\mathbf{u})=0$ is replaced by a collection of *parallel hyper-planes:* $\sigma=\mathbf{bu}^{\mathrm{T}}=t_k$, $k=1,..,tr$ where t_k is one of the "*tr*" real roots of the scalar non-linear equation $w(\sigma)=0$. The sign of the output function will remain invariant for any input vector, which lies between such planes, but changes to the *opposite* sign when an input vector crosses a separating hyper-plane. From this perspective, the class of linear separable functions is just a special case, when there is only one transition; namely, $tr=1$. In what follows we will show that for *any* Boolean functions there is at least a *default orientation* so that a solution based on "*tr*" separating hyper-planes always exists. Moreover, the number "*tr*" of parallel separating hyper-planes can be minimized by additional optimization of the *orientation vector* so that, if the function is linearly separable, the algorithm converges to *tr=1*. This approach not only allows for a simplification of the discriminant function design, but also gives a much more natural perspective over the distinction between a *linearly separable* and *a linearly not separable* Boolean function. In fact, the optimal *transition number "tr"* (i.e. the minimum number of separating hyperplanes) can be used to characterize the *complexity* of a Boolean function, or of any problem specified by a set of input-output samples.

Remark: Since any of the separating hyperplanes planes is given by the equation: $\sigma = \mathbf{bu}^{\mathrm{T}} = t_k$ it follows that the orientation vector **b** is perpendicular to these planes. Indeed, let us consider an arbitrary oriented line-segment $\mathbf{u}^{\alpha\beta} = \mathbf{u}^{\beta} - \mathbf{u}^{\alpha}$ lying on a separating hyperplane, joining the point $\mathbf{u}^{\alpha} = \left[u_1^{\alpha}, u_2^{\alpha}, ..., u_n^{\alpha}\right]$ with the point $\mathbf{u}^{\beta} = \left[u_1^{\beta}, u_2^{\beta}, ..., u_n^{\beta}\right]$. Since both $\mathbf{u}^{\alpha}$, and $\mathbf{u}^{\beta}$ belong to the same hyperplane, it follows that $\mathbf{b}(\mathbf{u}^{\alpha})^{\mathrm{T}} = t_k$, and $\mathbf{b}(\mathbf{u}^{\beta})^{\mathrm{T}} = t_k$. By subtracting the former equation from the latter equation it follows that $\mathbf{b}(\mathbf{u}^{\beta} - \mathbf{u}^{\alpha})^{\mathrm{T}} = t_k$, i.e. **b** is perpendicular to any line-segment in a separating hyperplane, and therefore **b** is an orientation vector perpendicular to all these planes.

Example 1:

Let us consider first a simple linearly non-separable example; namely, the "Parity2" function. Its graphical representation in the input space is shown in Fig.1. It consists of 4 vertices, 2 colored in blue and two colored in red.

Let us define a *projection axis* as a straight oriented line (parallel to the orientation vector **b**) passing through the origin of the input space and which is perpendicular to $tr \geq 1$ separating hyper-planes. In Figs 1-3 it is depicted as a bi-color "line-segment" intersecting the origin. The positive semi-axis (upper half of the line segment) is colored in light cyan, while the negative semi-axis (lower half line-segment) is colored in green. The *projection axis* is free to "rotate" around the origin of the *n*-dimensional input space.

Orientations

The orientation of the *projection axis* in the input space is specified by the *orientation vector* $\mathbf{b} = \left[b_1, b_2, ..., b_n\right]$.

Several different orientations, as well as their corresponding *orientation* vectors are shown in Fig.1 (a)-(c). Observe that the same vertices are projected onto *different* positions on the *projection axis* corresponding to different orientations. In what follow we will introduce several additional mathematical tools.

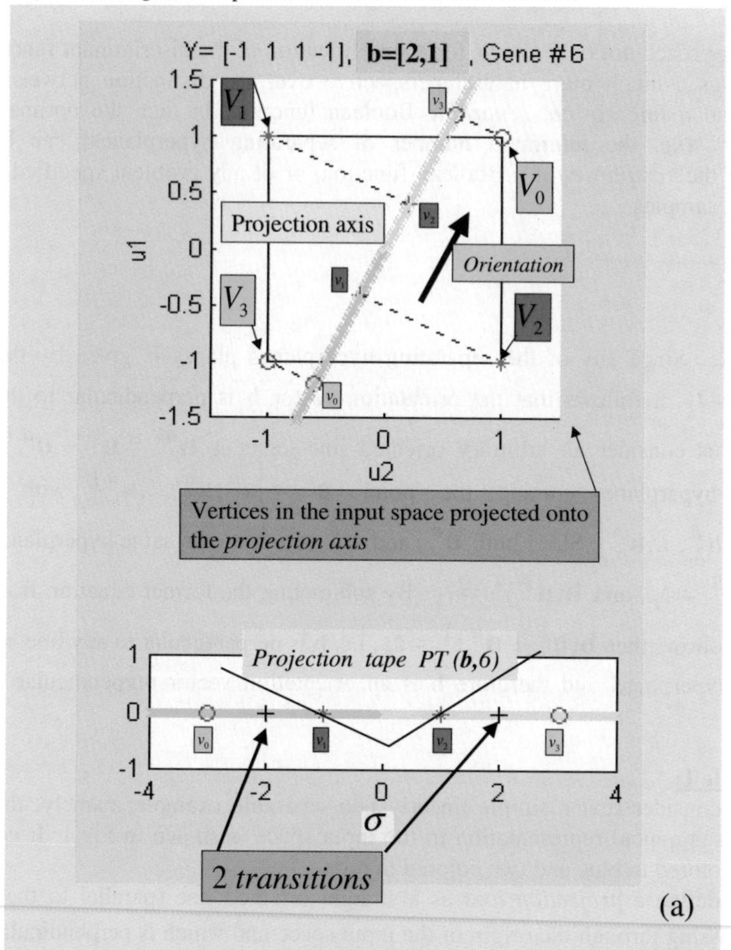

Fig.1:Piecewise-linear realizations of the Boolean function "Parity2" with 2 variables. The upper plot represents the input square, where each vertex represents a particular input vector and their color assignments define the output Boolean function. The projection tape associated with the orientation vector b and the orthogonal vertices projections of the 4 vertices V_0, V_1, V_2, and V_3 on the tape are transcribed into the lower plot. (a) A valid projection tape with 2 transitions for the Boolean function $Y = [-1,1,1,-1]$ (ID=6, or "Parity2"), defined by the *default orientation vector* $\mathbf{b} = [2,1]$.

Projection tapes

In the n-dimensional input space, a Boolean function is entirely specified by the set of pairs (vertex, color of vertex): $\{(V_0, Y(V_0)), (V_1, Y(V_1)), \ldots, (V_{N-1}, Y(V_{N-1}))\}$, where $N = 2^n$. The problem of realizing a given Boolean function is to find a n-dimensional discriminant function so that it will separate the "red" vertices from the "blue" ones. The problem can be dramatically simplified by simply projecting the vertices and their colors onto the *projection axis.* The result is called a *projection tape* (PT) where scalar projection of the vertices and their colors are transcribed onto the *projection axis,* in accordance with the following simple *projection transcription rule*:

Define
$$v_j = \sum_{i=1}^{n} b_i u_{ij} = \mathbf{b} \cdot \mathbf{V}_{\alpha_j}{}^T \quad , \quad j = 0,..N-1 \tag{7}$$
where $\alpha_j \in \{0,1,..N-1\}$ denotes the index of the vertex V_{α_j} where "j" is arranged in the same order as: $v_0 \le v_1 \le .. \le v_j \le .. \le v_{N-1}$. Therefore, the *projection tape PT* is specified by the set:
$$PT = \{ (v_j, Y(V_{\alpha_j})) \,|\, j = 0,..N-1 \} \tag{8}$$

Observe that a projection tape *PT* consists of *N* pairs, each pair being composed by a scalar representing the *excitation* associated with an input vertex and by a binary number representing the desired output (plus or minus) of the cell for that input vertex. Therefore, the *projection tape* $PT(B, ID)$ depends not only on the orientation of its associated projection axis but also on the associated Boolean function (specified simply by its identification number *ID*). In fact, the *projection tape* is a complete representation of a Boolean function of *n* variables in a *one-dimensional space* where the input variable for the discriminant function is the scalar *excitation* $\sigma = \sum_{i=1}^{n} b_i u_i = \mathbf{b} \cdot \mathbf{u}^{\mathrm{T}}$. Using this simple transformation, the task of finding a discriminant function to realize an arbitrary Boolean function reduces to that of finding a non-linear function $w(\sigma)$ of *one* variable σ, so that $\mathrm{sgn}(w(v_j)) = Y(V_j), \forall j$.

In each of Figs. 1-3, a specific Boolean function is chosen to illustrate the concept of a *projection tape,* which is represented both by a hypercube associated with the Boolean function, and by box containing the projection tape and an acceptable discriminant function, below each hypercube representation.

Default orientations

For the case of the "Parity2" function, let us start with the orientation $\mathbf{b} = [2,1]$, called a *default orientation,* which will be defined below for the general case for an arbitrary number of inputs. Observe in Fig.1 (a) that the resulting *projection tape* is characterized by four equidistant projected vertices: $PT(B,6) = \{(v_0,-1),(v_1,+1),(v_2,+1),(v_3,-1)\}$, where { $\alpha_0 = 3$, $v_0 = -3$ }, { $\alpha_1 = 2$, $v_1 = -1$ }, { $\alpha_2 = 1$, $v_2 = +1$ }, { $\alpha_3 = 0$, $v_3 = +3$ },

Let us first observe that:

$$\alpha_j = 2^n - 1 - j \tag{9}$$

and

$$v_j = -2^n + 1 + 2j \tag{10}$$

where $j = 0,..N-1$. This follows from our choice of the *default orientation* specified in (6), which in the general case of *n* inputs is defined by:

$$\mathbf{b} = \{2^{n-1}, 2^{n-2}, ..., 2^0\} \quad (11)$$

Remark: Since the default orientation will always project a vertex from the "hypercube" input space onto a *unique point* on the projection tape, there is no overlap with other projected vertices. Therefore the default orientation defined by (11) will always lead to a *valid projection tape* defined as one having no overlapping projection point.

On any valid projection tape one can identify *transitions* between intervals where consecutive projected vertices are labeled in "red", and intervals where consecutive projected vertices are labeled in "blue". We will see in the following, that this *transition map,* and its special case, called *a robust transition map,* are key concepts towards defining a discriminant function. The *default orientation* is not the only orientation leading to a *valid decoding tape* [Chua, 1998]; In fact most of the arbitrarily chosen orientations will lead to such a tape. For example, any permutation, and/or inversion (multiplication with –1), of elements in (11) will lead to a valid *projection tape,* as can be easily checked.

Valid and non-valid projection tapes

Note that in general, the number of distinct projected vertices on a *projection tape* can be smaller than *N.* For example, in Fig. 1(b), using the orientation $\mathbf{b} = [1,1]$ there are only three distinct points on the projection tape since $(v_1, 1) = (v_2, 1) = (0,1)$. Observe that in this case, the two vertices $Y(V_{\alpha_1}) = Y(V_{\alpha_2}) = 1$ have the same color and therefore projecting them from the hypercube onto *the same* point on the projection axis does not lead to a *conflicting situation.* However, in some cases, different vertices having *different* colors may project onto the same point on the *projection axis,* thereby leading to a *conflicting* situation, henceforth called a *non-valid projection tape.*

An example of a *non-valid projection tape* is shown in Fig. 1(d), where $v_0 = v_1 = -1$ and $v_2 = v_3 = +1$ but where $(Y(V_{\alpha_0}) = Y(V_1) = 1) \neq (Y(V_{\alpha_1}) = Y(V_3) = -1)$ and $Y(V_{\alpha_2}) \neq Y(V_{\alpha_3})$. Such situations are degenerate; they correspond to *failure boundaries*[3] in the parameter space of the orientation vector $\mathbf{b}$. Such failure boundaries are hyper-surfaces separating compact domains characterized by distinct *orderings* $\Omega = \{\alpha_0, \alpha_1, ..., \alpha_N\}$. A non-valid *projection tape* is *not robust,* and any small perturbation in the parameter space will lead to a valid *projection tape* corresponding to one of the domains bordering the failure boundary associated with the non-valid *projection tape.*

---

[3] In [Chua, 1998] a *failure boundary* was defined as a hyper-surface separating parameter domains of a cell which correspond to different Boolean function realisations. We extend this notion here to the case where each separated domain is associated with a class of realisable Boolean functions but where the same order $\Omega = \{\alpha_0, \alpha_1, ..., \alpha_N\}$ is preserved.

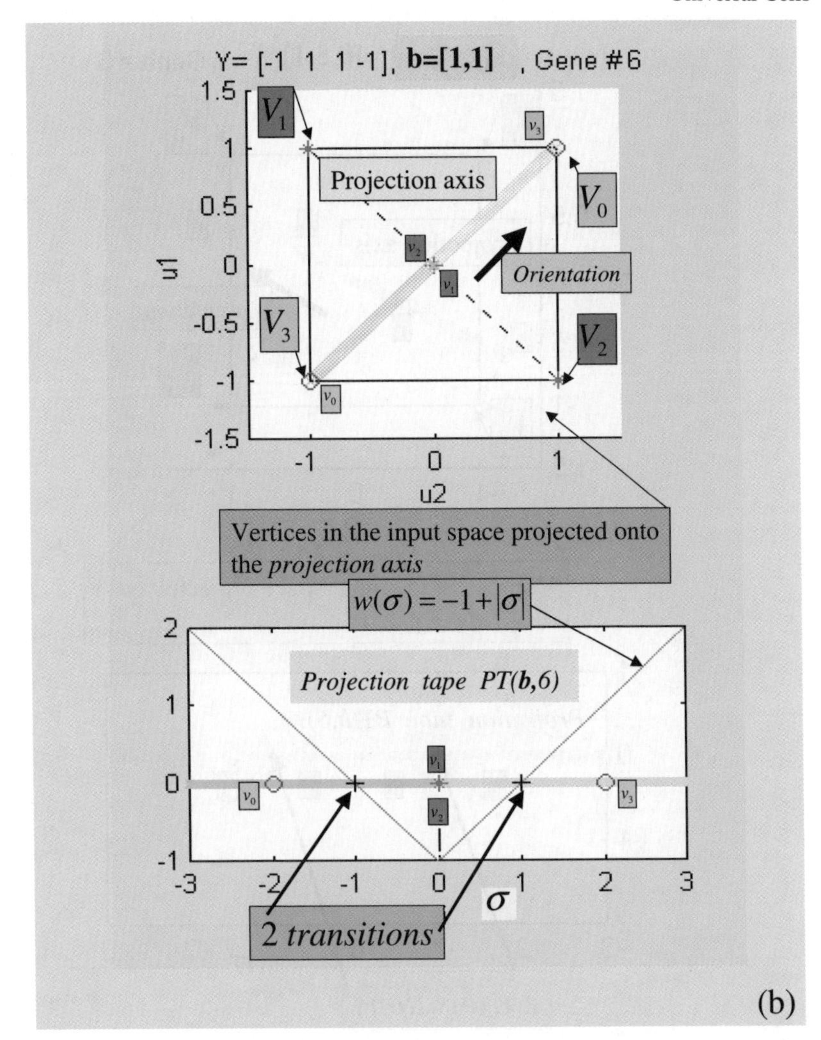

Fig. 1(b). Another valid projection tape with 2 transitions, defined by the *orientation vector* $\mathbf{b} = [1,1]$. Observe that in this case, two different vertices $(V_1, \text{and } V_2)$ having the same color in the input space are projected onto the same point on the projection tape.

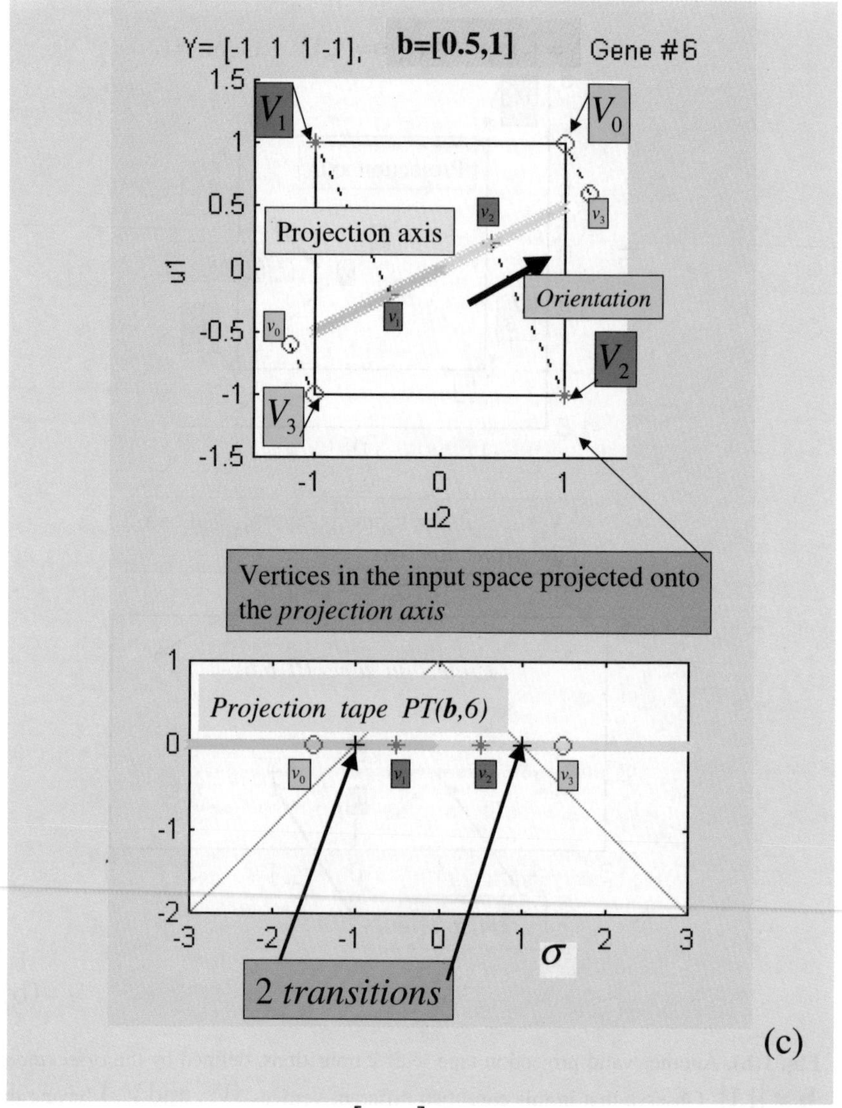

Fig. 1(c): The orientation vector $\mathbf{b} = [0.5,1]$ also leads to a valid canonical piecewise-linear realization of the "Parity2" Boolean function, using only one absolute value function.

The situation illustrated in Fig. 1(d), for **b**=[0 1] corresponds to a point on the failure boundary separating a domain for which $\Omega = \{3,1,2,0\}$ from a domain where $\Omega = \{1,3,0,2\}$. A point in the former domain corresponds to the valid *projection tape* shown in Fig. 1(c). It is interesting to observe that the valid *projection tape* in Fig.1(b) was obtained using the orientation vector $\mathbf{b} = [1,1]$ which lies on a separating boundary between the domains associated with $\Omega = \{3,1,2,0\}$ and $\Omega = \{3,2,1,0\}$, respectively. However, this failure boundary is not a failure boundary with respect to the objective function of interest; namely the *number of valid transitions*.

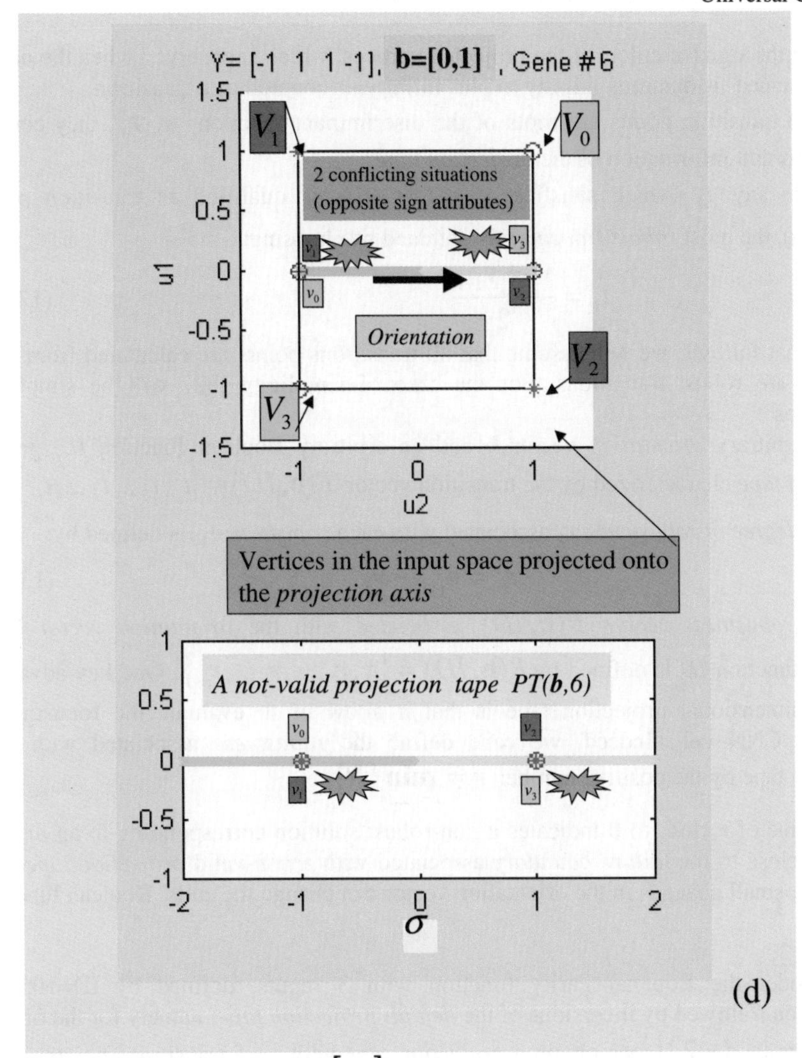

Fig. 1(d): The orientation vector $\mathbf{b} = [0,1]$ leads to a *conflicting situation*: Vertices from the input square having different colors are projected onto the same point on the projection tape. Such an orientation vector is not acceptable since it leads to a non-valid projection tape.

Indeed, even if the order of the projected vertices changes when the orientation vector is rotated and passed over the boundary $\mathbf{b} = [1,1]$ from Fig1(a) to Fig 1(c), via Fig 1(b), we found that the number of transitions remains constant (equal to 2) in this example and therefore with respect to the "*number of transitions*" objective function, the orientation vector $\mathbf{b} = [1,1]$ does not lie on a failure boundary.

Transitions and robust transitions

In a valid decoding tape, a *transition* exists if $Y(v_j) \neq Y(v_{j+1})$ and is defined as *any* real number t_k, lying between v_j and v_{j+1}; i.e., $v_j < t_k < v_{j+1}$. In other words, *every* change in the sign (or color) of the projected vertices, which is observed when the *projection tape* is scanned from minus infinity to plus infinity, is counted as a transition.

Since transition points are roots of the discriminant function $w(\sigma)$, they contain the most important information of the *projection tape*.

While any t_k which satisfies $v_j < t_k < v_{j+1}$ is qualified as transition point (by definition), the most *robust transition* is obtained by choosing:

$$t_k = \frac{v_j + v_{j+1}}{2} \tag{12}$$

In what follows, we will assume that all transitions points are calculated from (12) and therefore are robust transitions. For the sake of simplicity they will be simply called "transitions".

An arbitrary *orientation* vector **b** and an arbitrary Boolean function *ID,* generate a *projection tape* characterized by the transition vector $T(\mathbf{b}, ID) = \{t_1, t_2,.., t_k,..., t_{tr}\}$.

The *degree of robustness* r_k associated with each *transition* t_k is defined by:

$$r_k = v_{j+1} - v_j \tag{13}$$

The *robustness vector* $R(\mathbf{b}, ID)$ associated with the *orientation vector* **b** and a Boolean function *ID* is defined by $R(\mathbf{b}, ID) = \{r_1, r_2,..., r_k,..., r_{tr}\}$. One key advantage of the one-dimensional projection tape is that it allow us to evaluate the robustness of a particular CNN cell. Indeed, we cane define the *robustness* associated with a given projection tape by the positive number $r = \min_{k=1,...,tr}\{R\}$.

A value of *r* close to 0 indicates a non-robust solution corresponding to an *orientation* which is close to the *failure boundary* associated with a non-valid *projection tape.* In such situations, small changes in the orientation vector can change the cells' Boolean function.

Example 2:

Consider the Boolean Parity function with 3 inputs, defined by ID=105. For a permutation followed by inversions of the *default projection tape*; namely for the orientation vector $\mathbf{b} = [-4, -2, 1]$, we obtain a *projection tape* with *tr=5* transitions points, as shown in Fig. 2(a). In this case, the transition vector is given by $T = \{t_1, t_2, t_3, t_4, t_5\} = \{-6, -2, 0, 2, 6\}$. Together with the sign of the first projected vertex projection: $s = Y(v_0) = Y(V_{\alpha_0})$, the pair (s, T) constitutes a complete set of parameters required by the design algorithm to be developed in Section 3.2.3. Therefore the discriminant functions $w(\sigma)$ associated with a prescribed Boolean function, and a prescribed orientation vector **b** is completely specified by this set of parameters.

The simplest choice for a discriminant function, from a pure mathematical point of view, is a polynomial. Since the roots of the discriminant function are the transition points, the polynomial can be immediately written without any auxiliary design algorithm as follow:

$$w(\sigma) = s(-1)^{tr}(\sigma - t_1)(\sigma - t_2)...(\sigma - t_{tr}) \tag{14}$$

In our previous example, the color of v_0 is "blue" (corresponding to $s = -1$), and the there are 5 transitions for the orientation specified above. Since the robustness vector in this case is composed of 5 equal components, each one equals to 2, we have $r = 2$. Therefore, a valid discriminant function is represented by the polynomial: $w(\sigma)=(\sigma-6)(\sigma-2)\sigma(\sigma+2)(\sigma+6)$. Although useful as a mathematical construct, a polynomial discriminant function is not attractive in practical implementations since it requires many additional multipliers. Instead, a piecewise linear discriminant can perform the same task with the *addition* of only (*tr-1*) *absolute value function* units, but without any multiplication. Since an absolute value operator is much simpler to implement in current electronic technologies than multipliers, the piecewise linear discriminant function is a better choice from the perspective of electronic realizations. It also leads to a simpler mathematical analysis when used in CNN systems. A complete description for designing piecewise-linear discriminants via the theory of canonical piecewise-linear representations presented in [Chua & Kang, 1977] is given in the next section.

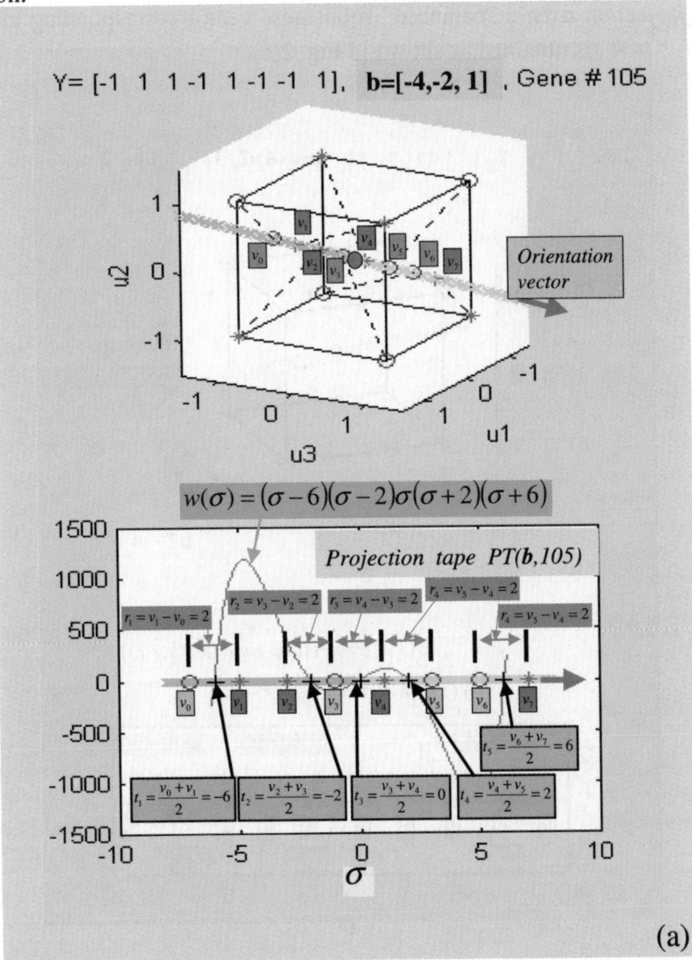

Fig. 2. Transitions, robustness and discriminant functions for the realization of Boolean logic with piecewise-linear CNN cells. The case of "Parity3" function (with 3 inputs), $Y=[-1,1,1,-1,1-1,-1,1]$ is considered. In Figs. 2(a)-(c), the upper plot represents the input (hyper)cube, where each vertex corresponds to a particular input vector and the color assignment of all vertices defines the Boolean function. The lower plot represents the projection tape, where all data required by the realization algorithm are clearly marked: (a) An inversion of the default orientation $\mathbf{b}=[-4,-2,1]$ leads to a projection tape with 5 transitions, as shown in the lower plot.

Finding the optimal orientation

Finding the optimal orientation vector $\mathbf{b}^*$, is geometrically equivalent to "rotating" the *projection axis* about its origin in the *n*-dimensional hyper-cube input space with the goal of minimizing the number of transitions *tr* (coarse optimization). This process may be followed by a "fine optimization", where the goal is to maximize the *robustness* parameter *r*. Such a process is illustrated in Fig.2 (a),(b),(c), where the "rotation" is generated by picking an arbitrary value for the *orientation* vector **b**. Observe that the number of transitions reduces from 5 in Fig2.(a) to 3 in Fig.2(b). The situation in Fig. 2(b), however, corresponds to an "unbalanced" robustness vector $R = [4,2,4]$, with a robustness of $r = 2$. By further rotating the *projection axis,* a "balanced" robustness vector corresponding to an optimized robustness of $r = 4$, is obtained as shown in Fig. 2(c).

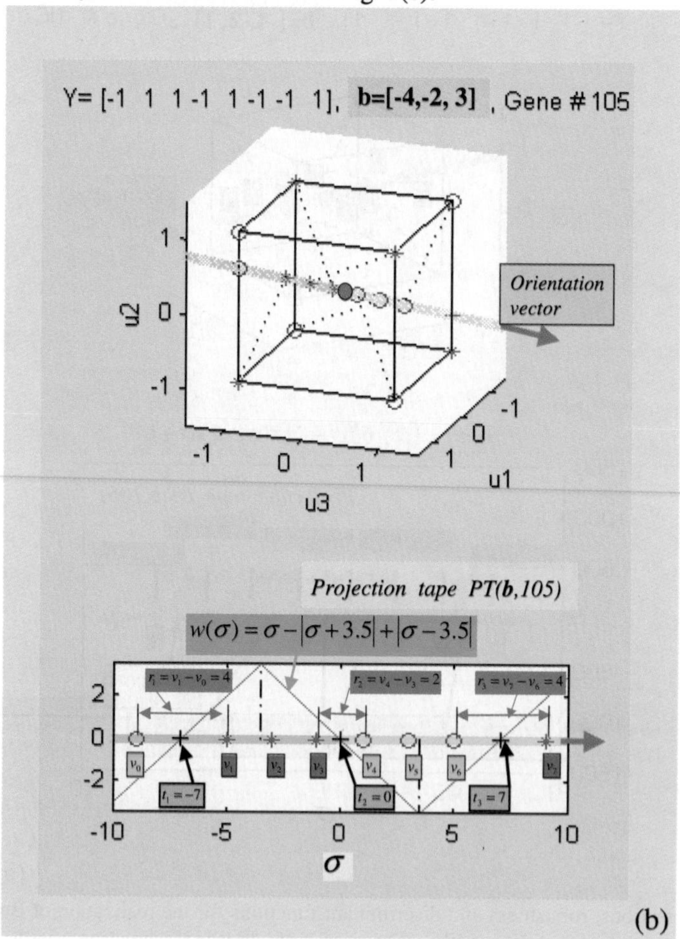

Fig.2(b). The number of transitions can be reduced, by rotating the projection axis: Only three transitions are found on the projection tape defined by the orientation vector $\mathbf{b} = [-4,-2,3]$. The associated piecewise-linear discriminant function is a special case of Eq. (22) obtained by applying the realization algorithm given by Eqs. (23) to (26). Observe that the robustness vector is unbalanced and hence further improvement of the orientation vector is still expected.

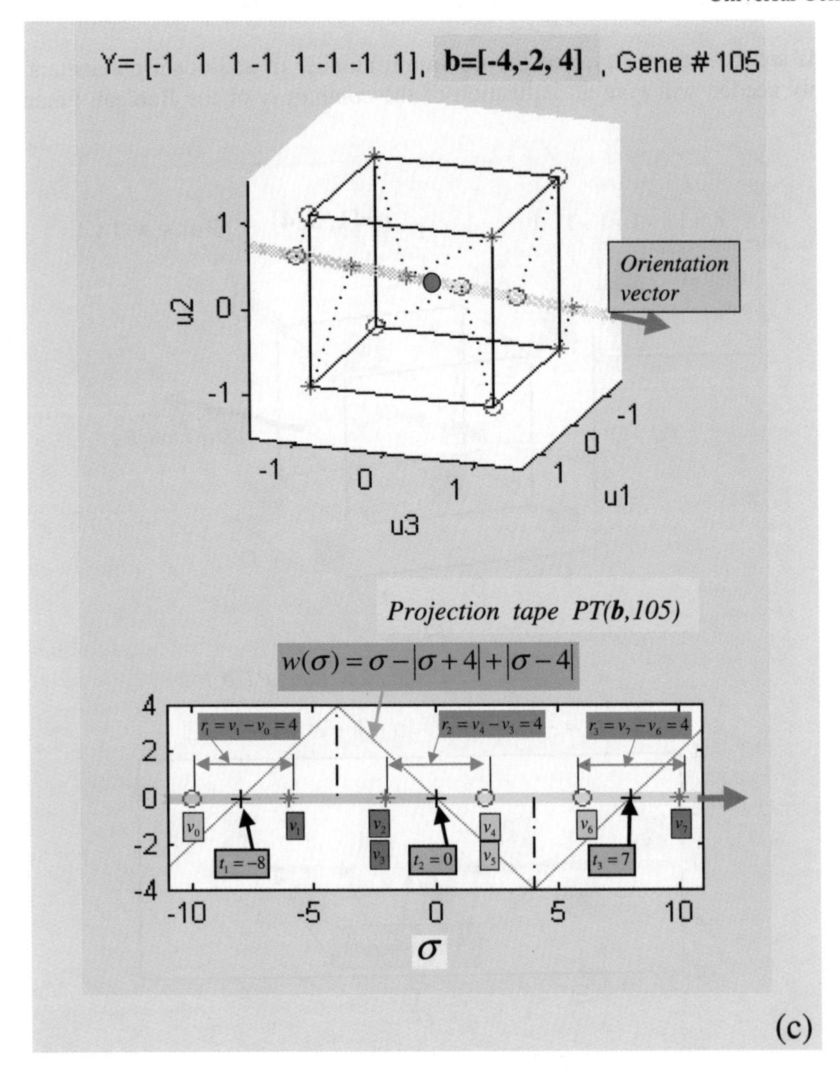

Fig.2(c) The near optimal orientation vector $\mathbf{b} = [-4,-2,4]$ leads to a balanced robustness vector while keeping the number of transitions minimal (equal to 3).

Example 3:

Consider next the case of a linearly separable function (ID=142). Starting from the arbitrary *default orientation vector* $\mathbf{b} = [1,2,4]$, the number of transitions is $tr = 5$, as shown in Fig.3 (a). Since we know in advance that this particular function is linearly separable, it is clear that this solution is not an optimal one. However, even for this not-optimized orientation, there exists a valid realization with the following discriminant function: $w(\sigma) = -\sigma + |\sigma + 3| - |\sigma + 1| + |\sigma - 1| + |\sigma - 3|$.

In fact, this is an important advantage of our approach; namely, that it offers a fast, valid solution even if it may not be optimal. Starting from this non-optimal solution and by using a global optimization algorithm, the orientation can be improved so that the number of additional parameters and absolute value functions is reduced as much as possible. Another aspect of our optimization approach is that both linearly separable and linearly *not* separable functions are treated equally, and the minimum number of additional parameters m that is ultimately needed will give an indication of the *complexity* of the Boolean function being realized.

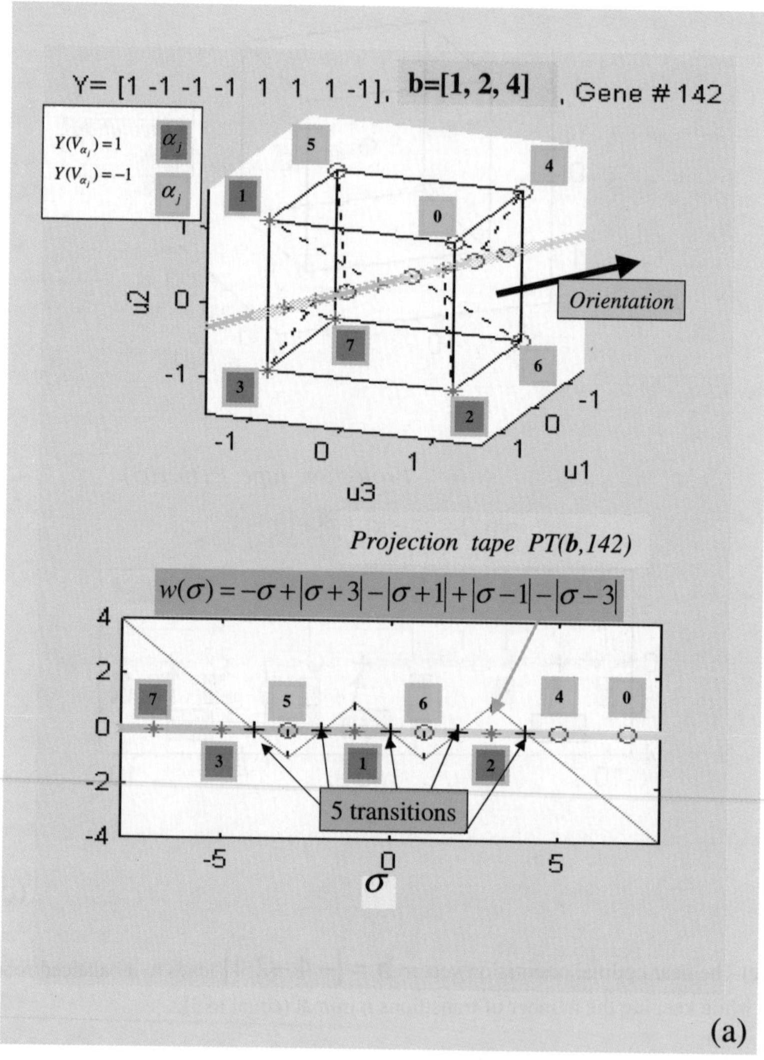

Fig. 3(a) The default orientation (non-optimal) leads to 5 transitions.

To optimize the *orientation vector,* one simple approach is to try all possible permutations and inversions of the initial (default) orientation vector. In this example, if we choose the orientation vector $\mathbf{b}=[-4,2,1]$, we would obtain a smaller number of transitions in the projection tape, in this case $tr=3$, as shown in Fig.3 (b). While in many cases this simple optimization procedure leads to a better solution, it does not necessarily lead to the optimal one, besides the time required to try all permutations is rather large, of $O(2^n n!)$. Experimentally was found that a near-optimal solution (in terms of *tr*) can be found much faster if *random* values are simply given to the orientation vector. Among a "population" of random mutations, the set of orientations minimizing *tr* is selected. This algorithm can be applied further using specific techniques of *genetic algorithms* [Koza, 1994], which are particularly well suited for this problem.

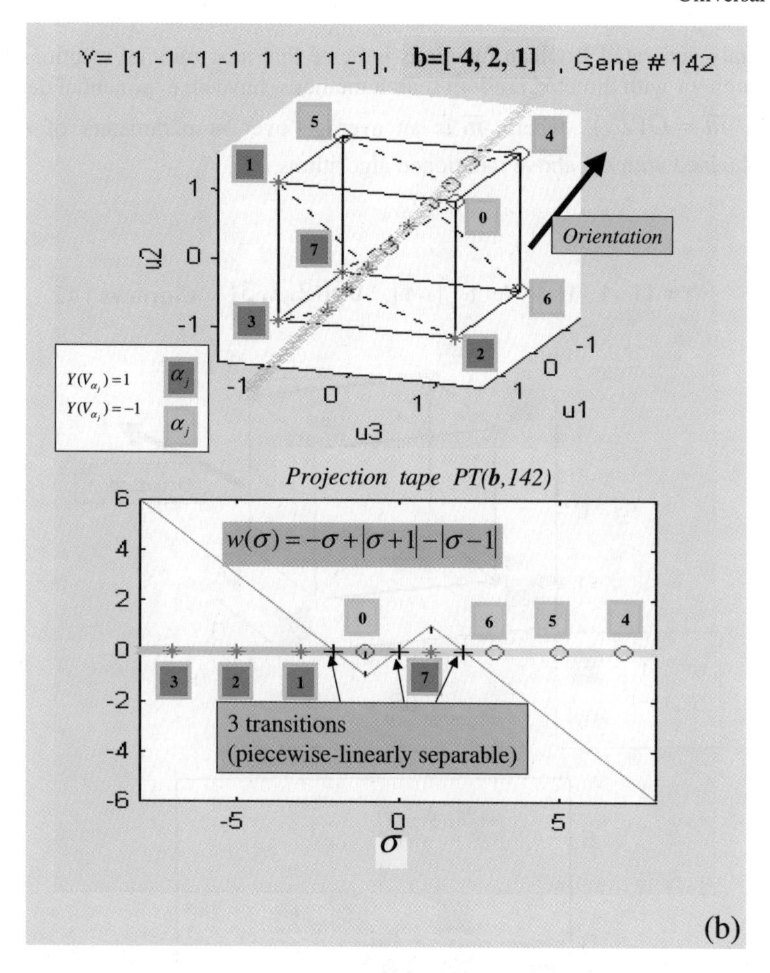

Fig. 3(b). A near-optimal orientation obtained by permutations and inversions reduces the number of transitions to three.

Applying such an approach to the above example, the optimal orientation vector $\mathbf{b}^* = [-2,2,3]$ is obtained. It corresponds to the *linearly separable* solution ($tr = 1$) shown in Fig3 (c). In this case the discriminant function is still a canonical piece-wise linear form, but with no absolute value terms; namely, $w(\sigma) = -\sigma$.

The problem of seeking an optimal *orientation* is a classical optimization problem where the goal function (the number of transitions *tr,* or a combination of *tr* and the *robustness r*) has an *unknown* dependence on the parameters to be optimized. In such cases, we can apply *genetic and evolutionary algorithms* [Koza, 1994], as well as techniques based on *directed random search* and *reinforcement* of the type described in [Harth & Pandya, 1988].

An open question still remains whether there is an analytical formula capable of producing the optimal orientation for an arbitrary Boolean function with *n* inputs. Another open question is to find a lower bound for the dependency between the number of transitions, (or the parameter *m* in (3)-(4)) and the number of inputs. Our estimative results with randomly generated Boolean functions indicate that near-optimal solutions found with random search, or with directed random search methods, have an exponential dependence on *n; namely,* $\overline{m} = O(2^n)$, where $\overline{m}$ is an average over *m* parameters of *near-optimal* solutions obtained with the above mentioned algorithms.

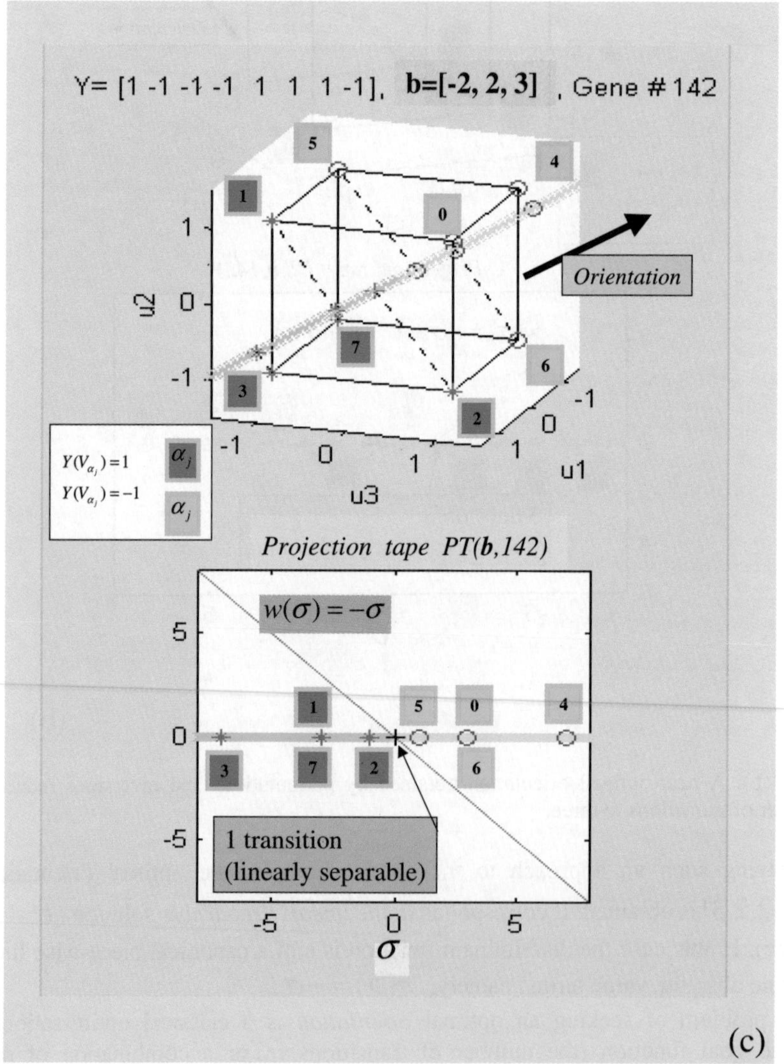

Fig. 3(c). The optimal orientation is found in this case using a stochastic optimization technique. For this orientation vector the minimum number of transitions (one) was reached indicating that the Boolean function is linearly separable.

Optimal orientations for totalistic and semi-totalistic Boolean functions

There are some special classes of Boolean functions, such that an optimal orientation can be easily found. We will discuss two such cases; namely, the *totalistic* and *the semi-totalistic functions.*

Totalistic functions were defined in [Wolfram, 1984] as Boolean functions where the output depends only on the *sum* of the input variables. While in his work Wolfram usually assumes linearly separable totalistic functions (his discriminant function being a linear one) we will extend this terminology in his paper to any piecewise linear discriminant function with an arbitrary number of transitions on its associated *projection tape.* By definition, any totalistic Boolean function has already optimized its *orientation vector* by restricting it to $\mathbf{b} = [1,1,..,1]$, i.e. all components are 1.

Example 4:
The cells implementing the Parity4 function[4] are known to posses interesting duplication properties when implemented on a Cellular Automata [Toffoli & Margulos, 1987]. The *default orientation vector* $\mathbf{b} = [8,4,2,1]$ yields a projection tape with $tr = 10$ transitions, as shown in Fig.4 (a).

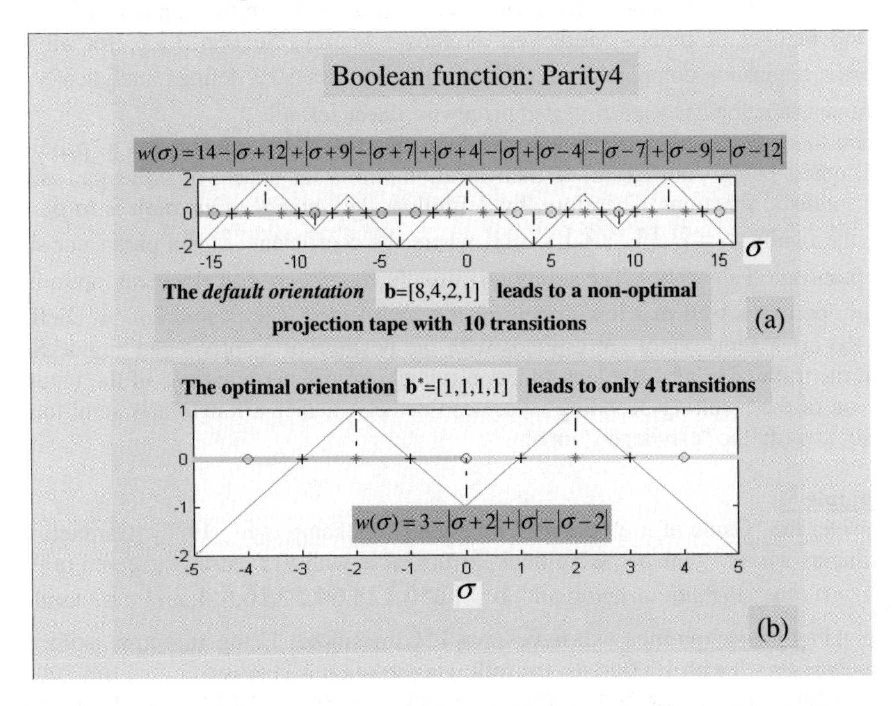

Fig. 4. Two realizations of the universal PWL CNN cells for implementing the Boolean function "Parity4" (Parity with four inputs). (a) The default orientation vector leads to a non-optimal solution with 10 transitions; (b) Since the Boolean function is totalistic the optimal realization is by definition $\mathbf{b}^* = [1,1,1,1]$. The resulting projection tape now has only 4 transitions.

No permutation and/or inversion of the *default orientation* vector can give a better result, as expected since in the case of totalistic functions, the output *does not depend on the* specific position of the inputs. However, since the function is totalistic (a simple test can be made, by allowing all inputs to be permuted and observing that the associated decoding tape

[4] The "Parity4" cell returns +1 if and only if there is an odd number of inputs (coming from north, south, east and west neighbouring cells) in the +1 state. Otherwise, the cell returns −1. This cell allows pattern replication in a generalized cellular automata [Chua, 1998] where the cell inputs are connected to the "east", "west", "north" and "south" neighboring cell outputs.

does not change), the *optimal orientation vector* is $\mathbf{b} = [1,1,1,1]$. Indeed, in this case, the resulting *projection tape* will have only $tr = 4$ transitions, as shown in Fig.4 (b) where realizations via the canonical PWL discriminant function corresponding to both optimal, and a not-optimal orientation are presented. The optimal canonical PWL discriminant function for implementing the Parity4 function is: $w(\sigma) = 3 - |\sigma + 2| + |\sigma| - |\sigma - 2|$.

It can be easily prove that for the general case of the "Parity" function with n inputs, the orientation vector associated with the definition of the totalistic function is an optimal one, with $tr = n$ transitions.

This result is very important because it contradicts some widely accepted reasoning, which considers that the class of Parity functions as one of the most "complicated" Boolean function. In fact, this class of functions yields a "complexity" of the associated piece-wise linear cell which grows only *linearly* with the number of inputs, while randomly selected Boolean functions were found to have an *exponential increase* in the number of transitions with the number of inputs. Moreover, as shown later in Section 3.2.4, for all Parity functions a realization complexity of $O(\log(n))$ can always be defined analytically if the discriminant function has a *multi-nested* piecewise linear formula.

Semi-totalistic functions are functions where only one of the n inputs has a "privileged" role, all others being "indifferent" to their location within the sphere of influence, as in the case of totalistic functions. Therefore, by definition, the *optimal orientation* is to be found among the family $\mathbf{b} = [1,1,1,..,\lambda,1,1,..,1]$, where the coefficient λ is a parameter subject to an optimization algorithm. The solution is usually found very fast since any optimization algorithm performs well in a low dimensional search space. The position of λ coefficient within the orientation vector corresponds with the position of a "privileged" input. Simple tests on the truth table of a Boolean function (which require permutations of the inputs and inspection of the resulting decoding tapes) can detect whether a function is semi-totalistic and if so, identify the "privileged" input.

Example 5:

Consider the "Game of life" Boolean function [Berlekamp *et al.*, 1982] [Gardner, 1970] with 9 inputs whose "gene decoding book" (truth table with 512 entries) is given in [Chua, 1998]. If the *default orientation* $\mathbf{b} = [256,128,64,32,16,8,4,2,1]$ is used, the corresponding *projection tape* will have $tr = 156$ transitions. Using an optimization based on a *random search* with 1000 trials, the following solution is obtained:

$$\mathbf{b} = [-15103,-14370,-6967,-10319,-13535,-5599,-14752,-5833,-14912],$$

leading to $tr = 79$ transitions, which is still far from being optimal. However, when we take advantage of the observation that the Boolean function "Life" is *semi-totalistic*, and take an arbitrary value $\lambda = -8$ a dramatically improvement is observed, the near-optimal orientation vector $\mathbf{b} = [1,1,1,1,-8,1,1,1,1]$ leading to a projection tape with only 4 transitions, as shown in Fig.5(a). This already good result can be further improved by carefully optimizing the parameter λ. In our case a simple search from –8.0 to +0.5with increments of size 0.5 was used. The resulting *optimal orientation* vector $\mathbf{b}^* = [1,1,1,1,0.5,1,1,1,1]$ has only $tr = 2$ transitions, which is the minimum possible[5], as shown in Fig. 5(b).

---

[5] We know that "Life" is *not* a linearly separable Boolean function, therefore the number of transitions on the *projection tape* must satisfy $tr > 1$. Since $tr = 2$ is the smallest value satisfying this property, the associated orientation vector is an optimal one with respect to the number of transitions.

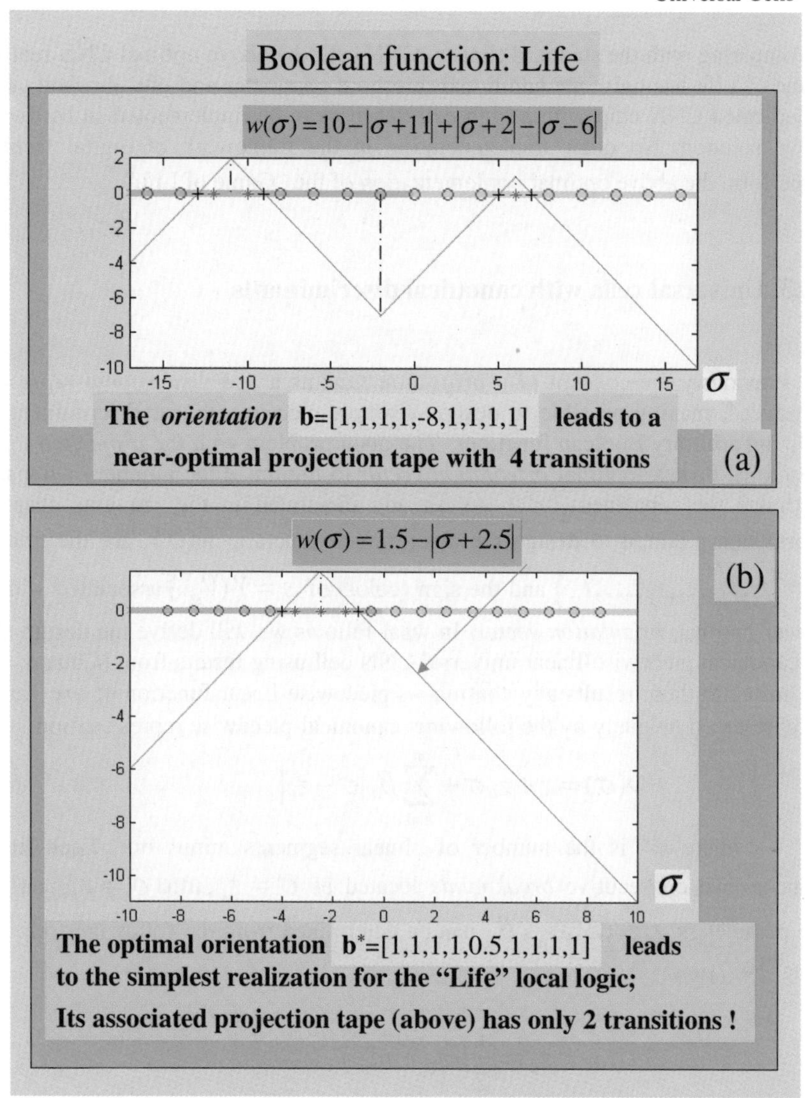

Fig. 5. Two realizations of the PWL CNN cells for implementing the *semi-totalistic* Boolean function "Life" with 9 inputs. (a) An orientation vector taking into account the *semi-totalistic* character of the "Life" function, but where the central element was arbitrarily chosen to be $b_5 = -8$, leads to a near-optimal projection tape with 4 transitions; (b) Further optimization of the central parameter leads to an optimal solution with only 2 transitions with $b_5 = 0.50$.

The discriminant function associated with this *projection tape* is simplified further to:

$$w(\sigma) = 1.5 - |\sigma + 2.5|$$

This is the most compact implementation found so far for a cell realizing the Boolean "Life" function.

Comparing with the standard (linear) CNN cell, the above optimal CNN realization of the Game of Life has only one additional threshold parameter and one absolute value function. A dedicated CNN chip with a high cell density can be implemented in hardware using the above solution. No other implementation in the framework of digital technologies can outperform the above optimal implementation of the "Game of Life".

3.2.3. Universal cells with canonical discriminants

Previously the concept of a *projection tape* as a *one-dimensional representation* was introduced, making possible to design a simple piecewise-linear discriminant function for realizing arbitrary Boolean functions. The main problem with the *projection tapes* is to find an optimal or near-optimal *orientation vector* to minimize the number of transitions and to maximize the robustness "r". As already illustrated in the previous chapter, the only information required to design the discriminant function $w(\sigma)$, are the transition vector $T=\{t_1,t_2,\ldots,t_k,\ldots,t_{tr}\}$ and the sign (color) of $s=Y(V_{\alpha_0})$ associated with an optimal, or near-optimal, *orientation vector*. In what follows we will derive the design algorithm for the canonical piecewise-linear universal CNN cell using results from [Chua & Kang, 1977]. According to these results any continuous piecewise-linear function of one variable σ can be represented uniquely by the following canonical-piecewise representation:

$$w(\sigma)=z+z_0\sigma+\sum_{k=1}^{m}\beta_k|\sigma-z_k| \tag{15}$$

where m is the number of linear segments minus one. Each linear segment connects two consecutive *breakpoints* located at $\sigma=z_k$, and $\sigma=z_{k+1}$ on the σ axis. The parameters $z,z_0,\ldots,z_m,\beta_k$ can be determined from the following formulas [Chua & Kang, 1977]:

$$z_0=\frac{1}{2}(m_0+m_m) \tag{16}$$

$$\beta_k=\frac{1}{2}(m_k-m_{k-1}),\quad k=1,\ldots,m \tag{17}$$

$$z=-z_0\omega-\sum_{k=2}^{m}\beta_k|\omega-z_k| \tag{18}$$

where m_k is the slope of the linear segment joining the breakpoint $(z_k,w(z_k))$ with the breakpoint $(z_{k+1},w(z_{k+1}))$, as shown in Fig. 6, and ω is any root of the discriminant function: $w(\omega)=0$. The slopes m_0 and m_m should be specified explicitly since they correspond to the leftmost and rightmost segments respectively , and therefore have only one breakpoint each.

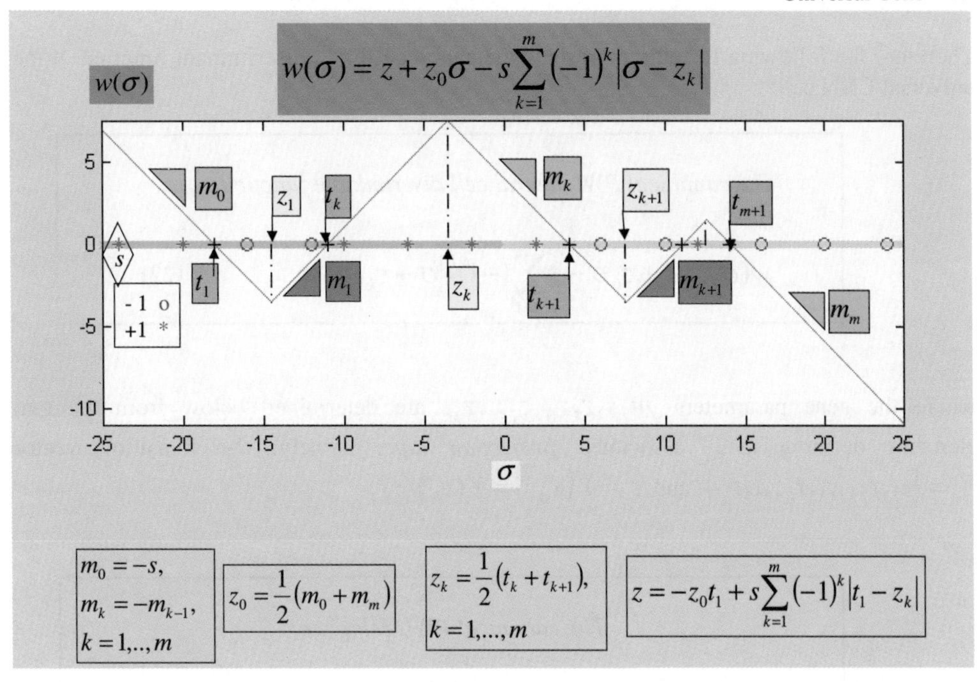

Fig. 6. Realization of the universal CNN cell via a canonical piecewise-linear discriminant function $w(\sigma)$. After a valid and eventually optimal orientation was found, the parameters $\{s, z, m, z_0, ..., z_m\}$ of the discriminant function can be simply determined from the associated *projection tape*.

In order to apply these formulas to our problem, we need to specify the breakpoints z_k as well as their associated slopes m_k. Given the transition points $T=\{t_1,t_2,...,t_k,..,t_{tr}\}$ and $s=Y(V_{\alpha_0})$, the following choice leads to a simple design solution;

$$m_0=-s \tag{19}$$

$$m_k=(-1)m_{k-1},\ \ k=1,...,m \tag{20}$$

$$z_k=\frac{t_k+t_{k+1}}{2},\ \ k=1,...,m \tag{21}$$

It can be easily checked that (19)-(21) leads to a valid discriminant function which intersects the σ axis exactly at the transition points $T=\{t_1,t_2,...,t_k,..,t_{tr}\}$ and having the prescribed sign (color) at the projected vertices. Moreover, it follows that $m=tr-1$.

Therefore the following formula represents the canonical-PWL discriminant function of the universal CNN cell:

The canonical PWL neural cell discriminant function

$$w(\sigma) = z + z_0\sigma - s\sum_{k=1}^{m}(-1)^k|\sigma - z_k| \qquad (22)$$

where the gene parameters $m, s, z, z_0, z_1, ..z_m$ are determined below from the two elements defining the associated *projection tape*; namely, the transition vector $T = \{t_1, t_2, ..., t_k, .., t_{tr}\}$ and $s = Y(V_{\alpha_0}) = Y(v_0)$:

The canonical PWL cell gene:

$$m = tr - 1, \qquad s = Y(V_{\alpha_0}) \qquad (23)$$

$$z_0 = \begin{cases} 0, \text{ if } m \text{ is odd} \\ -s, \text{ if } m \text{ is even} \end{cases} \qquad (24)$$

$$z_k = \frac{t_k + t_{k+1}}{2}, \quad k = 1, ..., m \qquad (25)$$

$$z = -z_0 t_1 + s\sum_{k=1}^{m}(-1)^k|t_1 - z_k| \qquad (26)$$

3.2.4. Compact universal cells with multi nested discriminants

Previously it was proved that any locally Boolean function can be realized with a CNN neural cell described by (3)-(4) where the discriminant function $w(\sigma)$ is defined by the canonical piecewise-linear formula (22). Compared to a standard CNN cell, this realization requires "m+2" *additional parameters* and "m" *absolute value functions*; both are determined by a *simple and exact* design procedure via (23)-(26). An *optimization algorithm* can be applied to find an optimal or near-optimal *orientation* vector **b**, which minimizes the number "m" of additional absolute value functions. It follows from Eq. (23) that the number m depends *linearly* on the number of transitions on the *projection tape*. In the worst case, of a poorly optimized *projection tape,* or of a complex Boolean function, the number of transitions can increase *exponentially* with the number of inputs n. An upper bound is $m \le 2^n$. Since a CNN cell has usually $n = 9$ inputs, the worst case would require up to $2^9 = 512$ additional absolute value functions.

In this section, I will present a method to overcome the disadvantage of a large number of transitions. This approach relies on a novel discriminant function $w(\sigma)$. The *multi-*

nested discriminant is a piecewise linear formula where the number m of absolute value functions and the additional parameters increases only with $\log_2(tr)$, where tr represents as before the number of transitions on the *projection tape.* As conjectured in [Dogaru & Chua, 1998f], any Boolean function can be realized with a multi-nested discriminant requiring only $m = n - 1$ absolute value functions and $m + 2$ additional parameters. The expression of the *multi-nested* discriminant function is given by:

$$w(\sigma) = s\left(z_m + \left|z_{m-1} + \left|\ldots + z_1 + \left|z_0 + \sigma\right|\right|\right|\right) \qquad (27)$$

By replacing the canonical piecewise-linear discriminant function (22) with (27) in the defining equations of the CNN cell (3)-(4) we obtain a cell with only $2n+1$ parameters but which still can realize *any* Boolean function! The payoff for this dramatic reduction in complexity is given by the fact that the 2^m roots of $w(\sigma) = 0$ are *not independent* and therefore, an additional restriction must be imposed on the *orientation* vector. This restriction stipulates that the transitions on the associated *projection tape* must match the root distribution structure specified by *a bifurcation tree* to be defined next. Therefore, unlike in the case of the canonical piecewise-linear CNN cell, for the multi-nested cell the *optimization* of the orientation vector should be tightly coupled with the process of determining the additional parameters $\{s, z_0, z_1, \ldots z_m\}$. Except for the case of totalistic functions, there is currently no explicit algorithm for designing this CNN cell. Instead, optimization algorithms based on directed random search and/or genetic mutations are used to solve this rather hard nonlinear optimization problem.

Bifurcation tree for multi-nested discriminant function

Equation (27) can be recast in the following recursive form:

$$\begin{aligned}
&w_0(\sigma) = z_0 + \sigma \\
&\text{FOR } k = 1,..m \\
&\qquad w_k(\sigma) = z_k + \left|w_{k-1}(\sigma)\right| \qquad (28)\\
&\text{END} \\
&w(\sigma) = s w_m(\sigma)
\end{aligned}$$

where k is an index associated with the *level of "nesting"*. Observe that at each level of nesting "k", the number of linear segments of the intermediate discriminant function $w_k(\sigma)$ is double to that of $w_{k-1}(\sigma)$. In particular, when $m = 0$, (27) and (28) correspond to the linear discriminant function (with one linear segment). After $k = m$ levels of "nesting", the resulting discriminant function $w(\sigma) = w_m(\sigma)$ will contain 2^m segments, as shown in Fig.7.

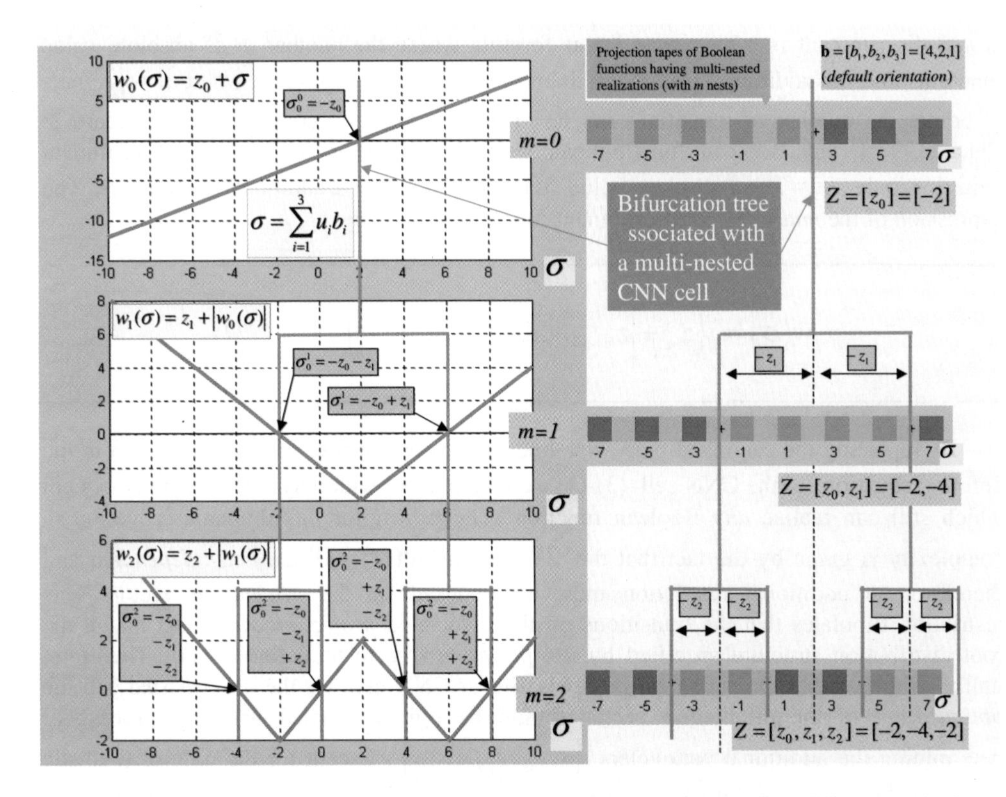

Fig. 7. Bifurcation tree for a universal multi-nested CNN cell. For each additional level of "nesting" the number of roots (transitions) of the discriminant function *doubles* instead of incrementing by one, as in the case of the canonical PWL cell. In this example we considered a Boolean function with 3 inputs and two levels of nesting. Therefore, given an appropriate orientation vector, the multi-nested discriminant leads to a more compact representation. However, the positions of the transitions on the projection tape are now restricted to lie within a configuration determined by a *bifurcation tree.* This leads to a more complicated procedure for finding the optimal orientation and the remaining cell parameters. Each level of nesting corresponds to a "branching" in the bifurcation tree corresponding to "root doubling" reminiscent of the "period doubling" scenario in nonlinear dynamics.

Since it is a piecewise-linear function, (27) admits always a canonical PWL representation. However, instead of 2^m absolute value function required by the canonical representation, due to the recursive form of (28), only m absolute value functions are now required. The converse is not true, i.e. not every function described by a canonical PWL representation admits a *multi-nested* piecewise linear form (28). This "irreversibility" is reflected by a special structure of the roots of the equation $w_m(\sigma) = 0$. In contrast to the canonical PWL representation where its roots are independent, in the case of (28) the roots are subject to the following constraints : By choosing the parameters $z_0, z_1, \ldots, z_m$ so that:

$$|z_m| < |z_{m-1}| < \ldots < |z_1|, \text{ and} \tag{29}$$
$$z_1, z_2, \ldots, z_m < 0$$

it can be easily proved that $w_m(\sigma) = 0$ has 2^m roots given by:

$$\sigma_q = -z_0 + \sum_{k=1}^{m} z_k \psi_k \tag{30}$$

where $\psi_k \in \{-1,1\}$, and q is the decimal equivalent of the binary number $\overline{\psi_1 \psi_2, \ldots, \psi_m}$ where the bar denotes changing each "-1" to "0". Equations (29) and (30) impose constraints on the parameters in (27) and on the positions of the roots on the *projection tape*. Therefore, in order to use (27) as a discriminant function, the orientation vector must be chosen so that (30) is satisfied and the cell parameters $\{z_1, z_2, \ldots, z_m\}$ obey Eq. (29). In fact (29) is not a strong restriction, since arbitrary values are allowed for the threshold parameters $\{z_1, z_2, \ldots, z_m\}$. However, in the case of (29) the number and the position of the roots are easier to control in developing the analytical design techniques to be presented next.

In a graphical representation, (30) can be associated with a *bifurcation tree* as shown in Fig. 7. for the particular case of $m = 2$ nests. The term "bifurcation" is used here because, for each additional level of nesting the number of roots is doubled in a manner reminiscent of the *period doubling* bifurcation observed in many nonlinear chaotic systems. The structure of the *bifurcation tree* is characterized by a *main trunk* positioned at $\sigma_0^0 = -z_0$ on the projection tape. This "trunk" corresponds to the unique root of the linear discriminant function, $w_0(\sigma_0^0) = 0$. For each additional level of nesting *k*, the roots σ_q^k of the intermediate discriminant function $w_k(\sigma) = 0$ can be calculated using the recursive definition (28). For example, at level 1, by imposing $w_1(\sigma) = 0$ in (28) and assuming that (29) is satisfied, it follows that the number of roots doubles and they can be determined by simply solving the piecewise linear equation $w_1(\sigma) = 0$ as follow:

$$z_1 + (z_0 + \sigma) = 0, \text{ if } \sigma > -z_0 \tag{31a}$$

or

$$z_1 - (z_0 + \sigma) = 0, \text{ if } \sigma < -z_0 \tag{31b}$$

If $z_1 < 0$ both alternatives are possible, leading to a doubling of the number of roots (in this case from 1 root at nesting level 0 to 2 roots at nesting level 1); namely:

$$\sigma_0^1 = -z_0 - z_1, \text{ and } \sigma_1^1 = -z_0 + z_1 \tag{32}$$

If $z_1 \geq 0$ (therefore, contradicting (29)), none of (31a) or (31b) has a solution. Indeed, (31a) has a solution $\sigma = -z_0 - z_1$ but since $z_1 \geq 0$ this solution does not satisfy the constraint $\sigma > -z_0$. A similar situation occurs for (31b). In terms of the *bifurcation tree* shown in Fig.7 this situation corresponds to a "collision" between the two resulting branches.

Observe that (32) is a special case of (30), when $m = 1$. Here we used the notation t_q^m to represent the "*q*" th root of the discriminant function associated with the nesting level "*m*".

In terms of the *bifurcation tree* (31a) and (31b) are equivalent to generating a set of *symmetrical* branches centered on the *trunk* associated with the previous nesting level, as shown in Fig. 7 for a particular case, when $Z = [z_0, z_1, z_2] = [-2, -4, -2]$. At each additional level of nesting, such branches will bifurcate into a double number of branches following the same scenario. Looking at the representation of the *bifurcation tree* in Fig. 7, it is clear that "branch collisions" can be avoided if and only if $|z_m| < \ldots < |z_2| < |z_1|$ and

$z_k < 0,\ k = 1,\ldots,m$, which is in fact the condition (29). Therefore, a "branch collision" leading to a decrease in the number of roots occurs whenever at least one of the inequalities in (29) is violated.

Uniform multi-nested cells and their bifurcation trees

Let us consider a special case of (29) leading always to a *uniform* distribution of roots on the *projection tape.* This case corresponds to the additional constraint:

$$z_k = -\eta 2^{-k}, \quad k = 1,\ldots,m \tag{33}$$

where η is an arbitrary scaling coefficient. Since (33) specifies explicitly the bias parameters $\{z_1, z_2, \ldots, z_m\}$, a dramatic simplification in our design method is expected. Indeed, in this case our design procedure reduces to that of finding a *uniform orientation,* i.e., an orientation vector **b** which will generate a *uniform projection tape* to match the *uniform multi-nested* discriminant function.

It follows from (30) that any root of the equation $w(\sigma) = 0$ is given by:

$$\sigma_q = -z_0 - \eta \sum_{k=1}^{m} 2^{-k} \psi_k \tag{34}$$

It can be easily checked that the distance between two consecutive roots on the σ axis is always equal to each other and it corresponds to a one bit change (from -1 to +1) in the least significant position ψ_m of the binary number $\overline{\psi_1 \psi_2, \ldots, \psi_m}$. Therefore, for *m* levels of nesting, the distance between two consecutive roots is given by:

$$\Delta^m = \eta 2^{-m}\left(+1 - (-1)\right) = \eta 2^{-m+1} \tag{35}$$

and the leftmost root σ_0 on the σ axis has its value:

$$\sigma_0 = -z_0 - \eta\left(1 - 2^{-m}\right) \tag{36}$$

For the multi-nested PWL function (with $m = 2$ nests) presented in Fig.7, the *bifurcation tree* is a uniform one, with $\eta = 2^{m+1} = 8$. As shown in the same figure, it splits the *projection tapes* into uniform segments containing an equal number of "red" and "blue" projected vertices, with a distance of 4 units between two consecutive roots (at nesting level $m = 2$).

The uniform multi-nested discriminant as an analog-to-digital converter

It is interesting to observe that the *multi-nested* discriminant function (27) where the threshold parameters are defined by (33) performs a very important function in signal processing; namely the conversion from analog values (in this case, the excitation σ) into binary ones. Indeed, if one consider the sign (color) $s_k = \mathbf{sgn}\left(w_k(\sigma)\right)$ of each intermediate discriminant function in (28), it is easy to check that the binary word $S = \overline{s_0 s_1 s_2, \ldots, s_m}$ is a Gray code[6] of the analog input σ. Such structures had been proposed for high-speed video signal analog-conversion [Fiedler & Seitzer, 1979] and with certain technological improvements they currently represent the most compact and fast analog-to-digital converters (e.g. [Liu & Liu, 2001]). There is also a nonlinear dynamics

[6] A Gray code is a binary code associated with a decimal number so that any increment or decrement of the decimal number will correspond to the change of only 1 bit in its associated Gray code.

interpretation for the recursive definition (28); namely, in the uniform case, each nesting level in (28) is equivalent to a "tent map" [Ott, 1993] transformation, as pointed out in [Kennedy, 1995].

Uniform orientations and projection tapes

A *uniform projection tape* associated with a Boolean function *ID* is obtained for a *uniform orientation* $\mathbf{b}^u$. As shown in Section 3.2.2, for a given orientation vector **b**, a *projection tape* is completely specified by its associated transition vector $T=\{t_1,t_2,..,t_{tr}\}$, and by the sign (color) of the leftmost projected vertex: $Y\left(V_{\alpha_0}\right)$. A *uniform projection tape* must satisfy the additional constraint:

$$t_{j+1}-t_j=\delta,\quad j=1,..,tr-1 \tag{37}$$

Experimental results on arbitrary Boolean functions indicate that the *robustness restriction* (12) imposed on the transition vector of a projection tape will dramatically reduce the chances of finding an associated uniform orientation. For example, in the case *n=3*, it was impossible to find *uniform orientations* which satisfy (12) for 24 out of all 256 possible functions. However, when this restriction was ignored, thereby allowing transitions to be located in arbitrary positions between projected vertices, uniform orientations were found for these 24 functions too. Therefore, in what follows we will remove restriction (12) from the definition of *projection tapes.* In effect, some Boolean functions will have less robust realizations than others. As shown next, by removing the restriction (33) which defines uniform multi-nested discriminant functions, the robustness will increase at the expense of storing the additional information represented by the parameters $\{z_1,z_2,\ldots,z_m\}$ for any prescribed Boolean function.

Boolean realizations: an analytic approach

If a Boolean function has a *uniform* projection tape with *tr* roots and the distance between consecutive roots is δ, a uniform multi-nested cell can be simply designed so that its associated bifurcation tree will match the projection tape.

Uniform multi-nested CNN universal cell realization procedure
Assumption: At least one uniform projection tape exists.

1. Choose $$m=\lceil \log_2(tr)\rceil, \tag{38}$$

where $\lceil x \rceil$ represents the first integer larger than *x* (also called a ceiling operator).

2. Following (35), determine η so that $\delta=\Delta^m$. It follows that:

$$\eta=\delta 2^{m-1} \tag{39}$$

3. Since the roots of a uniform multi-nested discriminant must satisfy (34), determine the bias z_0 so that (34) is satisfied. In the general case where *tr* is not a power of two, there are many valid choices[7]; However, we will always choose the first transition t_1 in the projection tape to be the first root σ_0 of the uniform multi-nested discriminant.

[7] Assuming that the $2^m - tr$ additional roots of the uniform multi-nested function lie in the "don't care" regions on the *projection tape,* i.e. regions where there is no input vertex projected.

Therefore: $\sigma_0 = -z_0 - \eta(1-2^{-m}) = t_1$, where t_1 is the leftmost transition on the projection tape. It follows that:

$$z_0 = -\eta(1-2^{-m}) - t_1 = -t_1 - \eta + \delta/2 \tag{40}$$

4. Determine the sign parameter s in (27). From (28) it follows that at any level of nesting k, with $k \geq 1$, the leftmost segment of the piecewise-linear discriminant $w_k(\sigma)$ has a slope equal to -1. This is true for the last level of nesting "m" as well. It follows from (27)' that the sign of the discriminant function when $\sigma < \sigma_0$ is equal to s. Since $\sigma_0 = t_1$ and there is only one projected vertex v_0 to the left of t_1 on the projection tape, it follows that:

$$s = \begin{cases} Y(v_0) = Y(V_{\alpha_0}), \text{ if } m \geq 1 \\ -Y(v_0) = -Y(V_{\alpha_0}), \text{ if } m = 0 \end{cases} \tag{41}$$

since for $m=0$ the slope of the linear discriminant function is +1.

Example 6

Let us consider again the Parity 4 function, for which we have obtained earlier a realization via the canonical PWL cell (Example 4). We know that this function is a *totalistic* Boolean function, so in a search for a *uniform orientation* (which leads to a *uniform projection tape*) the most natural choice would be $\mathbf{b} = [1,1,1,1]$. If the cell is considered from the perspective of having 9 inputs (e.g. , a parity 9 function in a 2-dimensional CNN grid), and knowing which are the "active" inputs, the orientation vector can be rewritten as $\mathbf{b} = [0,1,0,1,0,1,0,1,0]$.

Indeed, as shown in Fig. 8, its associated decoding tape is defined by the transition vector $T = \{-3,-1,1,3\}$ and it corresponds to a *uniform projection tape* with $\delta = 2$ and $t_1 = -3$.

Applying the above algorithm, a matching bifurcation tree is simply determined as follow:

$$m = \lceil \mathbf{log}_2(4) \rceil = 2 \ ,$$

$$\eta = \delta 2^{m-1} = 2 \cdot 2 = 4 \ ,$$

$$z_0 = -t_1 - \eta + \delta/2 = 3 - 4 + 1 = 0$$

Using the definition (33) for uniform multi-nested discriminant functions, the remaining values of the threshold parameters can be easily determined as follows:

$$z_1 = -\eta 2^{-1} = -2, \quad z_2 = -\eta 2^{-2} = -1$$

Similarly, the sign parameter is given by:

$$s = Y(v_0) = -1$$

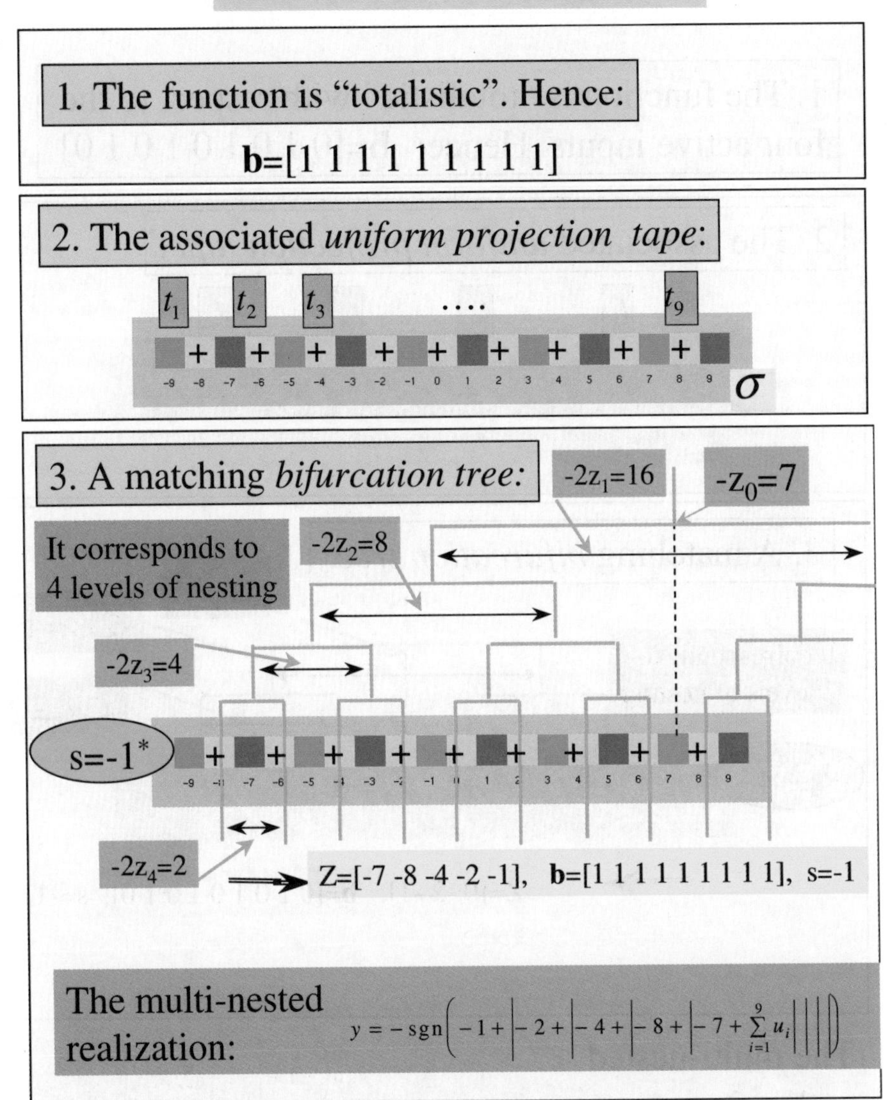

Fig. 8. A uniform multi-nested cell realization of "Parity 4"

Example 7

Let us now consider the Parity 9 Boolean function, a true valued function ($y = 1$) if the number of +1 inputs is even. An equivalent definition (within the binary formalism {-1,1}) of the Parity 9 function is that it computes the product of the 9 binary inputs. Since this Boolean function is *totalistic,* a good choice for the orientation vector is $\mathbf{b} = [1,1,1,1,1,1,1,1,1]$. Indeed, this choice leads to a *uniform projection tape* defined by the transition vector $T = \{-8,-6,-4,-2,0,+2,+4,+6,+8\}$ with $tr = 9$ and $\delta = 2$, as shown in Fig. 9.

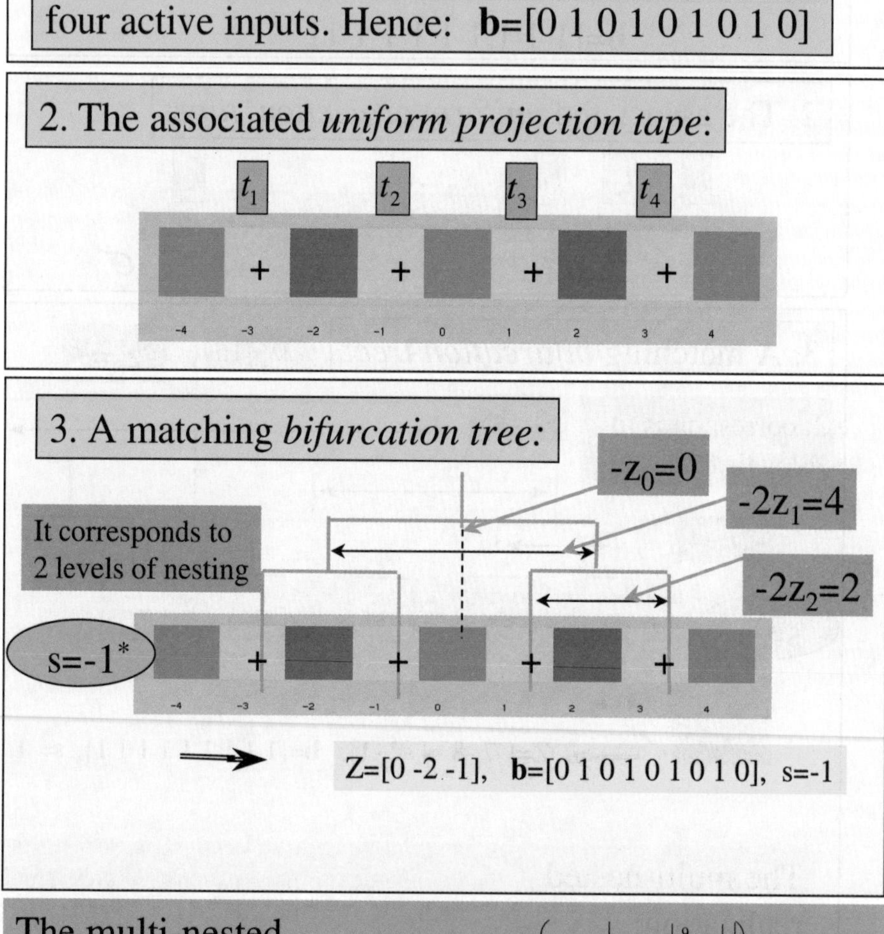
Boolean function: PARITY 4
1. The function is "totalistic" with respect to the four active inputs. Hence: b=[0 1 0 1 0 1 0 1 0]
2. The associated uniform projection tape:
t1 t2 t3 t4
-4 -3 -2 -1 0 1 2 3 4
3. A matching bifurcation tree:
-z0=0
-2z1=4
-2z2=2
It corresponds to 2 levels of nesting
s=-1*
Z=[0 -2 -1], b=[0 1 0 1 0 1 0 1 0], s=-1
The multi-nested realization:

$$y = -\mathrm{sgn}\left(-1 + \left|-2 + \left|\sum_{i=1}^{9} u_i\right|\right|\right)$$

Fig. 9. A uniform multi-nested cell realization of "Parity 9"

The simple design algorithm defined by (38)-(40) can be applied again, leading to the following realization:

$$m = \lceil \log_2(9) \rceil = 4\,,$$

$$\eta = \delta 2^{m-1} = 2 \cdot 8 = 16\,,$$

$$z_0 = -t_1 - \eta + \delta/2 = 8 - 16 + 1 = -7$$

Using the definition (33) for the uniform multi-nested discriminant functions, the remaining values of the threshold parameters can be easily determined as follows:

$$z_1 = -\eta 2^{-1} = -8, \quad z_2 = -\eta 2^{-2} = -4,$$
$$z_2 = -\eta 2^{-3} = -2, \quad z_2 = -\eta 2^{-4} = -1$$

Similarly, the sign parameter is given by:

$$s = Y(v_0) = -1$$

Observe that in this case, the rightmost 7 branches of the bifurcation tree remain unused, and they will intersect the *projection tape* in a "don't care" region void of any projected vertices. It is also important to observe that from a hardware realization perspective, only 4 nonlinear devices (absolute value operators) and 4 additional parameters are required to implement the Parity 9 function via the above uniform multi-nested cell.

Theorem 1. The Parity function with n inputs admits the simplest realization, having a complexity of $O(\log_2(n))$, via a uniform multi-nested PWL CNN cell.

Proof: The proof is constructive, and it consists of applying the design procedure defined by Eqns. (38)-(40):

It can be easily verified that the orientation vector $\mathbf{b} = [1,1,...,1]$ (n coefficients 1) is a uniform orientation vector which leads to a *uniform projection tape* with n transitions ($tr = n$) defined by the transition vector $T = \{-n+1, -n+3, ..., n-1\}$, and having $\delta = 2$. Therefore, by applying the above design algorithm, it follows that:

$$m = \lceil \log_2(n) \rceil,$$

$$\eta = \delta 2^{m-1} =]n[,$$

where $]n[$ is the first integer bigger than n which is a power of 2,

$$z_0 = -t_1 - \eta + \delta / 2 = n -]n[$$

Using the definition (33) for uniform multi-nested discriminant functions, the remaining values of the threshold parameters can be easily determined as follows:

$$z_k = -\eta 2^{-k} = -\frac{]n[}{2^k}, \quad k = 1, ..., m$$

Similarly, the sign parameter is given by:

$$s = Y(v_0) = -1$$

This result is extremely significant because the simplest realization of Parity functions with n inputs reported so far in the literature has a *linear* complexity, since it requires $m = O(n-1)$ Parity 2 (or XOR) gates. Since Parity functions are wide spread in modern information processing systems, their function being often associated with parity checksums for error detection and/or correction, the multi-nested CNN cell offers a dramatic reduction in complexity compared to any existing (digital) solution. Indeed, tasks like checking the parity of binary words of 64 bits are standard operations performed currently in computers. Observe that instead of the 63 XOR gates would be required to accomplish this task using a conventional approach, only 6 simple nonlinear devices (absolute value functions) and 7 additive parameters (biases) are required when using the multi-nested approach. The projection $\mathbf{bu}^{\mathrm{T}}$ is realized in this case by a simple summation of the inputs, and assuming that they are currents, it can be achieved with minimal hardware cost via the KCL (Kirkhoff's Current Law).

Finding the genes for arbitrary Boolean functions

Previously an *exact* design solution was presented, applicable only for the case of *uniform projection tapes,* assuming that a *uniform orientation* has been determined. In this section we will expand our view to different other methods of seeking a valid realization (described by its gene) for Boolean functions with arbitrary n inputs and no predetermined uniform orientation. For $n \leq 5$, we will show that several computable algorithms can be employed for this task.

Let remind that the defining equation of the multi-nested CNN cell is:

$$y(\mathbf{u}) = sign\left[s \times \left(z_m + \left| z_{m-1} + \left| ..z_1 + \left| z_0 + \sum_{i=1}^{n} b_i u_i \right| .. \right| \right| \right) \right] \tag{42}$$

where $\mathbf{G} = [\mathbf{s}, \mathbf{z}, \mathbf{b}] = [s, z_0, z_1, ..., z_m, b_1, b_2, .., b_n]$ is a *gene* i.e. a vector of $m+n+2$ parameters. The parameter s is either -1 or +1 while the rest of the parameters are chosen as integer numbers within a hypercube. Integer parameters are chosen and each parameter can be represented with a specified resolution r, which therefore will define the range of variation for each parameter: $-L \leq p \leq L$, where p is any of the z, or b parameters and the relationship of L with a specified representation resolution r (bits) is $L = 2^{r-1}$. For a given configuration of the *gene* vector the above equation will implement a specific Boolean function[8] specified by its ID:

$$ID(s, \mathbf{z}, \mathbf{b}) = [y(\mathbf{v}_{N-1}), y(\mathbf{v}_{N-2}), ..., y(\mathbf{v}_0)] \tag{43}$$

where $N = 2^n$, and $\mathbf{v}_j$ is the j-th binary vertex of a unit hypercube. In fact $\mathbf{v}_j$ is the binary code of j represented in a code where 0 is replaced by -1 and 1 by 1. For example, if $n=5$ there are $2^5 = 32$ vertices of the binary hypercube which correspond to the 32 entries in the truth table defining the function ID. Often, for compactness, the ID is represented as decimal number.

The design problem is to find a proper gene **G** for a specified ID, and the algorithm should be able to find a valid solution for an arbitrarily chosen ID. It turned out that the problem is not trivial at all and it demands high computational resource.

The following solutions are possible:

---

[8] Note that an alternative convention is $ID(s, \mathbf{z}, \mathbf{b}) = [y(\mathbf{v}_0), y(\mathbf{v}_1), ..., y(\mathbf{v}_{N-1})]$. It is straightforward to prove that the gene realisation in this case is that from the convention in (43) where only the sign of the b parameters is reverted.

(a) Systematic search

In this case the goal was to fill in a table containing all possible realizations for a given n. One possibility is to use a ***systematic*** search where each parameter is allowed to take values within its variation range i.e. $2L$ distinct values. For each combination of parameters the associated ID is evaluated according to (43). Then, if the ID is a newly discovered one a new record is added to a dynamic structure called a *Realizations Table* (RT). Each record contains the ID number and its realization (the gene parameters). A new set of parameters is then computed and the process stops when the table is filled with all possible Boolean functions. In addition one can also compute the robustness R for each ID function and consider filling a record even if a valid realization exists in the table if the robustness of the newly found function is better. In this second case, the process should stop only when all combinations of parameters were exhausted. The total number of iterations is $Nit = (2L)^{n+m+1}$. Observe that another problem is the choice of L. Since we want to minimize the number of iterations we are looking for a lower bound for L. Indeed, since L is related to the representation resolution r the following method can be used to estimate L.

A Boolean function with n inputs contains $Ib = 2^n$ bits of information. The information stored in the coefficients of a multi-nested realization is $Ir = 1 + (m+1+n)r$ bits where 1 bit comes from s and the remaining bits are coming from the representation using r bits of the $\{z, b\}$ parameters. It is clear that $Ir > Ib$, and in practice a coefficient β is considered, i.e. $Ir = \beta \cdot Ib$. Then, it follows that:

$$r = (\beta 2^n - 1)/(m+n+1) \qquad (44)$$

Example 8 Uniform multi-nested realizations for all Boolean functions with 3 inputs

If $n=3$ and we choose prescribed constant values for the parameters z_1, z_2 in (42) corresponding to *uniform* multi-nested realizations with 2 nests, $r = (8\beta - 1)/4$. Taking the lowest value possible $\beta = 1$, it follows that $r = 2$ is the smallest integer satisfying the above condition. It leads to $L = 4$, therefore the expected number of iterations $Nit = 8^4 = 4096$. Since 100,000 iterations take about 1 second on a 550Mhz Pentium processor, the table will be instantaneously filled. Table1 below lists the genes for all Boolean functions with 3 inputs. A similar table was presented in [Dogaru & Chua, 1998f] using a different approach, i.e. by learning parameters using a gradient descent procedure. The table in that case was derived for a non-uniform multi-nested formula (where each gene parameter has two additional parameters z_1, z_2). For the table presented below, these two parameters are uniquely specified for the entire table; namely, $z_1 = -3$, and $z_2 = -3/2$, corresponding to $\eta = 1.5 \cdot q = 6$. Fortunately, in this example $\beta = 1$ as a result of an optimization in choosing η, but in general, for larger number of bits, $\beta > 2$ is a better choice. Otherwise, it is possible that some of the Boolean functions will not be found as having a realization.

Table 1. Uniform multi-nested CNN cell realizations with 2 nests for the entire set of Boolean functions with 3 inputs. Observe that two of the bias parameters (z_1=-3 z_2=-1.5) are independent on the Boolean function, therefore only 5 parameters s, z_0, b_1, b_2, b_3 listed in the table will determine the "gene" associated with a particular function *ID*. For any function where $ID > 127$ the associated realization is the one listed in this table for the function $255 - ID$ but where the sign parameter *s* is inverted (multiplied by -1).

ID	s	z_0	b_1	b_2	b_3
0	-1	4	0	0	4
1	-1	-4	-4	3	3
2	-1	-3	4	4	3
3	-1	4	1	-3	0
4	-1	-4	3	-3	-4
5	1	-3	-3	0	-3
6	-1	-1	-1	4	4
7	-1	1	-4	3	3
8	-1	4	-3	3	-4
9	-1	1	1	-4	4
10	1	3	3	0	-3
11	-1	1	2	-3	3
12	1	3	3	-3	0
13	-1	-2	-2	-3	3
14	-1	-1	4	3	3
15	-1	-4	4	-3	0
16	-1	4	-1	-3	1
17	1	3	0	3	3
18	-1	-4	-2	1	-2
19	-1	2	-3	1	-3
20	-1	-1	4	4	-1
21	-1	2	-3	-3	1
22	-1	-4	4	4	2
23	-1	4	-2	-2	-2
24	1	0	-3	3	3
25	-1	3	-3	-2	-1
26	-1	-3	2	3	-1
27	-1	4	-1	-3	2
28	-1	3	-2	1	-3
29	1	4	3	-2	1
30	1	-2	-4	2	2
31	1	-1	2	-3	-3

ID	s	z_0	b_1	b_2	b_3
32	-1	3	-3	-4	4
33	-1	-1	4	-1	-4
34	-1	1	0	-3	-4
35	-1	2	-3	1	3
36	1	0	3	-3	3
37	-1	3	-2	-3	-1
38	-1	-3	3	1	-2
39	-1	3	-4	-2	1
40	-1	-2	-4	-1	-2
41	-1	-1	1	1	2
42	-1	-1	3	3	2
43	-1	2	-4	-1	1
44	-1	-3	1	-2	-3
45	-1	2	-4	-2	2
46	1	1	4	-3	2
47	1	2	-2	3	-3
48	1	3	-3	3	0
49	-1	-1	-3	-2	3
50	-1	2	3	1	3
51	-1	4	0	-1	0
52	-1	3	1	-2	-3
53	-1	1	3	2	-4
54	-1	4	-1	-2	-1
55	1	-1	-3	2	-3
56	-1	3	1	-2	3
57	-1	-2	2	4	-2
58	-1	-4	2	3	-1
59	1	2	3	-1	-3
60	1	0	4	-1	0
61	1	3	-1	4	3
62	1	-3	-2	-1	3
63	-1	3	-3	-3	0

ID	s	z_0	b_1	b_2	b_3
64	-1	-3	-4	3	4
65	-1	2	1	-4	-2
66	1	0	-3	-3	3
67	-1	-3	2	1	3
68	1	-3	0	3	-3
69	-1	-2	3	-3	-1
70	-1	-3	3	-2	1
71	1	-1	2	3	-4
72	-1	1	4	-1	4
73	-1	4	-1	4	-4
74	1	1	-2	-3	2
75	1	-2	-4	-2	2
76	-1	2	-3	-1	-3
77	1	4	2	-2	2
78	1	-1	2	4	-3
79	1	2	-2	-3	3
80	1	3	-3	0	3
81	-1	2	3	-3	1
82	-1	-3	-2	3	1
83	-1	-1	-3	4	-2
84	-1	1	3	3	2
85	-1	-1	0	0	4
86	1	4	-1	-1	2
87	1	-1	-3	-3	2
88	-1	-3	-2	-3	1
89	1	2	-2	2	4
90	1	0	4	3	-4
91	1	3	2	3	1
92	-1	-1	3	2	4
93	1	-1	-3	3	2
94	1	-3	-2	3	-1
95	-1	-3	3	0	3

ID	s	z_0	b_1	b_2	b_3
96	-1	-1	1	-4	-4
97	-1	1	-2	2	2
98	-1	3	3	-1	2
99	1	-2	-2	-4	2
100	-1	-3	-3	-1	2
101	1	1	1	-4	2
102	-1	0	-3	4	4
103	1	3	3	2	1
104	-1	-4	-4	-4	-1
105	1	-2	2	-2	2
106	-1	2	2	2	4
107	1	-1	-1	-1	4
108	1	-2	-2	4	-2
109	1	-4	-4	1	-4
110	-1	-3	1	-2	-2
111	1	-4	-4	1	1
112	-1	2	-1	-3	-3
113	-1	-2	-4	1	1
114	1	1	-4	-2	3
115	1	1	-3	-2	3
116	-1	1	-2	-3	-4
117	1	1	-3	3	-2
118	1	3	-3	2	1
119	1	-1	0	-3	4
120	-1	2	4	2	2
121	1	-1	4	-4	-4
122	1	-3	2	-3	1
123	1	-4	-1	-4	-1
124	1	3	-1	-2	3
125	1	4	2	-2	1
126	-1	0	3	3	3
127	1	4	-1	3	-1

A drawback of the systematic search method is that for larger *n* it may require a huge number of iterations, each iteration consisting in computing the ID using (43) and checking whether it is a newly discovered function or not. For example, let us consider the case when we are looking for the entire set of parameters in the case of uniform 3-nests realizations of 4-inputs Boolean functions. Taking $\beta = 2$, and applying (44) it follows that $r = 6$. Hence, $L = 32$, and $Nit = 64^5 = 1{,}073{,}741{,}824$ iterations which will take 10,737 seconds i.e. more than 3 hours on the same computer. It is easy to calculate that for larger number of inputs the resulting number of iterations will become completely unpractical. Another significant drawback of the systematic search method is its sensitivity to *L*.

In practice is difficult to estimate in advance the value of L because of the uncertainty regarding β. A larger value of β will reduce the possibility to skip some realizations but on the other hand will lead to a dramatic increase of the computation time.

(b) Random search

The second alternative **is** a ***random search*** where at each time step (iteration) t of the algorithm a gene G is not *enumerated* as above but rather *randomly generated* such that its parameters satisfy $-L \le p \le L$, where p is any of the z, or b parameters. The figure below gives an intuitive idea of the *random search* method.

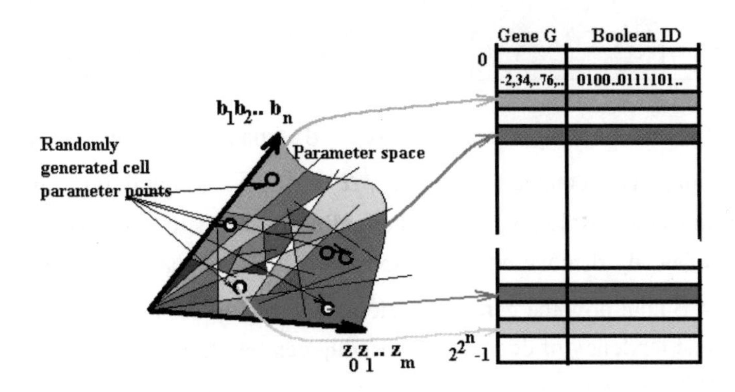

Within the iteration the same actions as for the systematic search are taken. It founds the ID associated with the current gene and if certain conditions are fulfilled, updates the *realization table* RT. The evolution of the number of newly discovered IDs as a function of iteration t has the characteristic of sigmoid function i.e. $DiscID = 2^{2^n} \tanh(\lambda t)$.

The shape of the sigmoid suggests that although most of the *realizations table* is relatively faster filled in, there still remain a lot of IDs for which quite a large amount of iterations T_{NewID} is needed to discover a new function. The parameter λ of the above function is proportional to the speed in discovering new realizations and depends among others, on the quality of the random number generator, and on the representation resolution of each parameter value.

It was experimentally found that the speed of discovering new realizations can be considerably increased if the following schedule is applied for the random number generators:

$$-Z \le p_z \le 0\text{; and } -B \le p_b \le B\text{; with } Z = L/\mu_z\text{, and } B = L/\mu_b \quad (45)$$

where p_z is any of the z parameters and p_b is any of the b parameters, and μ_a, μ_b should be carefully optimized as described below.

The first inequality is a consequence of (29), which confirm the negative sign for the z parameters. As shown next the scaling factors μ_z, μ_b allow one to reduce the representation resolution of the parameters to its lowest limits. It was also found that there is an optimal value of the ratio μ_b/μ_z for which the speedup of finding new realizations is maximized. Let us now consider the following example:

Example 9 Non uniform multi-nested realizations for *n=4* inputs

Let's take $L = 2^{14} = 16384$ and perform a random search as above. The table below shows the number of newly discovered functions after 10^6 iterations for different values of the scaling factors μ_z, μ_b

Found IDs	60322	59111	61228	61174	60074	58124	60190	52454	46388
μ_b	1	4	6	7	8	10	75	750	1500
μ_z	0.5	8	8	8	8	8	100	1000	2000

Observe that maintaining a constant $\mu_z = 8$ (shaded cells in the table above) a maximum speedup is obtained when $\mu_b/\mu_z \approx 0.75$. Maintaining this ratio while increasing 12 times the absolute values of the scaling factor there is still no dramatic change in the number of newly discovered realizations (60190 in the case $\mu_z = 100$ compared to 61228 for $\mu_z = 8$) but now the gene coefficients will be represented with less bits per each coefficient. Some tendency of decrease in speedup can be observed only when the absolute value $\mu_z = 2000$, corresponding to a limit in the representation resolution. In this latter case, the *realizations table* was completely filled on all its 65536 positions after 100 million of iterations (i.e. after about 1000 seconds) with the advantage that each parameter is represented on a minimal number of about 4 bits. A faster filling of the *realization table* can be obtained in the case corresponding to the third column in the above table. In this case, only about 16 million iterations were required to fill the table, but now each parameter of the realizations is represented typically on 14 bits. In any case the filling of the whole table lasts less than in the case of using the systematic search.

Using the random search algorithm as above, we are now interested to see what is the percentage of Boolean functions requiring *m=0* (or no nests), *m=1* nest, *m=2* nests or the maximum of *m=3* nests in the realization formula (42). The following table represents the results obtained for both *non-uniform* (N) and *uniform* (U) multi-nested cell realizations.

		0 nests	1 nest	2 nests	3 nests	Total
N	realizations	1882	14244	1358	48052	65536
	percentage	2.87%	21.73%	2.08%	73.32%	100%
U	realizations	1882	14244	7248	40002	63376
	percentage	2.87%	21.73%	11.06%	61.04%	96.7%

Observe that for the case of *uniform* multi-nested realizations, only for 63376 of the 65536 Boolean functions with 4 inputs, realizations were found. It still represents 96.7% of the whole set of Boolean functions. Also note that a relatively small percentage of the Boolean functions require the maximum value of *m* conjectured as *m=n-1*. This observation is consistent with that found for the multi-nested realizations of Boolean functions with 3 inputs presented in [Dogaru & Chua, 1998f]. In that case, only 9.37% of functions were found to require the maximal number of nests, ($m = 2$) while most of the functions (50%) were realized with only $m = 1$, as shown in Fig.10.

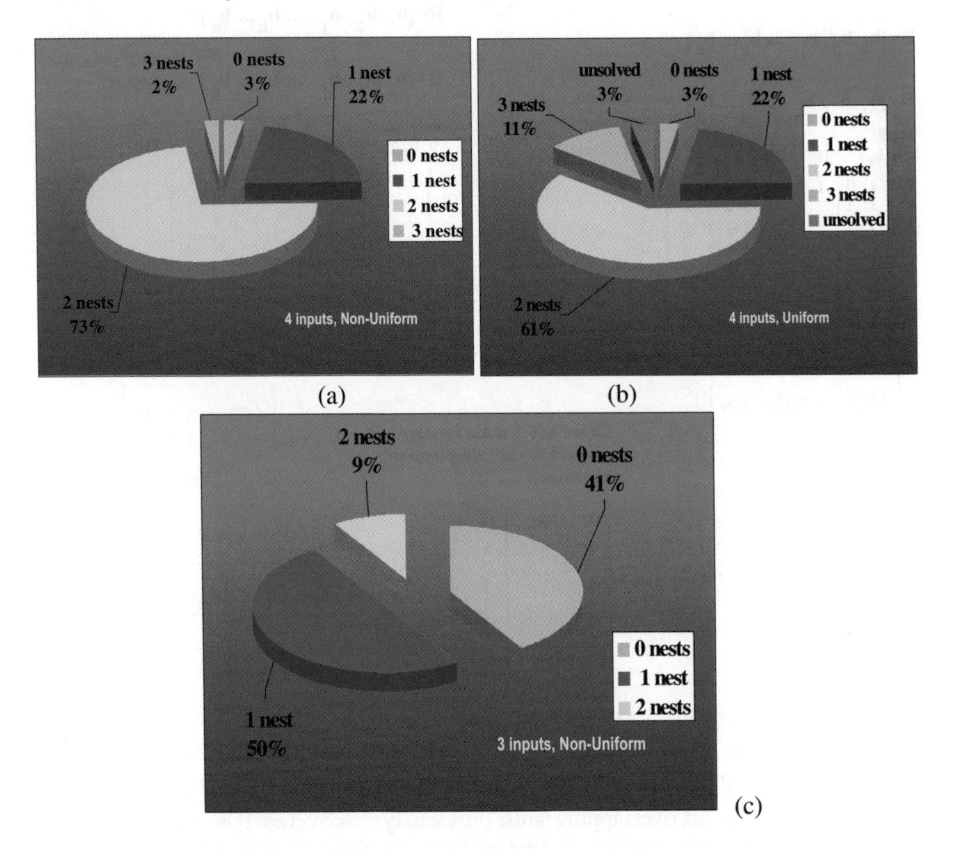

Fig.10. The distribution of gene realizations for different numbers of "nests": (a) The case of 4-inputs Boolean functions with *nonuniform* cells; (b) 4-inputs Boolean functions with *uniform* cell realizations; (c) 3-inputs Boolean functions.

The random or systematic search methods discusses herein can be applied using nowadays technologies for no more than *n=5* inputs. In this case, the computational demand is very high (in both storage and computing time) as a direct result of a dramatic increase in the number of Boolean functions. There are 2^{2^5} (about $4 \cdot 10^9$) distinct Boolean functions for which gene realizations should be found and stored. Several parallel and distributed computational solutions were recently proposed [Ionescu & Dogaru, 2002a][Julián *et al*, 2002b] but still required several months of computation time to list all realizations following the above procedures.

In [Dogaru *et al*, 2002b] a novel method was proposed based on the observation that Boolean functions can be grouped in classes (families). Instead of listing all Boolean functions one should now find a gene only for one (representative) member of each *familiy*. The realizations for all members in the family can be then obtained using very simple (linear) transformations of the *s* parameter and of the **b** (orientation) vector. Roughly speaking, there is an average of $(2^{n+1})n!$ members per family, which gives a reasonable number of *families*. For example, in the case *n=5* it follows that there should be about 4294,967,296 / 7680 = 559241 *families*. The speedup is expected to be of about three orders of magnitude since now the above search procedures are applied for a much limited number of representative Boolean functions corresponding to the number of families. The main three transformations generating the families of Boolean functions are depicted below:

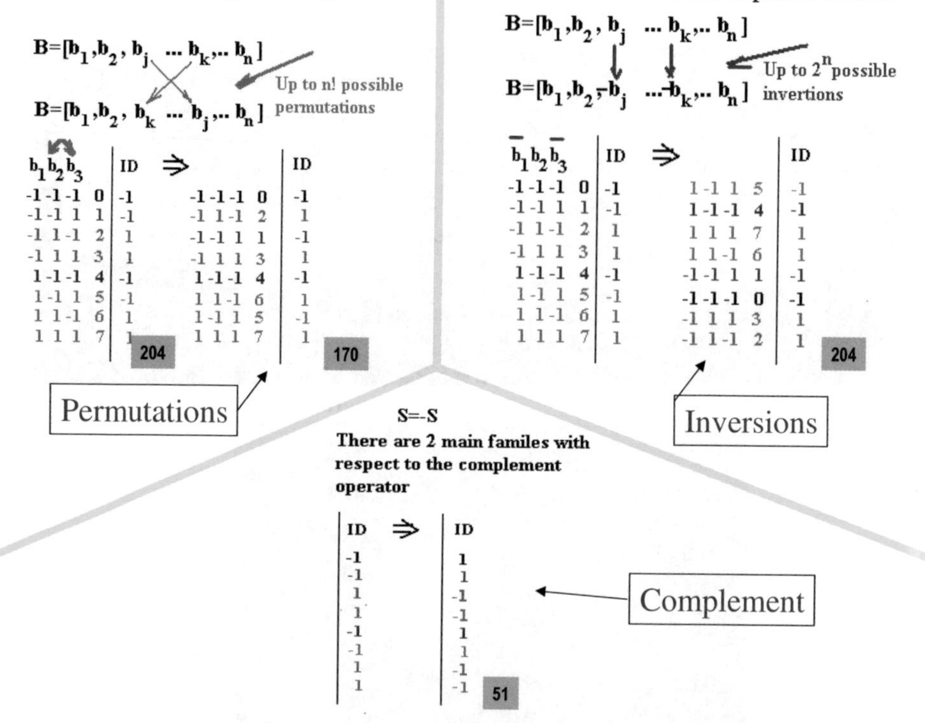

As shown in what follows, finding one ID realization is in fact equivalent to finding the realizations for at most $(2^{n+1})n!$ different IDs. While it is possible that some of these IDs are overlapping or there is an overlapping with previously discovered IDs, there is still a high potential in speeding up the process of finding new functions. Our experiments suggest that for random Boolean functions the consequent family of Boolean functions contains the maximum number (i.e. $(2^{n+1})n!$) while the number of members of a family is much smaller for "orderly" Boolean functions (e.g. PARITY, AND, OR, etc.).

Next we introduce two of the families and their corresponding transformations. The first family (***inverted inputs***) is defined by transformations of the type $b_i' = -b_i$ where index i can take up to n values. This leads to a maximum number of 2^n transformations, which can be easily listed. The effect of the above transformation over the ID is a change in the ordering of the vertices in (43). Thus the associated ID' can be easily computed by establishing the new ordering of its bits and making the corresponding permutations. The following example demonstrates this idea:

Example: Let us consider that b_3, b_2, b_1 is a realization of the $ID = \frac{01011011}{01234567}$ where the standard ordering is shown under the ID bits. The transformation $B' = [b_3, -b_2, -b_1]$ will transform the standard ordering 01234567 into 32107654. Hence the transformed $ID' = \frac{10101101}{32107654}$. The new ordering can be easily computed by representing the Boolean function in a truth table and complementing the columns associated to the corresponding inverted b parameter.

The second family (***permuted inputs***) is defined by the following transformation (permutation) in the B space: $b_j'' = b_k, b_k'' = b_j$ for a number of up to n pairs (j,k) of indices. The effect on the ID is again a reordering of its bits and there are $n!$ such possible transformations. The new reordering can be easily determined as above by writing a new truth table where the left side is composed of columns permuted according to the specified transformation. For the above example ID let us consider the permutation $B'' = [b_1, b_3, b_2]$. This will lead to the new ordering sequence: 04152637 and therefore to the transformed $ID'' = \frac{01100111}{04152637}$. Observe that in all these families of functions the number of 0/1 bits remains unchanged. One may eventually find additional families of transformations leading to a ordering change and this could be an interesting approach for an analytic solution of the multi-nested neuron design problem. Indeed, there are $(2^n)!$ possible re-orderings of the ID bits of which only $(2^n)n!$ permutations are covered by the above combined transformations in the *B* space. Although it is quite probable that some of the transformed ID will overlap, for the case *n=5* for each discovered ID there are potentially at most other 32*120=3840 new IDs that can be automatically listed thus enhancing the rate of discovery in a random search. Note that for a given *n* the table associated with all these 3840 orderings can be stored in memory and thus the process of identifying new transformed IDs become very fast. In addition to the above, by simply inverting the *s* parameter one gets a realization of the complement version of the actual ID. Thus the numbers of IDs in the above families could be further multiplied by 2 as an effect of this simple negate transformation.

(c) Finding realizations using global optimization techniques

For a relatively small number of inputs, exploratory search of the orientation vector parameter space can provide a reliable method for realizing multi-nested CNN cells with a uniform roots structure, as shown in the previous sections. However, this approach becomes inefficient for $n \geq 5$, as we discussed above. In such cases one should employ a focused search, where an optimization algorithm tries to find a valid gene for a multi-nested cell realization of a specified ID. The major challenge here is the speed of the algorithm and its guaranteed convergence. The figure below ilustrates the idea of a focused search.

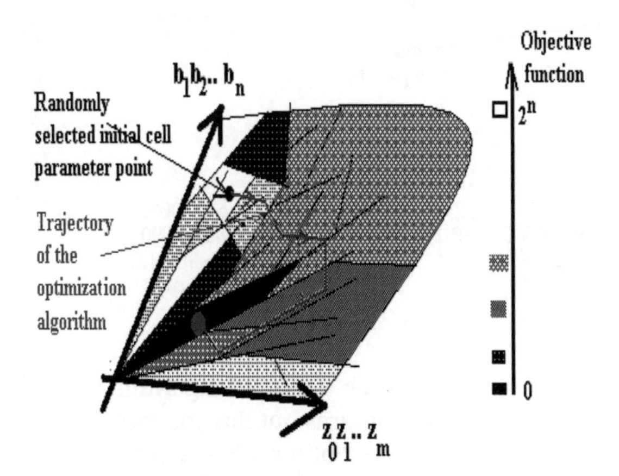

There are many global optimization algorithms that can be effectively used to find the gene parameters for our problem. In [Dogaru and Chua, 1999a] a modified version of the Alopex algorithm [Harth & Pandya, 1988] was sued. It is a form of "Simulated Annealing" [Kirkpatrick *et al.*, 1983] where some reinforcement information about the evolution of the optimization process is also included. Potentially useful are the *genetic algorithms* [Goldberg, 1989]. Next, a modified version of a *greedy-type* algorithm proposed by Matyas [Matyas, 1965] and later discussed in [Baba, 1989] is considered. The philosophy of this algorithm is rather simple and it can be expressed as follow (also illustrated in the figure below):

Pick a random gene. Then, in each iteration step:

(i) Do a random mutation (using a Gaussian random function generator with a specified dispersion σ) around the actual gene, and evaluate the new objective function (in our case the Hamming distance from our desired ID);

(ii) Accept the mutation only if there is a decrease in the objective function. Otherwise come back to (i). Stop the search process after a finite preset number of iterations (otherwise it may waste precious computing time).

One can do a number of repeats of the above algorithm starting with different initial conditions. After stop one can either have a valid solution or a message stating that a valid solution was not found. In this latter case one should run the algorithm again. The algorithm is theoretically convergent to the global solution if enough time is allowed. In addition to the above, it was found useful that at each iteration the dispersion σ will exponentially decrease according to the formula $\sigma_{new} = \alpha\sigma_{old}$ where α is an additional parameter of the optimization algorithm.

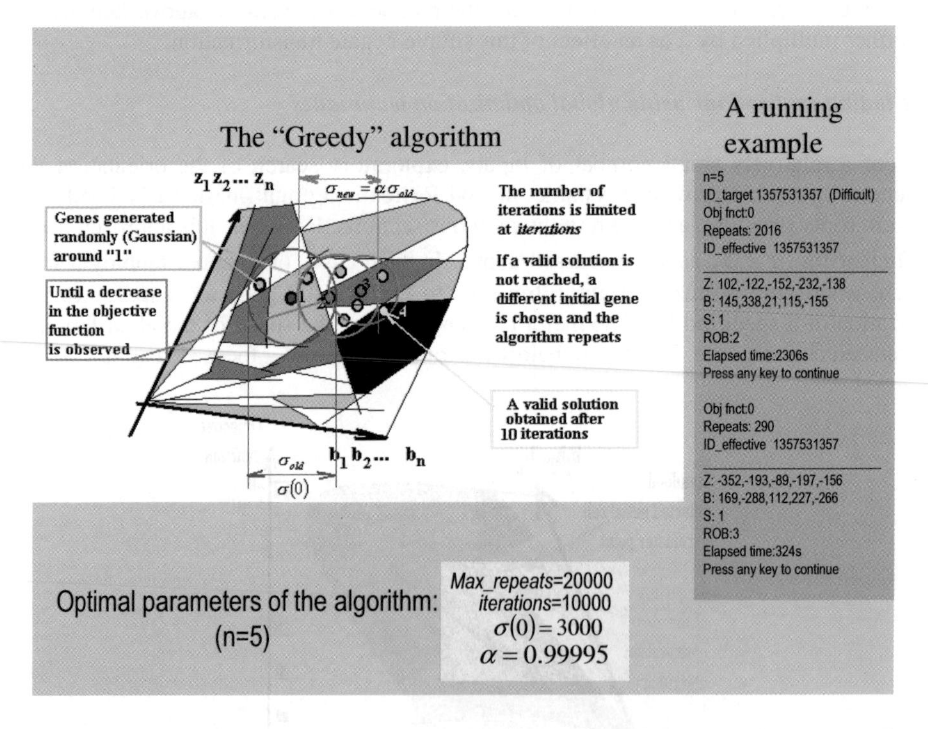

What are the reasons to choose such a simple algorithm? Because in my experience with stochastic and genetic algorithms it turns out that the most difficult part of the problem is the tuning of the parameters of the optimization algorithm. Since the processes are random different initial solutions for the same set of parameters will lead to completely different behaviors for the same set of parameters and thus is hard to evaluate their effect. The above algorithm has only four parameters {iterations, repeats, $\sigma(0)$, α} to tune. Next we describe several properties of algorithm and give some hints in using it. A pool of 942 randomly chosen IDs of 32 bits each (5 inputs) was considered, for which the above algorithm was applied to find 4-nests realizations. The optimal set of parameters was: {iterations=10000, repeats=20000, $\sigma(0) = 3000$, $\alpha = 0.99995$}. After running the algorithm the following were observed:

(1) The time until a solution is found: The ID=3569471485 was considered and the optimizer algorithm above was launched for 28 times. As seen in Fig.11(a), depending on the random initial condition, the number repeats until the final solution is reached can vary between 14 and 1483. On a Pentium-800Mhz it takes about *1 second per each repeat*. The total amount of repeats spent to solve 929 of the functions in the above pool was 831479. This represents an average of 895 seconds per function. The distribution of repeats until finding a solution for the pool of 929 functions is shown in Fig. 11(b).

(2) The probability of convergence towards the global optimum: In the pool of 942 randomly selected functions, for 13 of the functions the algorithm stopped without valid realizations. This is a consequence of limiting the number of repeats, in this case to 10000. Indeed, applying the optimization procedure again with the variable *repeats* set to 20000, valid realizations were found for all these 13 functions too. Again the average time for finding the gene realization was measured and it was of about 900 seconds (using the 800Mhz processor). This results has a very strong theoretical implication since it says on a statistical basis that any arbitrary Boolean function with 5 inputs admits a 4-nest universal CNN cell realization.

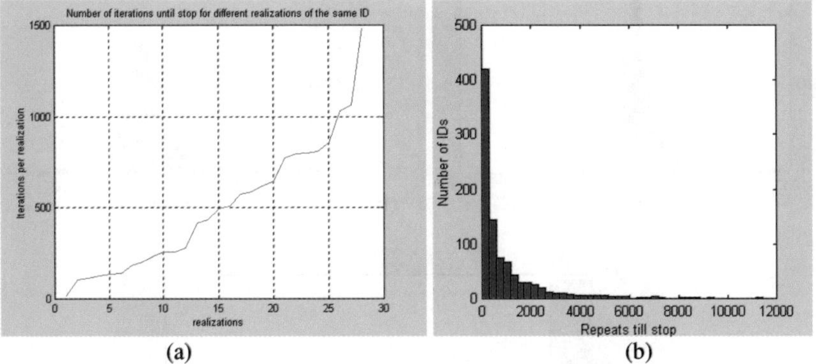

(a) (b)

Fig. 11. (a) Number of repeats till stop for the same ID; (b) Distribution of the *repeats* till stop for the entire set of 942 randomly chosen Boolean functions.

(3) The representation resolution and the robustness: For each realization *robustness* is defined as $rob = \min_{j=0,..,N-1}\left(\left|\tau(\mathbf{v}_j)\right|\right)$ where $\sigma(\mathbf{v}_j) = z_m + \left|z_{m-1} + \left|..z_1 + \left|z_0 + \sum_{i=1}^{n} b_i v^j{}_i\right|\right|..\right|$ using the notations in (42) and (43). In order to obtain a compact solution (i.e. a solution where parameters are represented as small integers), the raw result of the optimization algorithm is compacted using the following simple procedure:

Until *realized ID is altered* divide by 2 all parameters and check the newly realized ID.

The histograms in Fig.12 show the distribution of parameters obtained for the sample of 942 functions. The entropy of 9.55 bits/parameter was computed for the b parameters while the z parameters have the entropy of 9.85 bits. It follows that the average information stored in all parameters for an arbitrary ID is 1+5*9.55+5*9.85=98 bits. This is about 3 times 32 bits (the minimum information required) suggesting the effectiveness of our compression procedure. After running an additional compressing algorithm aimed to minimize the number of bits per parameter[9] better results were obtained, corresponding to variation ranges

[9] This algorithm actually computes the robustness *rob* for a given realization and maximizes this value by choosing the best *rob* obtained after each parameter of the gene was slightly perturbed (±1) without a change in the realized ID. Then, in a loop, all parameters are divided at 2 until the realization ID changes. The set of parameters obtained before the ID changed is the compact version of the original realization. The above compacting procedure can be repeated several times until no improvement is further observed.

between [-1061,806] for b and [-1106,655] for z and with an entropy of about 8 bits per parameter.

(4) Orderly versus random Boolean functions: It is interesting to note that *"orderly" Boolean functions* (e.g. Parity 5 defined by ID=1771476585) *require only a few bits per parameter* and valid realization are found 2 to 3 orders of magnitude faster than their randomly chosen counterparts. For example, z=[-1 -2 -4 -2 -1], b=[1 -3 -1 -1 3], s=1, is a valid realization of the Parity 5 function found after only one repeat. Note that no more than 3 bits per parameter are required in this case. For comparison remember that on average about 800 repeats are necessary to find the gene realization for randomly selected Boolean functions. Similar observations stand for other "useful" Boolean functions – by useful meaning functions that have certain significance.

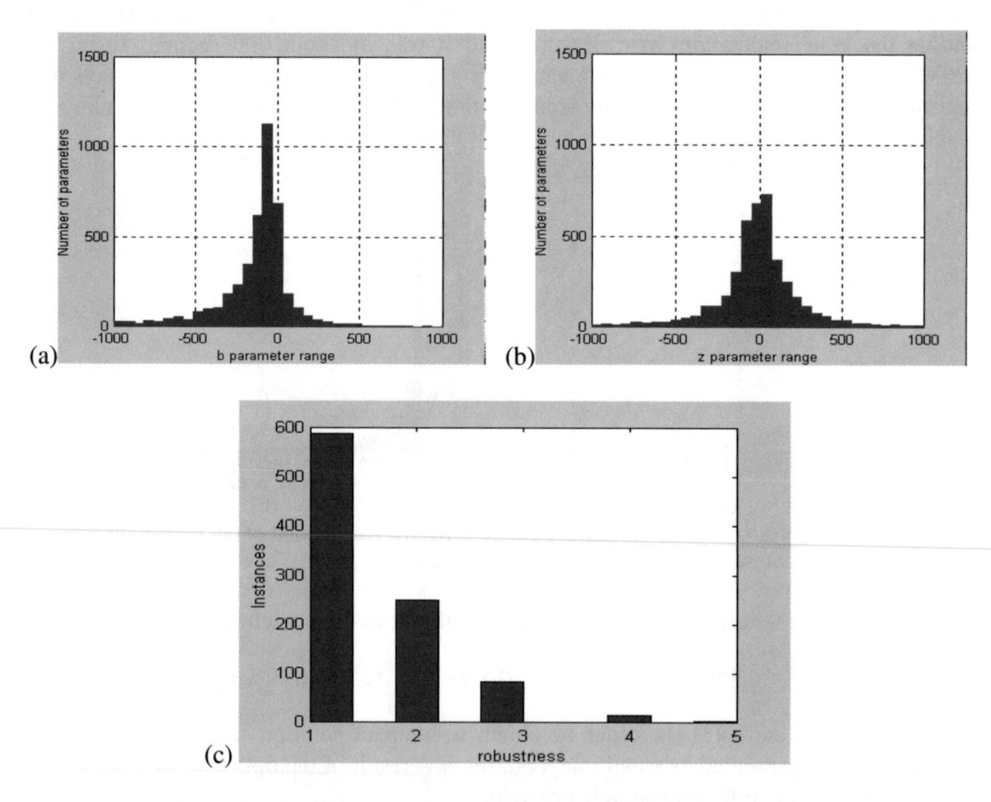

Fig. 12. (a)-(b)The distribution (histogram) of the b and z parameters for a pool of randomly selected 942 realizations of 5-inputs Boolean functions. Observe the Gaussian distribution with a maximum near the origin. The range of the b parameters was [-16698,13390] and of the z parameter was [-13757,16178] although for most of the IDs 9 bits suffices to represent the parameters. (c) Robustness profile: observe 5 levels of robustness with most Boolean functions having robustness equal to 1.

Other random search methods

In principle any global optimization method can be applied to solve the problem of finding a gene realization. An interesting solution was proposed recently in [Ionescu *et al.*, 2002b] where a genetic algorithm tailored to the problem was employed. Comparisons with the above method show an improvement of 10 times in terms of speed. Since the method is more complex that the Greedy-type method presented above, it may lead to more sophisticated and complex hardware implementation. However, by now it is the fastest procedure running on normal PCs to find the genes given a desired Boolean function.

Another method that gives good results is presented next. It is based on the Alopex algorithm [Harth & Pandya, 1988], and it represents an improvement over the Greedy method in that it replaces the random search around a candidate solution with a *directed* search. A directed search means that not all directions are equally privileged and the privileged direction is determined using a correlation between cause (mutation in the gene parameter) and effect (improvement of the objective function) in the previous iteration step. An annealing technique is also used, based on a *temperature* variable such that at the beginning when temperatures are high, the method resembles very much the Greedy algorithm above while during the run some more information is accumulated and the search direction can be more specifically defined. This situation corresponds to a decrease of the *temperature* parameter which tends to reach 0 at the end of the search cycle.

Given a multi-nested structure (27) with a well defined number of nests, assume the parameter s to be either +1 or -1, and generate a random gene. In this case the gene $G=[s,z_0,z_1,..z_m,b_1,..,b_n]$ is a vector with $m+n+1$ real-valued parameters. In the case of a completely random search, at each step a random mutation is generated and an objective function F is evaluated.

The function F, in our case, is defined to be the distance between the desired Boolean function Y and the effective Boolean function realized via (27) with the assumed gene. Whenever a gene is found to minimize F, its value is stored and the process continues until F reaches its global minimum value $F=0$.

The goal function in our case is defined as follow:

$$F=0.5\sum_{j=0}^{2^n-1}\left|Y(V_j)-\mathrm{sigm}\left(w\left(\sigma(V_j,G),G\right),\rho\right)\right| \tag{46}$$

where $w\left(\sigma\left(V_j,G\right),G\right)$ is given by (27) for inputs corresponding to the vertex V_j and having its parameters defined by the actual gene G. The steep "sign" function from equation (42) is replaced here with a smoother approximation; namely, the sigmoid function:

$$\mathrm{sigm}(x,\rho)=\frac{1-\mathbf{exp}(-\rho x)}{1+\mathbf{exp}(-\rho x)} \tag{47}$$

This replacement was found useful in order to achieve a faster convergence. By trial and error, the most convenient value for the sigmoid gain was found to be $\rho=0.5$ for our optimization problem. In this particular case we used floating point representations of the gene coefficients. If integer coefficients are used the gain of the sigmoid function should be adjusted accordingly. The idea of using the sigmoid is that the abrupt staircase-type error surface obtained when sign() function is used is now replaced with a smooth error surface where the local gradient may give important information on the search direction.

In a random search, the mutation ΔG is a binary vector with a number of components equal to the number of parameters composing the gene, and where the probability of each component being either -1 or +1 is equal; namely, 1/2 . Therefore any direction in the parameter space is equally privileged. Formally, this situation can be expressed as:

$$\Delta G=[g_1,g_2,..,g_k,..,g_{n+m+1}],\ \text{where}\ P(g_k=-1)=P(g_k=1)=0.5 \tag{48}$$

The mutation strength is determined by an additional parameter $\theta>0$. Independent of the type of search (random or directed), the resulting "mutant" gene is:

$$G = G + \theta \Delta G \tag{49}$$

In a *directed search*, the probabilities of the mutation vector are not necessarily equal to 1/2. Instead they are determined using the following steps:

(a) Determine the correlation between the last mutation g_k^- and its effect on changing the goal function from F^{--} (from two iterations earlier) to F^-:

$$c_k = g_k^-\left(F^- - F^{--}\right),\ k = 1,\ldots,m+n+1 \tag{50}$$

(b) Determine the following threshold parameter (which is influenced by the actual temperature T):

$$t_k = \frac{1}{1+\mathbf{exp}(c_k / T)} \tag{51}$$

Observe that $t_k = 0.5$ if the temperature is much higher than c_k, and it can range from 0 to 1 depending on c_k, when $T < |c_k|$

(c) Determine the new component g_k of the mutation vector

$$g_k = \mathbf{sgn}(t_k - \xi) \tag{52}$$

where ξ is a random variable uniformly distributed on the interval [0,1]. Such a variable may correspond to a chaotic signal in a practical realization.

The above Alopex algorithm for finding a multi-nested gene for a given Boolean function Y can be summarized as follow:

1. INITIALIZATION:
   1.1. Algorithm parameters:
   $n\_stps = 4000$, $T = 1$, $\theta = 1/n$, $\rho = 0.5$, $win\_temp = 10$
   1.2. Generate an arbitrary random gene[10] $G = \left[G_1, G_2, .., G_k, \ldots, G_{m+n+1}\right]$, where $G_k = \xi - 0.5$, i.e. a random variable uniformly distributed between -0.5 and 0.5.
   1.3. Generate a null vector of previous mutations and the maximum value possible for the previous goal functions:
   $\Delta G^- = \left[g_1^-, g_2^-, \ldots, g_k^-, \ldots g_{m+n+1}^-\right] = [0,0,\ldots 0]$, $F^- = F^{--} = 2^{n-1}$

2. FOR *step*=1 to *n_stps*
   2.1. Generate a "directed" mutation ΔG using (50)-(52) and update the gene with (49)
   2.2. Evaluate the objective function F using (46)
   2.3. IF $F < 0.45$ test if the actual gene is a valid realization by employing (42) where $w(\sigma)$ is given by (27) and its parameters are defined by the actual gene.
   IF (the realization is valid) EXIT
   END
   2.4. Prepare the next step: $\Delta G^- = \Delta G$, $F^{--} = F^-$, $F^- = F$

---

[10] Or a gene previously resulted from an unfinished optimisation process.

In addition, the Alopex algorithm has a schedule for updating the temperature, after each pool of *win_temp* steps. The new temperature is computed as the variance of a vector formed by all correlation coefficients c_k evaluated with (46) during the last *win_temp* steps. The algorithm parameters listed in the "Initialization" section of the algorithm were found to be the most convenient for solving Boolean functions with 4 inputs. They should be carefully tuned for other circumstances (e.g. larger number of inputs or other cell model).

Example 10

Consider the Boolean function with 4 inputs defined by $ID = 43567$. By choosing arbitrarily the sign parameter $s = -1$, the Alopex algorithm was first run for *m=3.* As shown in Fig. 13(a), the result is a valid gene, which was obtained after 4000 iterations. However, any attempt to find a realization with *m=2* failed, as shown in Fig13(b) where the algorithm saturates on a solution which differs by only 1 bit ($F = 1$) from the optimal one. Observe that the *goal function F* is not decreasing monotonically, as in the case of a well-tuned gradient algorithm; This is the effect of the partially random search which allows for a much better search strategy in the parameter space by avoiding being trapped in various local minims. It is also interesting to observe that allowing the biases z_1, z_2, z_3 to vary, the final solution does not usually satisfy the restriction (29).

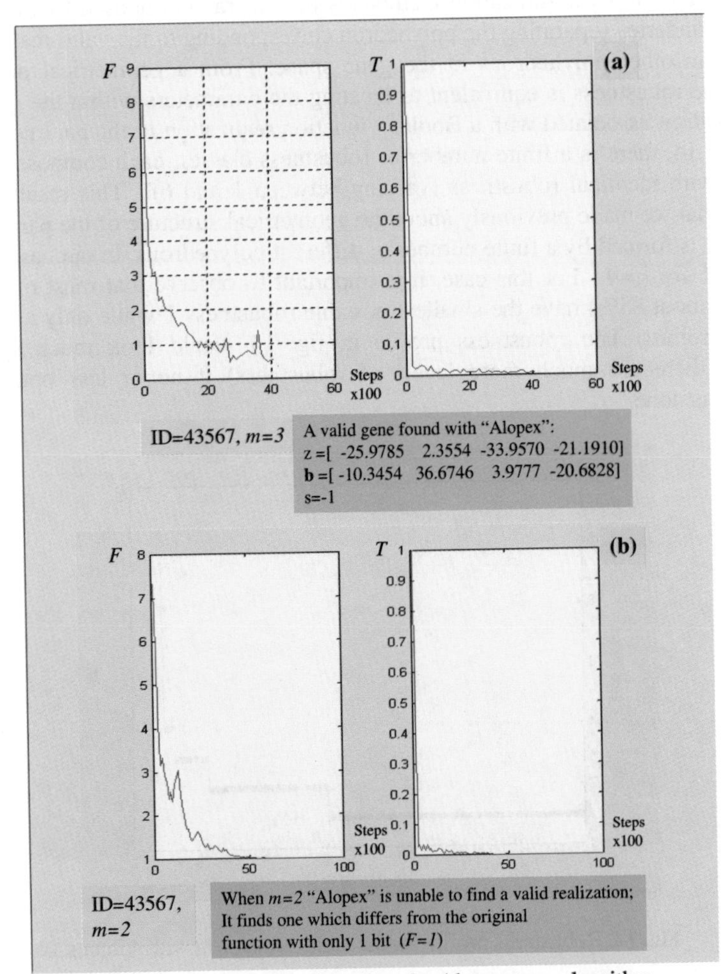

Fig. 13. An example of running the Alopex-type algorithm.

The above algorithm was first implemented in Matlab [Dogaru &Chua, 1999a] and then we used floating point representation for the coefficients. Under that configuration and using a 200Mhz processor, it took about 1 minute to find the realization for a 4-input function. In the light of the newer results presented herein we suggest that this algorithm would be much more effective if integer values in the ranges mentioned above will be used for the gene with a direct C or C++ implementation.

The Robustness of multi-nested realizations

For the realizations obtained using the methods discussed above it is useful to define a robustness measure for each function realization as follow:

$$\rho = \min_j \left(\left| w\left(s, z_0, \ldots, z_{m-1}, \sigma_j\right) \right| \right)$$

where σ_j is the sequence of excitations obtained for all possible combinations of binary inputs.

When a particular realization is found, it is likely that it is not the most robust. Thus, additional optimization procedures shall be employed to improve the robustness parameter above. These are global optimization methods with constraints imposed by the violation of the failure boundaries separating the polyhedron corresponding to the valid realization of the given ID from other polyhedrons in the gene space. From a geometrical point of view, optimizing the robustness is equivalent to locating the parameters within the gravity center of the polyhedron associated with a Boolean function realization in the parameter space. As shown in Fig,14, there is a finite number of robustness classes, each composed of Boolean realizations with identical robustness (varying between 1 and 6). This result sustains the observation that we made previously about the geometrical structure of the parameter space, namely that it is formed by a finite number of different polyhedrons. In our case this number seems to be 6 for $n=4$. For this case, it is important to observe that most of the Boolean realizations (about 85%) have the smallest possible robustness 1 while only a few can be 6 times more robust. The robustness profile in fig. 14 would look much smoother (or equivalently there are much more levels of robustness) if no or less optimization for robustness was done.

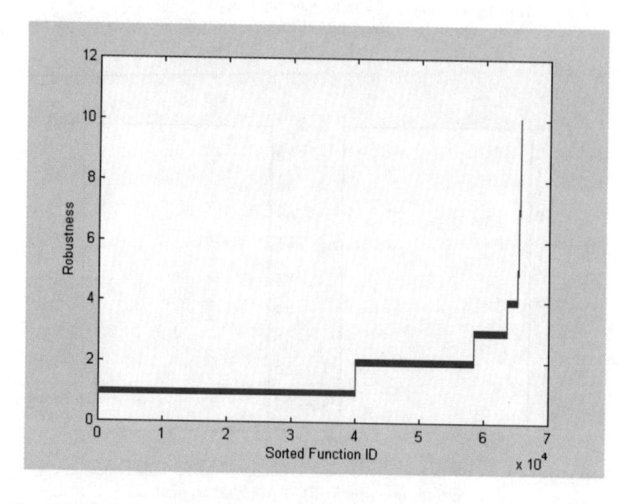

Fig. 14. Robustness profile for the multi-nested cell with 4 inputs.

3.3. Continuous state cells

In the previous section we introduced several models of universal CNN cells as simple piecewise-linear systems capable to represent arbitrary local Boolean functions. Although the inputs of such cells are allowed to vary within a continuous domain, the theory guarantees their universality only for the case when the inputs are binary. Although some exceptions were reported in [Dogaru et al., 1999b] where it was shown that universal CNN cells can also be used for some simple (and low dimensional) signal classification problems, in order to have an universal approximator for the continuous state domain one should develop different strategies. This chapter presents such an architecture, called a Simplicial neural cell [Dogaru et al., 2002a], [Julián et al, 2002a] which is capable of learning arbitrary input-output mappings when both inputs and outputs are allowed to vary within a bounded continuous domain. It is shown that this architecture is capable of similar performance to any of the "classic" neural paradigms while having a very simple and efficient mixed-signal implementation. The architecture and its realization circuit are described and the functional capabilities of the novel neural architecture called a simplicial neural cell are demonstrated for both regression and classification problems including nonlinear image filtering. The simplicial cell architecture can be applied only to discrete-time cellular systems such as the Generalized Cellular Automata (see Chapter 5).

3.3.1. Overview

The function of an universal approximator is often associated with neural network models such as multi-layer perceptrons (MLP), support vector machines (SVM) and so on. Each of these architectures has certain advantages and drawbacks with regards to the convergence and effectiveness of the learning algorithm, the structural complexity, and the functional representation capability. In a cellular system, where many similar cells of the same type have to be implemented on the same chip, compactness become an important issue as we already discussed in a previous section of this chapter.

For example, the SVM networks [Vapnik, 1998], deeply rooted in statistical learning theory offer very good solutions to complex problems while using a guaranteed convergent learning algorithm. However, its complexity is large and leads to non compact solutions. On the other side, the Adaline has the advantage of a simple and convergent training algorithm (the Widrow-Hoff's LMS) algorithm [Widrow & Stearns, 1985], but the functional representation capability is restricted to linearly separable problems only.

The above observations led to the following question: Is it possible to develop novel neural network models, with non-conflicting features? We hope that the reader will find a positive answer in what follows.

The theory of simplicial piecewise linear function approximations, developed by circuit theorists in the 70's, [Kuhnn, 1968], [Chien and Kuh, 1977], [Chua & Kang, 1977], was employed to develop the simplicial neural cell. The novel cell has several attractive properties such as:

- There are no multiplicative synapses an therefore the implementation complexity is low;
- The cell has a very simple and efficient mixed-signal implementation in common VLSI technologies but also in the promising nanotechnologies of the near future [Julián *et al*, 2002c];
- The functional performances for either regression and classification problems are found to be similar to those of classic yet more complicated neural network architectures;

- The central device around which the entire cell is built is a *digital RAM* which can be easily reprogrammed, or its emulation via a *compact neural network* described in Section 3.2.4.
- The learning process consists in a simple LMS algorithm, which has a guaranteed convergence and can be easily implemented in either software or hardware technologies.

The *simplicial neural cell* can be regarded as a linear perceptron operating in an expanded feature space. The expanded feature space with a dimension of 2^n is obtained by computing only *n+1* fuzzy-membership functions of the n-dimensional input vector. Therefore the computational complexity is only *O(n)* for both learning and retrieval. The novelty of our approach consists in exploiting the theory of simplicial subdivision (previously used in PWL approximations of nonlinear circuits) [Chien & Kuh, 1977], [Julián *et al.*, 1999] for defining the above fuzzy-membership functions.

In the previous section the *multi-nested CNN cell* was introduced. Since it can be viewed as a nonlinear extension of the Adaline, we will call it also a *nested Adaline* in the next. Let us recall that arbitrary Boolean functions can be represented using an *O(n)* complexity adaptive structure defined by:

$$c = sign\left(z_n + \left| z_{n-1} + \left| \ldots z_2 + \left| z_1 + \sum_{i=1}^{n} b_i v_i \right| \ldots \right| \right| \right) \tag{53}$$

where c is the binary output of the nested Adaline CNN cell, v_i are the n binary inputs and $\{z_i, b_i\}_{i=1,\ldots,n}$ is a set of trainable parameters. It is assumed that if $z_{k+1} = 0$ for $k \geq j$, there are only $j-1$ absolute value functions ("nests") in (1) , to the right of z_j. Hence, the Adaline or the linear threshold gate (LTG) is a special case of (1) for *j=1*.

From an image processing perspective, black-and-white image processing corresponds to binary CNN cells (where both inputs and outputs belong to the set {0,1}) while gray level image processing corresponds to inputs and outputs belonging to the interval [0,1][1].

A universal CNN cell implements any arbitrary nonlinear mapping $y = f(\mathbf{u}, \mathbf{G})$ of an input vector $\mathbf{u} = [u_1, u_2, \ldots, u_n]$ to a desired output y by properly choosing the *gene* [2] vector G. An image processing task, or in general a signal processing problem, can be formalized by an expert as a sequence of local rules, or it might be specified by a set of input-output samples (for example, an image filtering task can be specified by image samples such as those shown in Fig. 15). In the first case, a sequence of logical inferences of the type IF ($\mathbf{u} = \mathbf{v}_j$) THEN ($y = c_j$) is derived in the form of a *decoding book* [Chua, 1998], while in the second case a learning algorithm should be employed to determine **G**.

For the case of black-white image processing the *n*-dimensional input vector **u** associated with each cell (pixel), is confined to the vertices $\mathbf{v}_1, \mathbf{v}_2, \ldots, \mathbf{v}_j, \ldots, \mathbf{v}_N$ of an *n*-dimensional hypercube (Fig.16)., where a vertex is an *n*-dimensional vector $\mathbf{v}_j = \left[v_1^j, v_2^j, \ldots, v_n^j \right], v_i^j \in \{0,1\}, \ i = 1,..,n$. The number of vertices is equal to $N = 2^n$. The brightness of each vertex in Fig. 16 corresponds to the desired output value $c_j = y(\mathbf{v}_j)$.

---

[1] Without loss of generality and for the simplicity of exposition, the binary set used in this paper is {0,1} although most of the CNN literature uses the set {-1,1}.

[2] A cell *gene* was defined in [Chua, 1998] as a vector formed of all *m* parameters defining the cell model.

Conceptually, the simplest universal cell for the binary case is a ROM (Read Only Memory) where **u** is the address bus while the content of the ROM is a *N*-dimensional binary gene vector $C = [c_1, c_2, ..., c_j, ..., c_N]$. In order to achieve programmability, a RAM (Random Access memory) may replace the ROM. However this solution is rather expensive in terms of occupied area due to the RAM or ROM implementation complexity of $O(2^n)$.

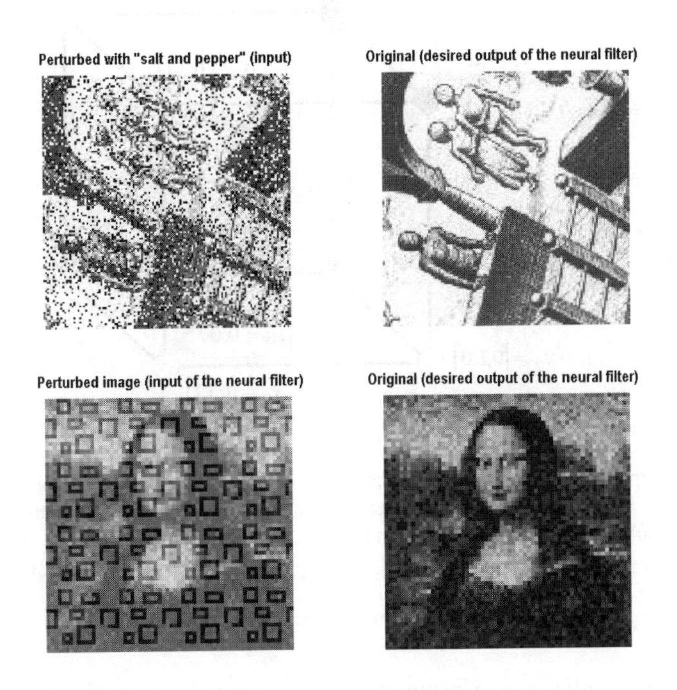

Fig.15. Image samples used to train the simplicial neural cell

As shown in Section 3.2.4 the multi-nested universal CNN cell (or *nested Adaline*) was proved to emulate the RAM while being a universal programmable logic (Boolean) operator with an implementation complexity of only *O(n)*. A circuit implementation of this concept exploits the non-monotone characteristic of resonant tunneling diodes (RTD) to achieve a very compact and fast realization in emerging nanotechnologies [Dogaru *et al.*, 2000a] although it can also have a convenient implementation in more usual CMOS technologies.

The transition from *binary* to *continuous* information processing is graphically represented by expanding the input space to the entire volume of the hypercube (see Fig.16) and allowing that $y = c_j \in [0,1]$.

Our main idea is to employ *simplicial piecewise-linear kernels* to expand the input feature space so that a simple Adaline can be trained in this expanded space. In order to decrease the approximation error each input can be further expanded to more inputs using a simple non-monotone transformation, as shown in Section 3.3.6. The resulting networks are able to approximate not only binary-to-binary mappings but any type of functions, thereby expanding the range of applications from Boolean function representation to arbitrary classification and regression tasks. All networks in this category have the major advantage of a guaranteed convergent and simple (gradient based) learning algorithm. As shown in the nest sections their circuit implementation is simple and convenient and its complexity can be reduced to *O(n)* when nested Adalines are employed to implement the embedded local logic.

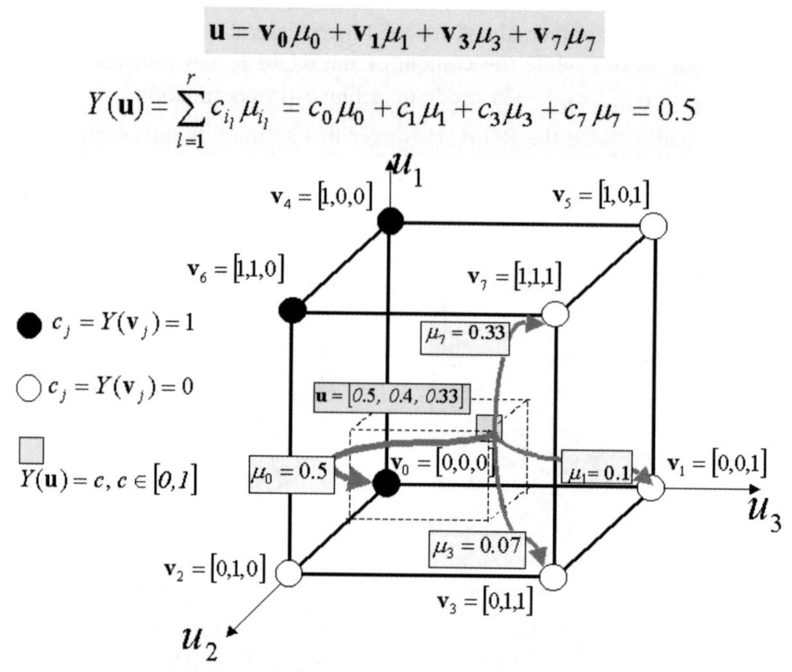

Fig.16. The input space of the simplicial cell. The vertices corresponding to binary processing are now basis vectors and any arbitrary continuous vector **u** can be decomposed (simplicial decomposition) in a weighted sum of no more than *n+1* such vertices. Moreover, any arbitrary input-output mapping $Y(\mathbf{u})$ can be also approximated as a weighted sum of certain coefficients c_j which are now representing the function.

Although the nonlinear expansion of the input space is not a new idea, dating back to Cover's results in [Cover, 1965], it proved to be highly effective, being adopted in modern approaches such as the Support Vector Machines and other types kernel-based networks. The general model of a neural net in this category is given by:

$$y = \sum_{j=1}^{ne} c_j \mu_j(\mathbf{u}) \tag{54}$$

where $\mathbf{u} = [u_1, u_2, ..., u_n]$ is the *n*-dimensional input vector, *ne* is the dimension of the expanded space, and c_j is a set of parameters composing the gene vector **G.** As long as the kernel functions $\mu_j(\mathbf{u})$ are predefined, learning from examples reduces to the task of training an Adaline in an input space formed by the outputs of the basis (or kernel) piecewise-linear functions $\mu_j(\mathbf{u})$. Instead of the LMS, modern training techniques capable to determine the optimal margin separation hyper-plane [Vapnik, 1998] can be employed leading to improved performances. The main issue for such kernel-networks is how to choose the kernel functions such that good functional representation performances are achieved while keeping the computational complexity as low as possible.

The *pyramidal cell* [Dogaru *et al.*, 1998e] was the first attempt to derive a kernel-based piecewise linear neural network, which was demonstrated to be *universal at least for the class of Boolean functions.* With the goal of binary universality in mind, a special set of

$ne = 2^n$ basis functions $\mu_j(\mathbf{u})$ was developed. Each basis was entirely determined by the vertex vector $\mathbf{v}_j$ which is nothing but the binary representation of the index "*j*". In addition to its Boolean universality, the cell was proved to posses some interesting properties in approximating continuous mappings such as those required by various nonlinear image filtering tasks. Although it requires 2^n parameters, the pyramidal cell model was the first for which was shown that each parameter can be quantified with a very small number of bits, i.e. it is very robust. The same interesting property stands for the case of the *simplicial neural cell* as discussed later in Sections 3.3.2 to 3.3.5.

Despite its interesting features, the pyramidal universal cell still has two major drawbacks: First, there is no proof that it represents a universal approximator for continuous mappings (the proof stands only for binary mappings, i.e. Boolean functions). Second, although it has a small number of basic building blocks, its implementation complexity is of $O(2^n)$ and the only way of reducing it is by sacrificing computation time which then becomes $O(2^n)$ for an $O(n)$ implementation complexity. In order to overcome the above drawbacks, the *simplicial neural circuit* was first proposed [Dogaru *et al.*, 2001a] following a theoretical motivation exposed in [Julián et al., 2001]. Although it still requires 2^n coefficients c_j in the model given by (54), it was shown that an elegant mixed-signal circuit implementation exists so that the implementation complexity can be effectively reduced to $O(n)$ while the computation time is of the same order. Moreover, there are theorems in [Chien & Kuh, 1977] showing that the resulting network is a universal approximator for any continuous input-output mapping.

The mathematical model of the simplicial neural cell is a particular case of (54) for $ne = n + 1$ and can be written as:

$$y = \sum_{l=1}^{n+1} c_{i_l} \mu_{i_l}(\mathbf{u}) \qquad (55)$$

where i_l is a vertex index ranging between 0 and $2^n - 1$. It is equivalent to index j in the general kernel based architecture described by (54).

The *simplicial neural cell circuit*, presented in more detail in the next sections, implements (55) as follows (see also Fig.17): A ramp signal increasing from 0 to 1 (assuming that all inputs $u_i \in [0,1]$) is generated during a predefined period of time T (the systems' clock period). During this time the correct sequence of *n+1* binary vertices $\mathbf{v}_{i_l}$ (also defined by the binary code i_l) in (55) is provided at the outputs of the *n* comparators. The comparators are connected such that their inverting inputs are all connected to the unique ramp signal while their non-inverting inputs are the actual inputs of the simplicial neuron. Each binary code of i_l is present at the output of the comparators for a period of time, which is a fraction of T equal to $\mu_{i_l}(\mathbf{u}) \cdot T$. A universal binary device (e.g. a Random Access Memory addressed by $\mathbf{v}_{i_l}$)[3] maps the input binary code $\mathbf{v}_{i_l}$ to its corresponding coefficient c_{i_l} and it is assumed that this mapping is done almost instantaneously. Thus a current signal c_{i_l} is generated at the output of the RAM for the

---

[3] $\mathbf{v}_{i_l}$ is the binary vertex (code) representing the index i_l as an unsigned binary word.

same period of time $\mu_{i_l}(\mathbf{u})\cdot T$. It is obvious that employing a simple capacitor to integrate the output of the RAM, at the end of the entire computation cycle T the voltage on the capacitor is proportional to y in (55). The essential component of the *simplicial neural network* is the device mapping a binary input vector $\mathbf{v}_{i_l}$ (representing the index i_l in a binary form) to its associated coefficient c_{i_l} in (55). A detailed discussion on implementing this device is provided later in Section 3.3.3.

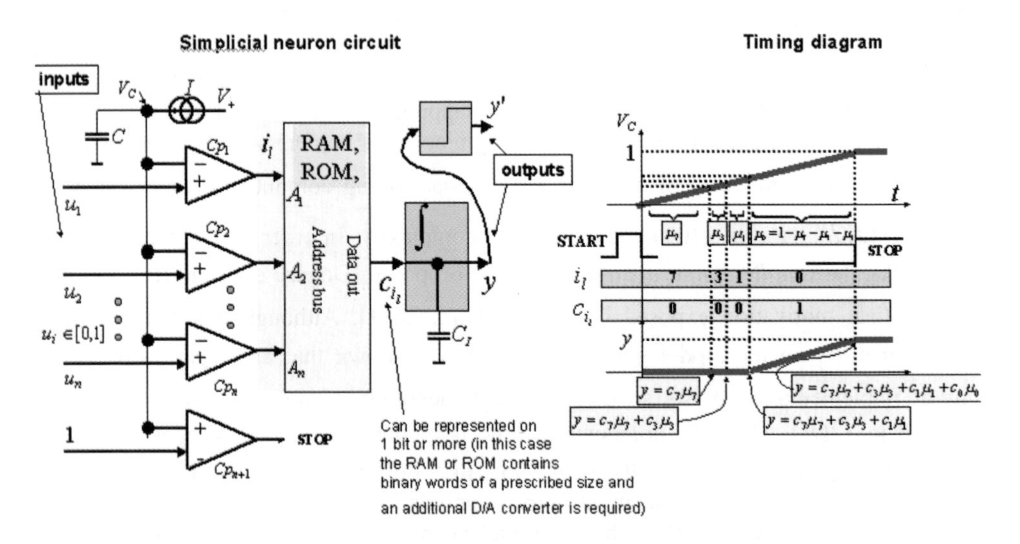

Fig. 17. The circuit architecture of the simplicial cell. The output is available at the end of the cycle as shown in the timing diagram. In a cellular system there is only one comparator per cell and the logic controlling the reset of the two capacitors is the same for all cells.

In order to learn a specific input-output mapping, the c_{i_l} coefficients are subject to LMS (Widrow-Hoff) or other similar training algorithms. It is important to point out that the learning process has the same computational complexity as that of a simple linear perceptron with n inputs although there are 2^n distinct coefficients. The reason is that once an input vector is presented at the input only $n+1$ of the basis functions are non-zero therefore making unnecessary the learning process for other than the corresponding coefficients.

In what follows the theory of the simplicial subdivision is exploited showing that the inference performed by the simplicial neural network can be recast conveniently in a neuro-fuzzy formalism, as detailed in Section 3.3.2. Several implementation issues are then discussed in Section 3.3.3, and details on the training procedure are presented in Section 3.3.4. Section 3.3.5. deals with the regression functional capabilities of the simplicial neuron, evaluated for several nonlinear image filtering problems. Several results in employing the simplicial cell in classification problems are also briefly discussed. Next, in Section 3.3.6 the issue of improving the approximation capability by nonlinearly expanding the input space is discussed. A brief analysis of the hardware efficiency compared to classic multi-layer perceptron solutions is given in Section 3.3.7.

3.3.2. Some theoretical issues on simplicial neural cells

Relationships with fuzzy logic

Let us consider an arbitrary point **u** belonging to the input hypercube, and defined by its associated vertices $\mathbf{v}_j$ (see the example in Fig.16). From the fuzzy theory point of view, each vertex represents a crisp IF-THEN rule, more precisely it codes the rule IF $(\mathbf{u} = \mathbf{v}_j)$ THEN $(y = c_j)$. Such a list of rules can be loaded in a RAM (or ROM) with c_j corresponding to the data bus[4]. A crisp (Boolean) inference cell will provide an output only when **u** is one of the vertices. What happens when **u** represents an arbitrary vector coding a gray-scale portion of an image or a set of continuous signal samples? The crisp (Boolean) table stored in a RAM *cannot* provide a solution. However the fuzzy logic theory provides an elegant solution for determining the output. Indeed, the premises $(\mathbf{u} = \mathbf{v}_j)$ are no longer FALSE or TRUE, instead they are assigned a truth value $0 \le \mu_j(\mathbf{u}) \le 1$, where $\mu_j(\mathbf{u})$ denotes the fuzzy membership of vector **u** to the premise $(\mathbf{u} = \mathbf{v}_j)$. The choice of the fuzzy membership is subjective in nature, thereby giving rise to criticisms from many adversaries of fuzzy-logic. By establishing an equivalence between the nonlinear mapping (55) and a fuzzy inference system, this criticism can be responded. Indeed, the theory of simplicial subdivision provides an effective method for *evaluating* $r \le n+1$ nonzero membership functions instead of 2^n arbitrarily predefined membership functions, as in a typical fuzzy or neuro-fuzzy inference system. An efficient and simple algorithm for computing the simplicial membership functions can be defined via a mixed-signal circuit, as shown already in Fig.17.

The resulting *simplicial neural cell* can be regarded as the equivalent of a Tsukamoto's fuzzy inference system [Lin & Lee, 1996] (page 154) and therefore its output is computed following the "center of gravity rule" as $y = \sum_{j \in J} c_j \mu_j \Big/ \sum_{j \in J} \mu_j = \sum_{j \in J} c_j \mu_j$, since $\sum_{j \in J} \mu_j = 1$ by definition in the simplicial decomposition [Chien and Kuh, 1977]. The above relation is identical to (55). For the example in Fig.16, the $r = 4$ vertices (rules) used to infer the output y are given by $J = \{i_l\} = [i_1, i_2, i_3, i_4] = [7,3,1,0]$, which can be obtained using the simplicial decomposition algorithm in [Chien and Kuh, 1977], or its circuit implementation described herein. Consequently, the corresponding membership functions are evaluated and the output can be computed for a given choice of the gene vector.

Training and testing samples

In order to use the simplicial neural cell for regression or classification, one has to determine the *gene,* i.e. the vector of coefficients $\mathbf{C} = [c_1, c_2, ..., c_j, ..., c_N]$ such that $\left\| d^p - \sum_{l=1}^{r} c_{i_l} \mu_{i_l}(\mathbf{u}^p) \right\| < \varepsilon$ for all pairs $\lfloor \mathbf{u}^p, d^p \rfloor$, $p = 1, ...P$ of input vectors and desired outputs available from the P training samples. To solve this problem, the simplicial neural cell (55) can be regarded as a linear adaptive neuron (Adaline) in an expanded N-dimensional feature space defined by the sparse outputs $\mu_{i_l}(\mathbf{u}) = \mu_j(\mathbf{u})$. Consequently, the

---

[4] If c_j is not binary, a wider binary word coding it can still be stored in a digital RAM (or its multi-nested universal CNN cell emulation).

simple LMS (Widrow-Hoff) rule [Hassoun, 1995] (page 66) can be employed to determine the gene parameters c_j after sufficient epochs of presenting the training samples.

Therefore, in each step p of the learning process, the coefficients c_j are updated using the following algorithm:

1. **Given $\mathbf{u}^p$, compute the sequence of vertices $\mathbf{v}_{i_l}$, and non-zero membership functions μ_{i_l}, as well as the $r < n+1$ indexes i_l (see Section 3.3 for an example).**
2. **Compute the effective output of the simplicial CNN cell:** $y^p = \sum_{l=1}^{r} c_{i_l} \mu_{i_l}(\mathbf{u}^p)$
3. **For all indexes i_l ($l = 1,\ldots,n+1$) update the coefficients:** $c_{i_l} = c_{i_l} + \eta(d^p - y^p)\mu_{i_l}$

To evaluate the efficiency of the training process a test set with Q samples is used and the training stops when $\left\| d^q - \sum_{l=1}^{r} c_{i_l} \mu_{i_l}(\mathbf{u}^q) \right\| < \varepsilon$, where ε is a prescribed minimal error, and $q = 1,\ldots,Q$. At the beginning of the algorithm all parameters are initialized to 0, i.e. $c_j = 0,\ j = 1,\ldots,2^n$. The training rate η is chosen large enough to ensure fast convergence of the algorithm. Theoretical bounds for choosing an optimal value of η are given in [Hassoun, 1995].

Quantization of gene's coefficients

The results obtained using various training sets for either regression (nonlinear filtering) and classification problems led to the conclusion that under a proper choice of the dimension n of the input space[5], the overall performance (i.e., the error or the misclassification rate) does not change significantly when the coefficients c_j are quantified with a small number of bits. Particularly, the case of only one bit is important since it leads to a lower implementation complexity. The resulting cell in this case is called a *compressed simplicial neural cell.* The following two algorithms can be used to perform the gene coefficients quantization:

1. ***Direct quantization:*** In this case, c_j is replaced by $\hat{c}_j = \alpha + \beta \operatorname{sgn}(c_j - \tau)$, where α, β, and τ are parameters which should be optimized independently for a maximum of performance (i.e. minimization of misclassification error on the test samples).
2. ***Direct quantization followed by evolutionary programming:*** In this case, a better performance of the overall system is obtained. This optimization procedure begins with an initial solution given by a direct quantization as above and continues with a set of evolutionary programming techniques applied to the string of binary genes c_j until the overall performance of the system reaches a global optimum.

---

[5] A large dimension n is preferred because it leads to a large number of gene parameters. Consequently the information allocated to each of these parameters will necessarily be small, therefore, at the limit one bit per parameter may suffice to encode the problem learned by the neural network. Although the number of inputs is in general fixed by the nature of the problem, the input space dimension can be easily expanded by some simple linear or nonlinear transforms (see section 3.3.6 for an example).

3.3.3. Circuit implementation issues

The key issue in computing the output of the *simplicial neural cell* for a given input vector $\mathbf{u}$ is the evaluation of all membership functions μ_{i_l} and their associated indexes i_l and vertices $\mathbf{v}_{i_l}$. Recall that the vertices can be used as a unique base in which any arbitrary $\mathbf{u}$ can be decomposed (Step 1 in the algorithm in Section 3.3.2). The theoretical foundations of this decomposition are given in [Chien and Kuh, 1977] and [Julián et al, 2002a]. In order to exemplify the *simplicial decomposition* process let us consider the example in Fig.16, which corresponds to $\mathbf{u}=(0.5, 0.4, 0.33)$. As we are assuming that all components satisfy: $0 \le u_i \le 1$ all available data should be first scaled to [0,1]. The algorithm follows:

1. **Pick the minimum non-zero component of $\mathbf{u}^{(1)}$, i.e. 0.33, and decompose it as:** $\mathbf{u}^{(1)} = 0.33(1,1,1)+\mathbf{u}^{(2)}$**, where** $\mathbf{u}^{(2)}=(0.17, 0.07, 0)$.
   Consequently: $\mathbf{v}_{i_1}=(1,1,1)$, $i_1 = 111_2 = 7$, $\mu_{i_1}=\mu_7=0.33$
2. **Pick the minimum non-zero component of $\mathbf{u}^{(2)}$ i.e. 0.07 and decompose it as:** $\mathbf{u}^{(2)} = 0.07(1,1,0)+\mathbf{u}^{(3)}$**, where** $\mathbf{u}^{(3)}=(0.1, 0, 0)$.
   Consequently: $\mathbf{v}_{i_2}=(1,1,0)$, $i_2 = 011_2 = 3$, $\mu_{i_2}=\mu_3=0.07$
3. **Pick the minimum non-zero component of $\mathbf{u}^{(3)}$ i.e. 0.1 and decompose it as:** $\mathbf{u}^{(3)} = 0.1(1,0,0)+0$**, where this step marks the end (no further decomposition is possible)**
   Consequently: $\mathbf{v}_{i_3}=(1,0,0)$, $i_3 = 001_2 = 1$, $\mu_{i_3}=\mu_1=0.1$
4. **Choose** $v_{i_4}=(0,0,0)$, $i_4 = 000_2 = 0$ **and compute** $\mu_{i_4}=\mu_0=1-(\mu_7+\mu_3+\mu_1)=0.5$

It is easy to check that the computation above is done straightforwardly by the comparators in Fig.17. The essential element in the circuit implementation is the RAM (or its replacement) which memorizes the gene coefficients c_{i_l} and therefore the input-output mapping in a compressed form. The output integrator is also a simple capacitor but it realizes the important function of summing in time the weighted contribution of all coefficients, thereby implementing (55) with a minimum of electronic devices. Unlike most neuro-fuzzy networks, no multiplier circuit is required by our simplicial neuron, the output scheme being thus similar to pulse-stream neural networks reported revently in the literature [Ota & Wilamowski, 1999].

Note that if the simplicial neural cell is used for image processing, in a fully parallel CNN implementation there is no need to implement all 9 comparators per cell. Instead one cell contains only the comparator associated with the central pixel, the digital RAM (or ROM in the case of a predefined function), and the output integrator. The current source I, the capacitor C and the comparator Cp_{n+1} are *common* to the entire chip. Since dynamic RAM densities in actual commercial technologies can easily reach 10^9 bits/chip, a rough calculation indicates that with minor modifications, on the same chip one can easily integrate about 10^6 simplicial neural cells. Each cell is composed of a 512 bits memory[6] plus a few components implementing the input comparator and the output integrator.

---

[6] For a homogeneous CNN, the RAM in each cell can be replaced by a 2^n to 1 multiplexer which has

This corresponds to transforming an obsolete DRAM chip into a 1024x1024 pixel resolution programmable CNN which allows for up to $2^{512} = 1.3 \cdot 10^{134}$ different gray level and binary image processing functions !

The implementation density of the circuit can be increased furthermore by emulating the RAM with a *nested Adaline* defined by (53).

Considerations regarding the implementation of the local Boolean logic

The mapping $c_{i_l} = C(\mathbf{v}_{i_l})$ corresponding to the RAM or other local Boolean logic device in Fig.15 can be implemented as follows:

- Using a RAM (or ROM) where the address bus receives the code $\mathbf{v}_{i_l}$ and the corresponding coefficient (using a limited number of bits, say *p* bits) is stored at the associated address. As shown next for many practical problems the one-bit quantization (*p*=1) may suffice. In such cases, the implementation complexity is $O(2^n)$. Although it represents an exponential complexity this result should be interpreted optimistically since the RAM technology with increased densities is readily available and represents the driving force of the VLSI industry. On the other let us consider the following question: What is easiest to implement? A system where all parameters are *concentrated* in a single device (i.e. the $p \cdot 2^n$ one bit RAM cells) or a system such as the Multi Layer Perceptron (MLP) where one has to build several tens or hundreds of neurons, each with its own storage devices (e.g. analog RAMs or similar) for their synaptic weights? We suggest that for many applications, particularly those characterized by a small number of inputs (e.g. $n < 16$) the use of a digital RAM in a simplicial neural cell architecture is the most efficient and convenient solution.

- Emulating the RAM with a neural network solution. Indeed, one can train a convenient neural network architecture with binary inputs and outputs to learn the mapping $c_{i_l} = C(\mathbf{v}_{i_l})$ instead of storing the coefficients in a RAM memory table. It is expected that a dramatic reduction in complexity will occur, particularly for the cases with large number of inputs *n,* when a problem is defined by a number of training samples $M << 2^n$. Indeed, if we consider the fixed point representation of c_{i_l} using *p* bits, each of these binary outputs represents a Boolean function with *n* inputs (the binary code $\mathbf{v}_{i_l}$) which can be conveniently learned by a *nested* Adaline described by (53). This architecture solution is represented in Fig.22. Therefore, the overall complexity in terms of number of devices and parameters is now only $O(n)$. And this result stands while the resulting simplicial neural network is a universal approximator, therefore being capable to learn arbitrary problems! We should consider this result with caution because each parameter of the multi-nested cell may require a representation resolution of up to $2^n / n$ bits per parameter [Dogaru & Chua, 1999a]. At this point, the question mentioned above becomes more evident.

---

the same implementation complexity as the RAM. Apparently simpler, this solution is unpractical since it requires a 2^n lines gene bus connecting all cells. Instead, a single 1 bit data bus suffices to program all cells, when the RAM is used.

Indeed, which device is more convenient to implement: The RAM which is a readily available device containing an array of 2^n binary cells, or the $O(n)$ complexity multi-nested neural cell where each of the $2n$ parameters require some storage mechanism with a resolution of $O(2^n / n)$? Put in this way, the answer to the question will tend to be: The RAM. However, there might be numerous practical situations for which the emulation of the RAM with a neural network is far more convenient. Let us consider the example of a Median filter, which will be discussed further in Section 3.3.5. This example demonstrates another viewpoint of the simplicial neural cell. According to this viewpoint, the functional capability of a neural network (the one emulating the RAM) is highly increased by embedding it within the additional circuitry formed by the input comparators and the output integrator. Let us first consider the simple linear threshold gate defined by: $y = sign\left(\sum_{i=1}^{9} u_i\right)$. If we apply this LTG as a local filter for a 3x3 neighborhood in a noisy (salt and pepper type) gray level image, the result of processing will be a black-and-white image without any meaning. However if we employ the same simple neural structure as a mapping $c_{i_l} = C(\mathbf{v}_{i_l})$ in the simplicial neural cell architecture, the result is a highly efficient Median filter which restores the original gray scale image from the corrupted one. The implementation complexity of the resulting simplicial cell is just a bit larger than that of the linear threshold gate, leading to the simplest mixed-signal implementation known so far for a Median filter.

Concluding, we suggest that the solution of emulating the RAM with nested Adalines is in fact highly desirable and can lead to dramatic reduction in complexity for many real world problems.

The method of RAM emulation becomes practical for situations where a large number of inputs are required, making thus impossible the use of very high capacity RAMs. In most cases such situations correspond to a number of samples $M << 2^n$, therefore the RAMs will be inefficiently used (many locations will be never accessed). For example, consider a problem with $n = 30$ inputs and 3000 training samples. The implementation solution using a RAM will require several RAM chips of 1024 Mbits each (still not available on the market) while at most 30x3000=90.000 of their locations will be effectively used (assuming that in the worst case, each of the input vectors will generate a set of vertices which has a void intersection with the sets generated by any other input). It is clear that in this case, independently of how complex the problem to be learned is, the RAM emulation represents a much more reasonable solution. In most cases it might be possible that a simple linear threshold gate or a multi-nested cell (nested Adaline) with a reasonably small number of "nests" (i.e. absolute value functions in (53)) will emulate well enough the RAM so that the overall simplicial cell will have good generalization performances.

Software implementations

Although the simplicial neuron architecture is tailored to the mixed signal implementation, it has also a convenient implementation in software. Particularly, the main advantage of a software implementation is the possibility to replace the simple RAM structure with a dynamically allocated one. Indeed, for a large number of inputs (e.g. n=30) the circuit implementation in Fig. 17 becomes prohibitive since it requires a RAM with a capacity of 2^{30} bits. However, in such cases only a small fraction of the RAM locations are actually accessed, since for most of the practical problems the number of training samples would be $M << 2^n$. Therefore, a dynamically allocated RAM, easily to program in a software implementation, would replace the linear RAM removing the memory space problems as long as M has reasonable values. It is easy to verify that in the worst case a maximum of $M(n+1)$ memory cells would be required to store the gene.

3.3.4. A general procedure for training the simplicial cell

In order to use the above architecture and its circuit realization for different tasks, the following steps should be considered:

1. Prepare a set of training samples using representative signals (images) for a given task. Also prepare a different set of test signals (images) to evaluate the generalization performance of the simplicial neural cell.

2. Use the above training samples to determine the gene coefficients $c_j = c_{i_l}$ in the simplicial decomposition (55). As the structure is linear with respect to the outputs μ_{i_l}, the gene coefficients can be learned in a straightforward manner, for example using the LMS algorithm. As a result of learning, the resulting gene coefficients are *not binary*.

3. Quantify the coefficients $c_j = c_{i_l}$ using *one bit per coefficient*. Quantization ensures that each coefficient can be computed as the output of a nested Adaline (53). For example, the direct quantization scheme described above was often employed with acceptable results. One can also choose $p>1$ bits per coefficient, leading to a more accurate representation but also requiring more devices in the cells' circuit implementation.

4. Determine the parameters of the nested Adaline emulating the RAM using one of the methods described in Section 3.2.4 such that it can represent the Boolean function defined by the pairs of samples $(\mathbf{v}_j, c_j)_{j=1,..,N}$ where $\mathbf{v}_j$ is a binary input vector and c_j is the desired output. Further iterations trough steps 3 and 4 can improve the quality and the compactness of the solution. Step 4 will be removed if the choice is to implement the local Boolean logic in the simplicial neuron with a RAM. Consequently, the RAM has to be loaded with the quantified values of the gene coefficients c_j, as they resulted from step 3.

3.3.5. Functional capabilities and applications

Within the framework of the Cellular Neural Network (CNN), the adaptive nature of the simplical neural cell makes it suitable for image filtering, signal classification and other applications where the problem is specified by samples. Next some representative examples are considered.

Square scratch removal

The problem of square scratch removal is defined by a set of training image samples (such as in Fig.15(lower row)). Other test image samples (see Fig.18) are used to evaluate the performance of the resulting neural filter. An ideal filter should reconstruct any other original image when the CNN input is provided with a perturbed input image.

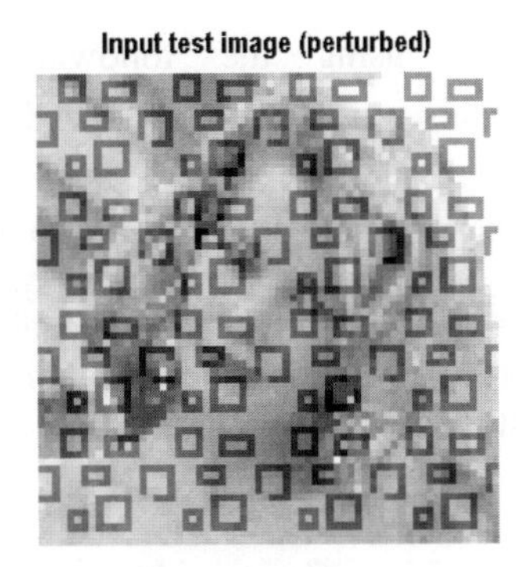

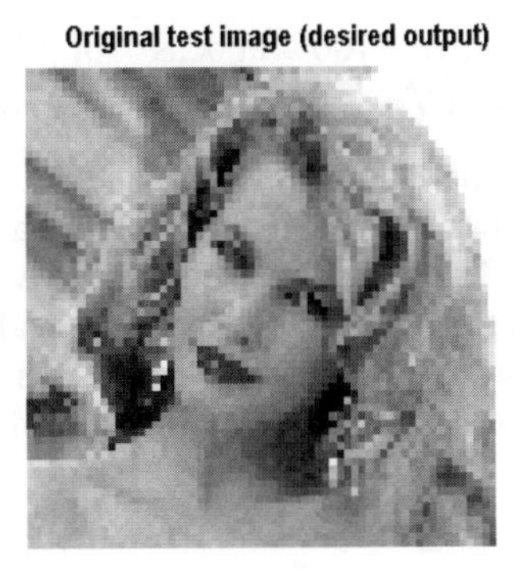

Fig. 18. Test images to evaluate the functional capability of the simplicial cell as a noise removal filter.

After 100 epochs of LMS training of the simplicial neural cell, a good quality reconstructed image is obtained (Fig. 19(a)). In this case no quantization was performed after learning. For comparison, Fig.19(b) presents the reconstructed image using the standard (linear) CNN cell with no feedback.

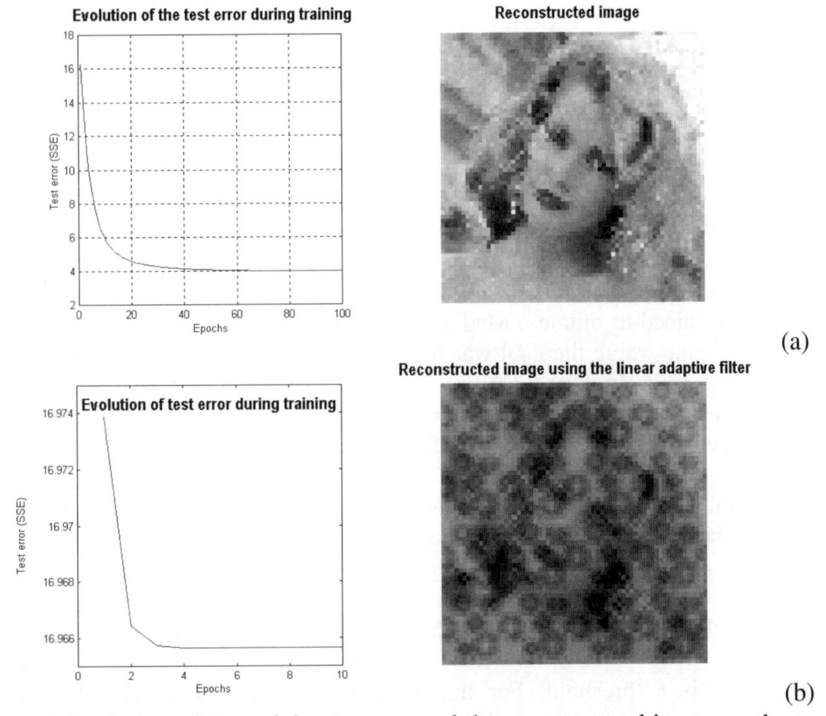

Fig.19. (a) Evolution of the training process and the reconstructed image on the test set after 100 epochs of training; (b) For comparison, using a simple linear filter to do the same task the quality of the reconstructed image is rather poor although the learning speed is about one order of magnitude larger.

Let us now consider the direct binary quantization of the gene (i.e. a Boolean gene) obtained after 100 epochs of learning by using the following transform: $c_j = 0.3 + 0.7\,\mathrm{sgn}(c_j - 0.2)$ where the parameters were ad-hoc optimized after several trial-and-error experiments.

As shown in Fig. 20, the output image is not significantly altered when compared to the image in Fig. 19(a) obtained using real valued coefficients. It resembles quite well the original image, thereby providing evidence that the compressed simplicial neural cell learned correctly the task of square scratch removal. Additional optimization of the gene using genetic algorithms may provide an even better result.

Fig.20. The reconstructed image when one-bit coefficients are used in the local RAM memory of the simplicial cell (their distribution is shown on the left).

Median Filters

The problem of “Salt and Pepper” noise removal is usually solved by employing Median filters [Castleman, 1996]. However the same problem can be defined by a set of training image samples such as those presented in Fig. 15(upper row). Such samples were used to train the simplicial neural cell for the above task. In other words, the simplicial neural cell was trained to mimic a Median filter. The learning algorithm was followed by a quantization scheme where the goal was to find the optimal gene which can be represented by a linearly separable Boolean function. The results shown in Fig. 21 indicate a very good performance on a different data set (test set) and comparisons with results obtained by employing a median filter indicates a perfect similarity between the two filters in terms of functionality.

In addition, our simplicial neural filter leads to a much more compact implementation than that of the traditional Median filter. Indeed, the binary gene (Boolean function) represented in Fig. 21 as a 16x32 matrix can be emulated using a very simple linear threshold gate defined by $c = \mathrm{sgn}\left(\sum_{i=1}^{n} b_i x_i + z_1\right)$, where $\mathbf{b} = [b_1, b_2, \ldots, b_n]$ is a vector of weights and z_1 is a threshold. For the particular case of the emulated median filter, $b_i = 1, \forall i$ and $z_1 = 0$, leading to an implementation similar to the one depicted in Fig. 22 but with no “nest” in the *nested Adaline*. No more than ten to twenty transistors are necessary to implement this cell. For comparison, an advanced Xilinx FPGA implementation of a cell performing median filtering requires several thousands transistors per cell [Maheshwari *et al.*, 1997]. It is straightforward to generalize the *simplicial neural cell* to Median filters with larger neighbourhoods by simply employing a neural cell with the corresponding number n of inputs, $b_i = 1,\ i = 1,..,n$ and $z_1 = 0$.

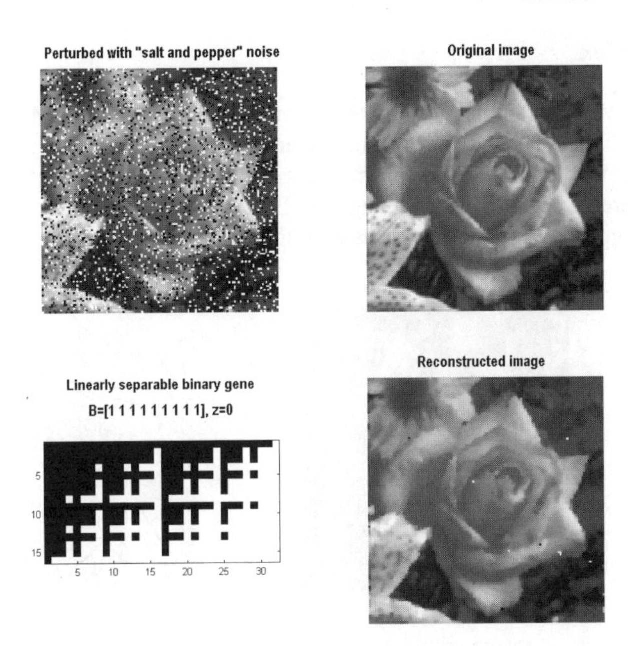

Fig.21. A set of 512 binary coefficients, represented here as a 16x32 matrix allow the simplicial cell to function as a Median filter. Note that the above binary configuration corresponds to a very simple linearly separable threshold gate which can be used instead of the RAM for this very specific application.

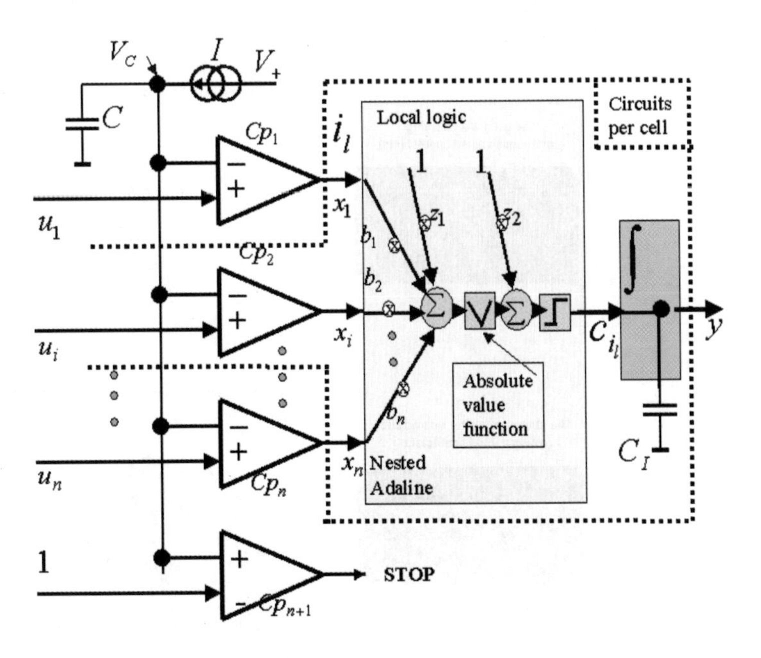

Fig. 22. A compact simplicial cell where the RAM is replaced with a nested Adaline to reduce complexity

Edge detection

The previous examples led to RAM emulations using *nested Adalines* without "nests" i.e. "classic" Adalines with a very simple hardware implementation. In this example we will consider the task of edge detection from a gray level (continuous level) image. The training samples were obtained from the images in Fig.23, where a "Canny" edge detector from the Matlab image processing toolbox was used to process the original.

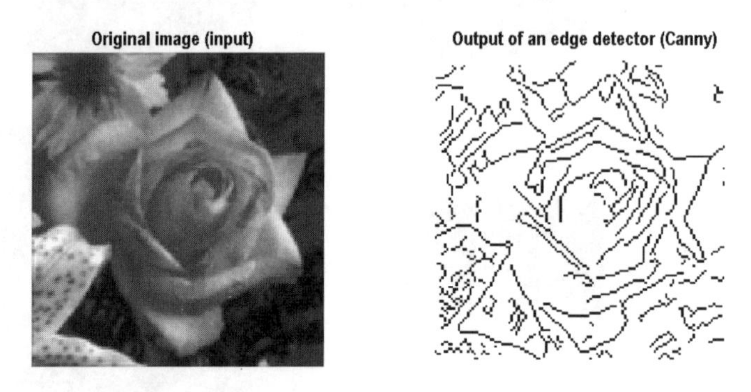

Fig.23. Training images to learn the "edge detection" task

After direct quantization (using $\alpha = 0$, $\beta = 1$, $\tau = 0.4$) it is seen that no major difference occur in the processed image, in fact the image obtained after quantization has better edges. A simple two nested Adaline with $\mathbf{B} = [1,1,1,1,1,1,1,1,1]$ and $\mathbf{z} = [z_1, z_2, z_3] = [0,-4,-4]$ is capable to represent the table of coefficients composing the binary gene in Figure 24(down).

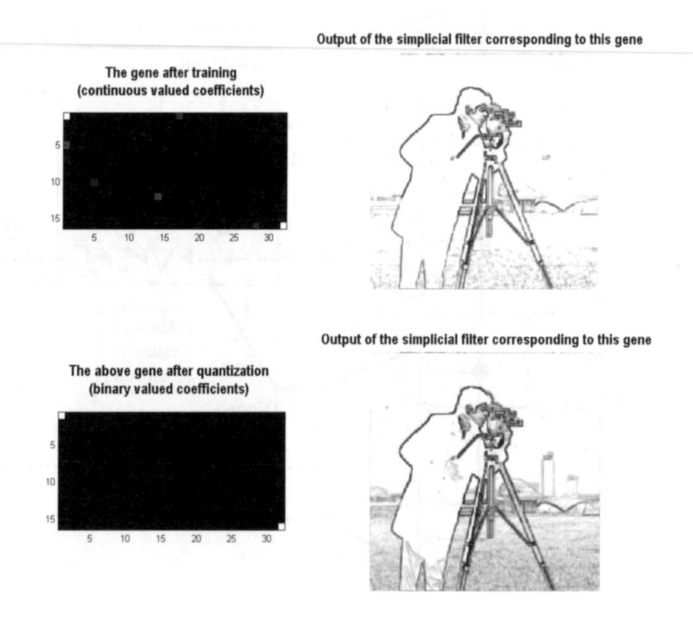

Fig. 24. The output of a cellular with simplicial cells trained to detect edges. Observe that no significant change occurs when the c_j coefficients are one-bit quantified (lower row) compared to the case of using continuous valued coefficients (upper row).

Observe the existence of gray levels in the output image of the simplicial nested Adaline which can be removed while emphasizing various details by an additional thresholding as seen in Fig. 25 where the original image and the image filtered with the "Canny" edge detector are simultaneously presented for comparison.

Output of the simplicial edge detector (binary coefficients)

Fig.25. Outputs of the CNN using simplicial cells without and with thresholding (lower row) compared to the original image and the output of the Canny edge detector (used to generate the training images)

Pattern classification

With the addition of a hard limiter at the output (providing a binary output y' instead of the continuous output y) the simplicial neural cell (55) can be trained for various pattern classification tasks. For problems where more than two classes are involved, a simplicial cell (except the input stage based on comparators) will be used for each output associated with a class assuming that an –1 output corresponds to a pattern which does not belong to that class while an output equal to +1 signify the membership of the input vector to that class. The hard limiter is introduced only during the retrieval phase, and the same learning algorithms as for the regression problems is employed.

To evaluate the functional capabilities of the simplicial neural cell to perform classification tasks several benchmark problems described in [ELENA, 1995] were used. The problems were previously learned using different neural classifiers and the comparative results are readily available in a technical report [ELENA, 1995]. A detailed analysis is given in [Dogaru *et al.*, 2002a]. Here we will briefly discuss the conclusions.

First, it was found that if the number of inputs is relatively small, the classification performance can be improved by increasing the dimension of the input space using a simple linear transform. For example replacing the input vector $\mathbf{u} = [u_1, u_2, u_3, u_4]$ with $\mathbf{u} = [u_1, u_2, u_3, u_4, u_1 - u_2, u_2 - u_3, u_2 - u_3]$ let to 0% misclassification error for the IRIS problem while a percentage of 1.35% is obtained when no particular transform is used. Second, it was found that by expanding the input space as discussed above, one bit quantization of the c_j coefficients could be done without a significant loss of performance. Moreover, the best classification performance obtained using the simplicial neural cell is almost equal with the value obtained when other neural architecture is trained for the same problem.

3.3.6. Nonlinear expansion of the input space

In the above examples it was shown that a linear expansion of the input space before using the simplicial neural network can significantly improve the quality of the classification, especially when combined with the one-bit quantization of the gene coefficients. However, there are situations where a linear expansion of the type considered above cannot be applied. Such situations correspond for example to problems characterized by a single input. For example, let us consider the case of *the simplicial neural cell* trained to learn the function $f(u_1) = \sin(2\pi u_1)$ with $u_1 \in [0,1]$.

In such cases, the preprocessing method in [Dogaru *et al.*, 1999b] can be applied. This method is based on a very simple nonlinear transform inspired from the multi-nested nonlinearity employed in (53). This method of expanding the input space assumes that instead of applying the unique input to the simplicial neural circuit (which will result in a very poor performance since only 2 gene parameters will be available to learn the problem) one will generate *m* additional inputs using the following scheme:

$$u_k = 1 - |2u_{k-1} - 1|, \quad k = 2, \ldots, m$$

Typically, the performance of the simplicial neural cell improves when *m* is increased up to a value m_{opt}. For $m > m_{opt}$ no significant improvement in performance is observed. Using the above observation one can easily determine m_{opt} for a given problem. The same idea of nonlinear preprocessing can be also applied to problems with $n>1$. Using this method with $m_{opt} = 3$ for the PHONEME classification problem led to a better result than for any of the linear input space expansion models used previously. For the SIN problem considered in our example choosing $m_{opt} = 5$ results in convergence of the LMS learning algorithm towards solution characterized by a very small sum of square error (SSE=0.037) as shown in Fig.26(a). The whole problem is therefore "learned" in the 32 RAM coefficients and further investigations indicate that the above approximation error does not change significantly even if each coefficient is quantified with at least 6 bits.

For comparison, the case of *m*=3 is presented in Fig. 26(b). In this case, the SSE error is much larger, of about SSE=1.

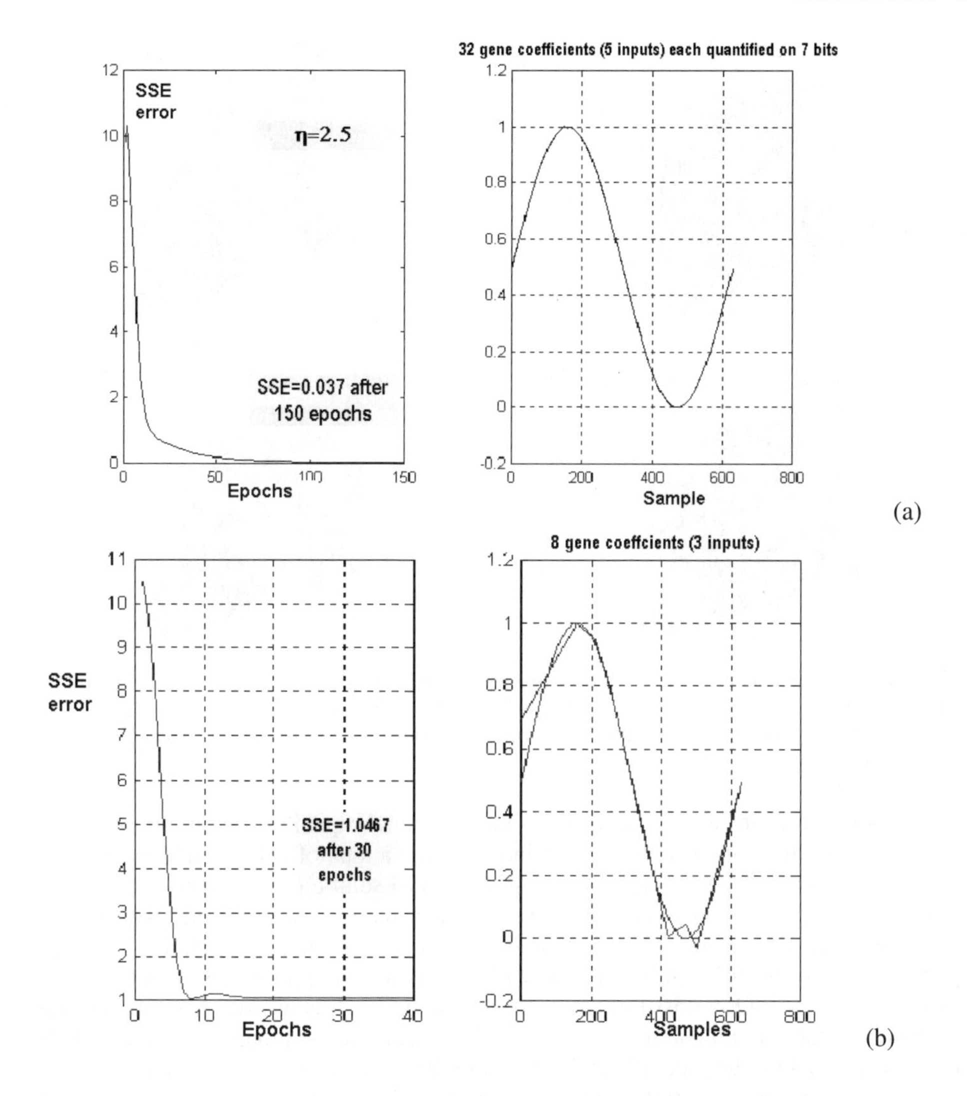

Fig. 26. (a) Using enough "nests", in this case 5, in the nonlinear expansion a very good learning of the sinusoid curve is obtained; (b) A lower quality of the approximation but with a more compressed representation (8 coefficients) is obtained using less recursions – in this case 3.

3.3.7 Comparison with multi-layer perceptrons

In a hardware realization of a neural system the goal is to achieve a proper functionality while minimizing the area of silicon (by reducing the number of electronic devices) and the power consumption. The easiness of programmability and reconfigurability is also an important issue.

To evaluate the efficiency of the simplicial neural cell against other solutions, let us consider again the problem of square scratch removal with a direct one bit quantization and a RAM-based implementation of the simplicial neuron. In order to achieve the same level of performance a standard Multi-Layer Perceptron (MLP) was trained and the results are displayed in Fig. 27.

Recovered using a 9:15:1 MLP, SSE=3.89

Fig. 27. A comparison between the output generated by multilayer perceptrons with different nodes per hidden layer (lower row) and the simplicial cell with one bit per coefficient. Both neural architectures were trained using the same training samples from Fig. 15 (lower row).

The multi-layer structure which best matches the functional capabilities of the simplicial cell is one with 15 neurons on the hidden layer and therefore a total of 150+16=166 synapses. In order to achieve programmability, the mixed-signal VLSI solution in [Cardarilli *et al.*, 1994] can be used to implement the MLP. Since each synapse has to be programmed with 5 bits, a RAM of at least 166x5=803 bits is required to store the weights. In addition, the VLSI implementation of the MLP in [Cardarilli *et al.*, 1994] would require 166 synaptic circuits (each containing 24 MOS transistors) and 16 neuron units. It is obvious from Table 1 that the simplicial neuron require almost 63% of the RAM cells and less than 2.5% of the CMOS transistors required by the programmable MLP in [Cardarilli *et al.*, 1994] (assuming that 10 CMOS transistors suffice to implement a comparator in the simplicial neuron).

Table 1. A hardware efficiency comparison between the simplicial neuron implementation and a mixed-signal multilayer perceptron realization in [Cardarilli *et al.*, 1994]

	MLP realization [Cardarilli *et al.*, 1994]	Simplicial cell	Devices in the simplicial neuron (% of the MLP)
RAM bits	803	512	63%
Transistors	≅ 4016	≅ 102	2.5%
Flexibility and reconfigurability	Good	Very good	

Moreover, the simplicial cell is much more flexible since any new application may require only the loading of the RAM with a new string of 512 bits. Instead, the MLP in [Cardarilli *et al.*, 1994] requires additional wiring circuitry to accommodate a variable number of hidden nodes as required by a certain application. Also, while the RAM storage in the MLP is distributed in many registers of 5 bits each, one single compact RAM is used by the simplicial neuron thus making the silicon use more effective. Similar comparisons stand for any other of the problems considered, leading always to a much more efficient implementation for the simplicial neuron architecture. In fact, although the requirement of RAM cells is quite similar (as expected, since the same amount of knowledge should be stored), the simplicial neuron architecture leads to a dramatic reduction in the number of devices otherwise associated to the synapses and neurons of the equivalent MLPs or other type of neural network.

Another important advantage of employing the simplicial cell instead of a multi-layer perceptron or some other neural network is of theoretical nature. Indeed, since the simplicial cell is a piecewise-linear function the tasks of analysis and understanding of dynamical phenomena in systems made of such cells are much more easy to attack.

3.4. Concluding remarks

In this chapter we focused on the possibility of defining universal computing structures at the cell level while using as the simple mathematical formalism of piecewise-linear functions. The solutions presented herein lead also to very compact microelectronic implementations, a topic which is essential for a cellular system made of many similar cells.

First the universality in the binary domain was investigated. It was shown that for a cell system with n inputs binary universality can be achieved by combining two blocks. The first is a linear weighted summation which corresponds geometrically to a projection of the input vertices in the n-dimensional space on to a 1-dimensional projection tape. Using proper optimization programs one can tune the weights of the linear weighted block to minimize the number of transitions (from positive to negative desired outputs) on the projection tape. The linearly separable Boolean functions are those functions for which there exist an orientation such that only one transition is observed on the projection tape. The second block is a one-dimensional discriminant function, which effectively implements the desired function transition map. Piecewise-linear functions in either canonical or multi-nested form can be used efficiently for this task. In particular, it was shown that the multi-nested approach allows one to match any arbitrary Boolean function to a system having an O(n) complexity, a result which is far beyond in terms of simplicity to any other similar systems previously described (e.g. neural networks made of linear threshold gates or polinomial networks). Several hints of determining the gene parameters as well as examples of optimization algorithms were given in Section 3.2.4.

Second, we investigated the possibility to expand the universality issue to the bounded but continuous domain [0,1] this corresponding to a much wider area of practical applications. Employing the theory of simplicial decomposition and elements of multi-nested structure a novel type of neural cell called a simplicial neuron is described in Section 3.3. It is shown there that using pulse width modulation and therefore allowing a finite time for processing a very simple circuit can be defined which implements the simplicial neural cell. The continuous input-output mapping is now approximated with the discrete content of a memory and the memory itself can be furthermore replaced with the binary universal cell described in Section 3.2.4. The result is a O(n) complexity structure where the gene coefficients are compactly stored in a RAM and which was demonstrated to perform regression and classification tasks with the same success as traditional functional approximators (e.g. multi-layer neural networks) . The accuracy of approximation can be controlled by using a nonlinear expansion based on the multi-nested idea introduced in Section 3.3.6 and it essentially leads to more gene coefficients to code the mapping.

Chapter 4
Emergence in Continuous-Time Systems: Reaction-Diffusion Cellular Neural Networks

The local activity theory [Chua, 1998], [Chua, 1999] offers a constructive analytical tool for predicting whether a non-linear system composed of coupled cells, such as *reaction-diffusion* and *lattice dynamical systems*, can exhibit complexity. The fundamental result of the local activity theory asserts that a system cannot exhibit emergence and complexity unless its cells are *locally active*. This chapter presents an application of this new theory to several Cellular Nonlinear Network (CNN) models of some very basic physical and biological phenomena. The first is a model of nerve membrane due to FitzHugh-Nagumo, the second is the Brusselator model used by Nobel laureate Ilya Prigogine to substantiate on his "far from equilibrium" systems theory and the last is the morphogenesis model of Gierer and Meinhardt to explain pattern formation in living systems.

Explicit inequalities defining uniquely the *local activity parameter domain* for each model are presented. It is shown that when the cell parameters are chosen within a subset of the local activity parameter domain where at least one of the equilibrium states of the decoupled cells is stable, the probability of emergence increases substantially. This precisely defined parameter domain is called the "edge of chaos", a terminology previously used loosely in the literature to define a related but much more ambiguous concept.

Numerical simulations of the CNN dynamics corresponding to a large variety of cell parameters chosen on, or nearby, the "edge of chaos" confirm the existence of a wide spectrum of complex behaviors, many of them with computational potentials in image processing and other applications. Several examples are selected in this chapter to demonstrate the potential of the local activity theory as a novel tool in nonlinear dynamics. It is not only an instrument for understanding the *genesis and emergence* of *complexity*, but also as an efficient tool for choosing cell parameters in such a way that the resulting CNN is endowed with a brain-like information processing capability.

4.1. The theory of local activity as a tool for locating emergent behaviors

Many processes observed in nature can be described as a hierarchy of homogeneous interactions between many identical cells. Such cells may consist of molecules, physical devices, electronic circuits, biological cells, dynamical systems, artificial life-like cells, and other abstract entities defined by rules, algorithms, semi groups etc. The common characteristic of such systems is that under certain circumstances, collective *complexity* may emerge, i.e. the function of the entire system is more than simply summing the functions of its parts. Life itself is a supreme manifestation of complexity, achieved by mechanisms that are not yet completely understood Such complex behaviors are of great interest not only for understanding various natural phenomena but also from the perspective of artificial intelligence and information processing. The term "emergent computation" [Forrest, 1990] was recently proposed to describe all complex behaviors which can be of interest from the computational perspective. Electronic circuits, for example, currently provide an ideal vehicle for such applications of complexity. Indeed, the recent explosion of information technologies has provided a striking example of the benefits of exploiting various dynamical behaviors in nonlinear electronic circuits.

For systems made of arrays of interconnected cells, called *Cellular Nonlinear Networks* (CNN) [Chua, 1998], an extensive theory and many effective tools have been developed. For such systems, it is logical and useful to partition the system into two parts; the *cells* and their *couplings*. By simply connecting many identical locally active cells through appropriate coupling devices, one can achieve self-organization and brain-like information processing. An important example is the Cellular Neural Network [Chua & Yang 1988], [Chua 1998], inspired by the organization of biological visual systems. Such a system, when used as the kernel of a CNN Universal machine [Roska & Chua 1993], is capable of universal computations [Crounse & Chua, 1996]. In particular, an important class of computationally useful dynamical systems, the binary Cellular Automata [Toffoli & Margolis, 1987], had recently been shown to be just a simple special case of a CNN, called *generalized cellular automata* (see Chapter 5).

While it is generally recognized that nonlinearity is a condition for complexity [Schrödinger, 1967], [Prigogine 1980], [Nicolis, G. & Prigogine, I. 1989],[Haken 1994], it was shown in [Chua 1998], for the first time, that nonlinearity *is too crude a condition* for complexity. A complete theory for "Reaction-Diffusion" CNNs reveals that the necessary condition for a *non-conservative* system to exhibit complexity is to have its cells *locally active* in a precise sense to be defined in Section 4.3.3. In other words, *unless the cells are locally active* (or equivalently, are *not locally passive*) no complexity can emerge. This theory offers a constructive *analytical method* for uncovering local activity. In particular, given the mathematical model of the *cells* (the *kinetic* term in reaction-diffusion equations), one can determine the domain of the cell parameters in order for the cells to be locally active, and thus potentially capable of exhibiting complexity. The theory of local activity provides a definitive answer to the fundamental question: What are the values of the cell parameters for which the interconnected system may exhibit complexity?

A large variety of complex dynamical phenomena from reaction-diffusion systems, including chaos, spiral waves, Turing patterns, etc., have been described in the literature [Murray 1989]. The CNN paradigm provides a unified treatment of all such systems in terms of the associated *cells,* and *couplings.* Until the recent development of the local activity theory [Chua, 1998], to obtain a desired dynamical behavior, the cell parameters are usually determined empirically by trial-and error. By restricting the cell parameter space to the local activity domain, a major reduction in the computing time required by the parameter search algorithm is achieved. Another important aspect of the local activity theory for reaction-diffusion systems is that it reduces the complexity determining procedure to only the cell parameter domain, thereby ignoring the "couplings"; namely, the diffusion coefficients.

4.2. Narrowing the search, "Edge of chaos" domains

The name "edge of chaos" had been used "loosely" in the literature, [Packard, 1988], [Langton, 1990] for the domain of parameters in a complex system where interesting and computationally useful complex dynamics may emerge. For example, based on various computer experiments [Kauffman, 1993] [Kauffman, 1995], suggested that in order to achieve optimality, *evolution* tunes the parameters of a complex system such that it is operating in an ordered regime near the "edge of chaos". The current idea of "edge of chaos" is based on the empirical observations that interesting complex behaviors seem to occur on, or near the boundary, of a domain separating the *ordered* regime from the *chaotic* regime in the parameter space. However, no precise *definition* of *edge of chaos,* let alone algorithms for finding the associated parameter domain, has been advanced.

In this chapter, the edge of chaos will be defined *precisely* in Section 4.3 to be a specific and *relatively small* subset of the local activity domain, for which the additional constraint that the uncoupled cell is also stable in all equilibrium points is added.

Examples of such sub-domains are presented in various two-dimensional cross-sections of the parameter space in Sections 4.4. to 4.6. By arbitrarily choosing parameters within the *edge of chaos* domain, or near its boundaries, it was found that complex behaviors become abundant, even with only a very coarse tuning of the coupling parameters.

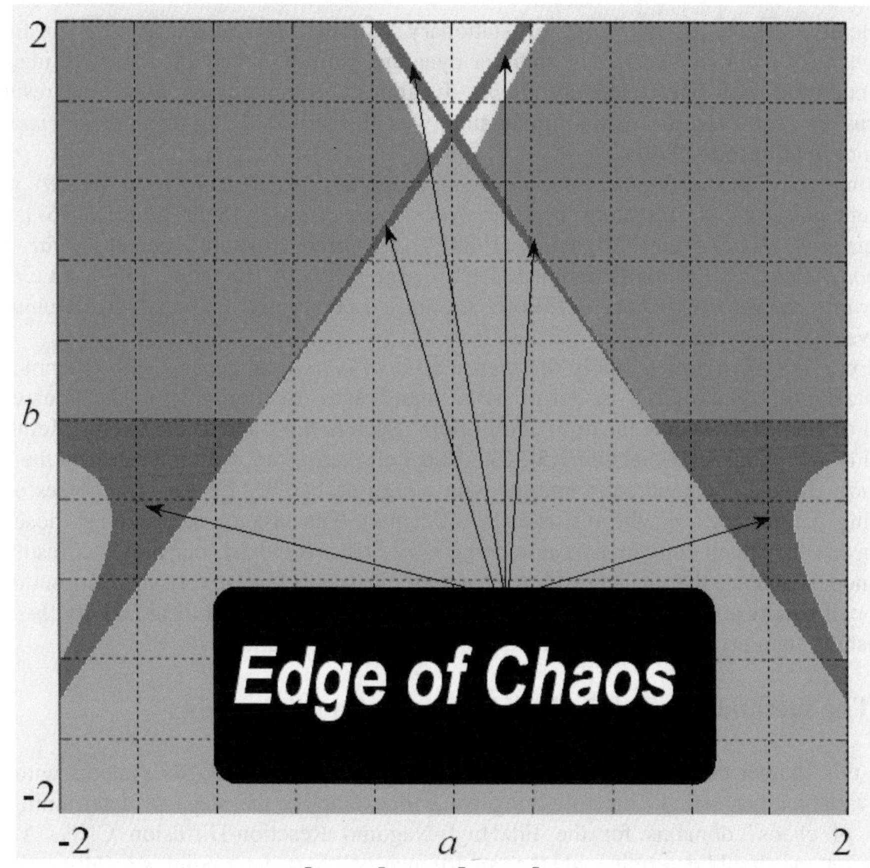

Fig. 1. A cross section ($a \in [-2,2]$, $b \in [-2,2]$) identifying the *edge of chaos* domain (coded with color red) of the FitzHugh-Nagumo Reaction-Diffusion Equation, with one diffusion coefficient $(D_1 > 0,\ D_2 = 0)$, for $c = 1$ and $\varepsilon = -0.10$. For a detailed color code, see Fig. 10.

The recent theory of *local activity* [Chua, 1998] offers an effective and unified framework for determining whether a lattice dynamical system made of coupled cells can exhibit emergent behaviors. This theory is not only fundamental in concept but also provides a practical and explicit analytical tool for determining a subset of the cell parameter space where *complexity* may emerge. In particular, a rigorous analytical test can be performed at the cell level in order to determine whether a cell is *locally passive* or not (i.e. *locally active*). If a cell is locally passive, i.e. if there is no equilibrium point where it is locally active, *independent* of the passive coupling between cells, then no complexity can emerge in a system made up of such cells. A region in the cell parameter space, called the *local passivity domain,* can thus be discarded as being not interesting in so far as the *synergetic* phenomenon of *emergence* is concerned. The remaining region, called the *local activity* domain, includes cell parameters for which both unstable and stable dynamics of the cell can spawn emergent behaviors (static and dynamic patterns) in the corresponding CNN with appropriate choice of *couplings*. However, if the coupling is simple, as in reaction-diffusion equations, only a relatively small subset of the local activity domain, called the *edge of chaos* [Dogaru & Chua, 1998b], [Chua, 1999], has a high probability of exhibiting complexity, usually in the form of stationary patterns. By *complexity* we mean the emergence of a *non-homogeneous*, static or dynamic, pattern in a coupled array of identical cells endowed with only relatively *simple* dynamics. In some cases where the resulting patterns are useful for information processing, we call the underlying phenomena *emergent computation* [Forrest, 1990].

Complexity and emergence are universal properties which have been studied under different guises in the literature, such as *order from disorder* [Schrödinger, 1967], *self-organization* [Nicolis and Prigogine, 1989], *dissipative* structures operating *far from thermodynamic equilibrium* [Prigogine, 1980], *edge of chaos* [Langton, 1990], e.t.c. The *theory of local activity* [Chua, 1998] offers a unified perspective and analytical techniques for investigating complexity.

Next it is shown how to apply the *theory of local activity* in general with examples for the FitzHugh-Nagumo CNN model of nerve membranes (Section 4.3), to the Brusselator model of chemical systems far from equilibrium (Section 4.4), and to the Gierer-Meinhardt model of morphogenesis (Section 4.5). Choosing cell parameters within, or nearby the *edge of chaos*, complex behaviors did emerge with a high probability for various choices of the coupling parameters (i.e., the diffusion coefficients). The task of determining those cell parameters for which complexity can emerge is *not trivial*. Indeed, our previous results on both models indicate that the *edge of chaos domain* is usually a relatively small subset of the entire cell parameter space. Finding this subset without applying the local activity theory is like fishing for a needle in a haystack and would have been impractical.

4.3. The methodology of finding "edge of chaos" domains

In this chapter our study is focused on the Reaction-Diffusion CNN systems with two state variables per cell. In the following we will exemplify the steps in determining the "edge of chaos" domains for the FitzHugh-Nagumo Reaction-Diffusion CNN. At the beginning one should determine whether a particular cell parameter point is locally active or not.

4.3.1. Four steps precluding the local activity testing

A Reaction-Diffusion system is mathematically described by a system of Partial Differential Equations (PDE) with one or two positive diffusion coefficients, respectively:

1. *PDE of the Reaction-Diffusion system with one diffusion coefficient* D_1

$$\boxed{\begin{aligned} \frac{\partial V_1(x,y)}{\partial t} &= f_1(V_1(x,y),V_2(x,y)) + D_1\nabla^2 V_1 \\ \frac{\partial V_2(x,y)}{\partial t} &= f_2(V_1(x,y),V_2(x,y)) \end{aligned}} \tag{1}$$

2. *PDE of the Reaction-Diffusion system with two diffusion coefficients* D_1 and D_2

$$\boxed{\begin{aligned} \frac{\partial V_1(x,y)}{\partial t} &= f_1(V_1(x,y),V_2(x,y)) + D_1\nabla^2 V_1 \\ \frac{\partial V_2(x,y)}{\partial t} &= f_2(V_1(x,y),V_2(x,y)) + D_2\nabla^2 V_2 \end{aligned}} \tag{2}$$

In both cases, $f_1(\cdot)$ and $f_2(\cdot)$ specify the particular CNN cell. For example, in the case of the FitzHugh Nagumo model they are defined by :

$$\boxed{\begin{aligned} f_1(V_1,V_2) &= c\, V_1 - \frac{(V_1)^3}{3} - V_2 \\ f_2(V_1,V_2) &= -\varepsilon\left(V_1 - bV_2 + a\right) \end{aligned}} \tag{3}$$

where a, b, c and ε are the cell parameters. Although c is usually assumed to be unity in the literature, it will be more illuminating in understanding the new concept of local activity by allowing the parameter c to assume any real number. In keeping with the physical meaning of diffusion we will assume $D_1 \geq 0$ and $D_2 \geq 0$ throughout this chapter.

Step ❶

To apply the concept of *local activity*, the *first step* is to map the above partial differential equations (PDE) into the following associated *discrete-space version*, called a Reaction-Diffusion Cellular Neural Network (CNN) in [Chua, 1998] :

Reaction-Diffusion CNN with one diffusion coefficient

$$\boxed{\begin{aligned} \frac{dV_1(j,k)}{dt} &= f_1(V_1(j,k),V_2(j,k)) + I_1(j,k) \\ \frac{dV_2(j,k)}{dt} &= f_2(V_1(j,k),V_2(j,k)) \end{aligned}} \tag{1'}$$

Reaction-Diffusion CNN with two diffusion coefficients

$$\boxed{\begin{aligned} \frac{dV_1(j,k)}{dt} &= f_1(V_1(j,k),V_2(j,k)) + I_1(j,k) \\ \frac{dV_2(j,k)}{dt} &= f_2(V_1(j,k),V_2(j,k)) + I_2(j,k) \end{aligned}} \tag{2'}$$

where (j,k) denotes the "discretized" spatial coordinates in the $x-y$ plane , $j,k = 1,2,..N$, and

$$I_1(j,k) = D_1\left(V_1(j-1,k) + V_1(j+1,k) + V_1(j,k-1) + V_1(j,k+1) - 4V_1(j,k)\right)$$
$$I_2(j,k) = D_2\left(V_2(j-1,k) + V_2(j+1,k) + V_2(j,k-1) + V_2(j,k+1) - 4V_2(j,k)\right)$$

denote the "discretized" 2-dimensional Laplacian associated with the diffusion coefficients D_1 and D_2, respectively.

Step ❷

The *second step* in applying the theory of local activity is to find the *equilibrium points* of Eqs. (1') and (2') (we will henceforth *delete* the spatial coordinates (j,k) to avoid clutter) :

Equilibrium Points for one diffusion case:

$$\begin{aligned}\dot{V}_1 &= f_1(V_1,V_2)+I_1=0\\ \dot{V}_2 &= f_2(V_1,V_2)=0\end{aligned} \tag{4}$$

Substituting (3) into (4), we obtain:

$$c\,V_1-\frac{(V_1)^3}{3}-V_2+I_1=0 \tag{5}$$

$$V_1-bV_2+a=0 \tag{6}$$

Solving (6) for V_2 and substituting the result into (5), we obtain:

$$\boxed{\frac{1}{3}V_1^3+\left(\frac{1}{b}-c\right)V_1+\left(\frac{a}{b}-I_1\right)=0} \tag{7}$$

Equation (7) may have one, two, or three real roots, henceforth denoted by $V_1(Q_1)$, $V_1(Q_2)$ and $V_1(Q_3)$, respectively. They can be found numerically, or by explicit formulas [Bronshtein & Semendyayev, 1985]. In general, these roots are functions of the 3 cell parameters a, b, c and the "coupling" input I_1 associated with the diffusion coefficient D_1, namely,

$$V_1(Q_i)=V_1\big(Q_i(a,b,c,I_1)\big),\quad i=1,\ 2,\ 3. \tag{8}$$

Equations (4)-(8) are derived by assuming I_1 is the coupling input (i.e. $D_1>0,\ D_2=0$). A corresponding set of equations with the subscript "1" changed to "2" can be derived, *mutatis mutandis* , when I_2 is the coupling input (i.e. $D_1=0,\ D_2>0$).

Equilibrium Points for the two diffusion coefficients case:

$$\begin{aligned}\dot{V}_1 &= f_1(V_1,V_2)+I_1=0\\ \dot{V}_2 &= f_2(V_1,V_2)+I_2=0\end{aligned} \tag{9}$$

Substituting (3) into (9), we obtain:

$$c\,V_1-\frac{(V_1)^3}{3}-V_2+I_1=0 \tag{10}$$

$$-\varepsilon\left(V_1-bV_2+a\right)+I_2=0 \tag{11}$$

Solving (11) for V_2 and substituting the result into (10), we obtain

$$\boxed{\frac{1}{3}V_1^3 + \left(\frac{1}{b} - c\right)V_1 + \left(\frac{a - I_2/\varepsilon}{b} - I_1\right) = 0} \tag{12}$$

Equation (12) may have one, two, or three real roots $V_1(Q_1)$, $V_1(Q_2)$ and $V_1(Q_3)$, respectively. In general, these roots are functions of the 4 cell parameters a, b, c, ε and the *two* "coupling" inputs I_1 and I_2 associated with the two diffusion coefficients D_1 and D_2 respectively, namely, $V_1(Q_i) = V_1(Q_i(a,b,c,\varepsilon,I_1,I_2))$, $i = 1, 2, 3.$

Step ❸

The *third step* in applying the theory of local activity is to calculate the 4 *coefficients* $a_{11}(Q_i)$, $a_{12}(Q_i)$, $a_{21}(Q_i)$ and $a_{22}(Q_i)$ of the Jacobian matrix of (1') and (2'), respectively, about *each* equilibrium point Q_1, Q_2 and Q_3 (for the case of a general CNN cell, there might be more or less than 3 equilibrium points):

$$\begin{bmatrix} a_{11}(Q_i) & a_{12}(Q_i) \\ a_{21}(Q_i) & a_{22}(Q_i) \end{bmatrix} = \left.\begin{bmatrix} \dfrac{\partial f_1(V_1,V_2)}{\partial V_1} & \dfrac{\partial f_1(V_1,V_2)}{\partial V_2} \\ \dfrac{\partial f_2(V_1,V_2)}{\partial V_1} & \dfrac{\partial f_2(V_1,V_2)}{\partial V_2} \end{bmatrix}\right|_{\substack{V_1=V_1(Q_i) \\ V_2=V_2(Q_i)}} \tag{13}$$

where $V_2(Q_i) = \frac{1}{b}(V_1(Q_i) + a)$ for the one-diffusion coefficient case, and $V_2(Q_i) = \frac{1}{b}\left(V_1(Q_i) + a - \frac{I_2}{\varepsilon}\right)$ for the two-diffusion coefficient case. For example, in the case of the FitzHugh-Nagumo cell, we have:

$$\boxed{\begin{aligned} a_{11}(Q_i) &= c - V_1^2(Q_i) \\ a_{12}(Q_i) &= -1 \\ a_{21}(Q_i) &= -\varepsilon \\ a_{22}(Q_i) &= b\varepsilon \end{aligned}} \tag{13'}$$

Step ❹

The *fourth step* in applying the theory of local activity is to calculate the trace $T(Q_i)$ and the determinant $\Delta(Q_i)$ of the Jacobian matrix (13) about *each* equilibrium point:

$$\begin{aligned} T(Q_i) &\overset{\Delta}{=} a_{11}(Q_i) + a_{22}(Q_i) \\ \Delta(Q_i) &\overset{\Delta}{=} a_{11}(Q_i)a_{22}(Q_i) - a_{12}(Q_i)a_{21}(Q_i) \end{aligned} \tag{14}$$

For the example considered herein (the FitzHugh-Nagumo model), we have:

$$T(Q_i) = c + b\varepsilon - V_1^2(Q_i) \tag{15}$$

$$\Delta(Q_i) = b\varepsilon\left(c - V_1^2(Q_i)\right) - \varepsilon \tag{16}$$

4.3.2. The concept of local activity

To provide an intuitive understanding, as well as a physical basis for the concept of *local activity*, it is instructive to associate the variables ($V_1(j,k)$, $I_1(j,k)$) in (1') and (2') with the voltage and current of a *2-terminal* electronic circuit "cell" described by the same equations.

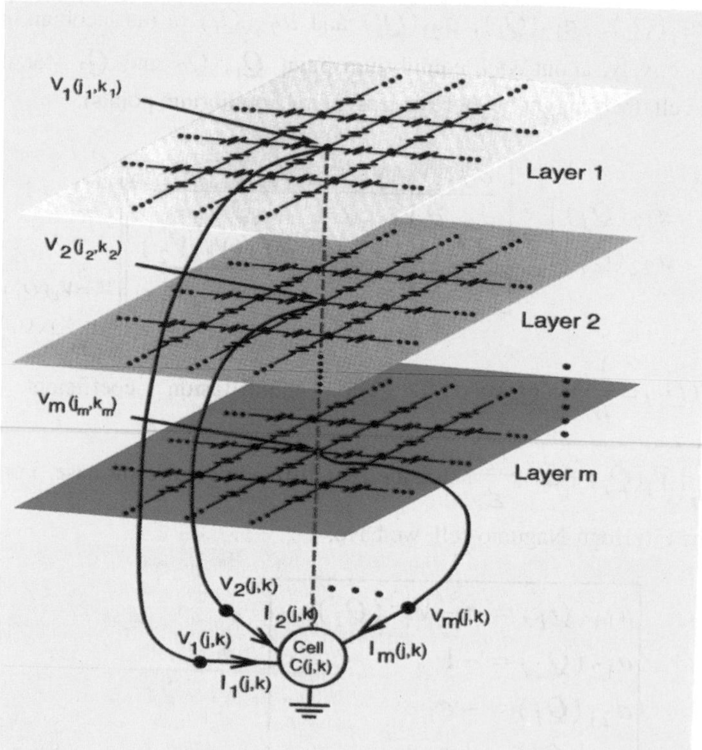

Fig.2. A two-dimensional Reaction-Diffusion CNN depicted as a restive grid circuit. The state values of the cells are associated with voltages V_1 to V_m while the currents flowing through the m-port cell correspond to the diffusion terms in equations (1') and (2'). A m-th order nonlinear cell is coupled with its 4 closest neighbors via m layers of resistive grids. All resistors in the grids have the same conductance which correspond to the diffusion coefficient D_m in the same equations.

In particular, let us identify the cell state variable $V_1(j,k)$ with the *node-to-datum* voltage at node (j,k) of a 2-dimensional *grid* G_1 of linear resistors. Similarly, let us identify the *coupling input* $I_1(j,k)$ with the current *leaving* node (j,k), i.e. *entering*[1] the *cell* connected to node (j,k). For the 2 diffusion coefficient case, let us likewise identify $V_2(j,k)$ and $I_2(j,k)$ with the node-to-datum voltage and current associated with a corresponding node (j,k) of a *second*[2] resistive grid G_2. In this case, the associated electronic circuit cell has 3 terminals with 2 node-to-datum voltages (V_1,V_2) and two terminal currents (I_1,I_2) [Chua, 1998].

The positive diffusion coefficients D_1 and D_2 correspond to conductance of the linear resistors in the two resistive grids, G_1 and G_2 respectively. The importance of the circuit model lies not only in the fact that it offers a convenient physical implementation, but also the fact that well-known results from classic circuit theory can be used to test the cells for local activity. From this perspective, each cell is assumed to be operating near an equilibrium point $Q_i = (\bar{V}_1, \bar{V}_2, \bar{I}_1, \bar{I}_2)$ where the bar over the variables denotes the time average (d.c.) voltages and currents, respectively.

If there is *at least one* equilibrium point where the cell acts like a source of "small signal" power, then the cell is said to be *locally active*. The small signal power source is defined in a precise sense [Chua, 1998], as a cell capable of injecting a net small-signal *average power* into the *passive* (i.e. $D_1 \geq 0$, $D_2 \geq 0$) resistive grids,.

On the contrary, if *there is no* equilibrium point such that the cell can act like a source of "small-signal" power, then the cell is said to be *locally passive*. It can be shown [Chua, 1980] that in this case all cells must generically tend to the same unique hence *homogenous* steady state voltage at all nodes, thereby precluding the possibility of any complex phenomena.

4.3.3. Testing for stable and unstable local activity

In Section 4.3.1, a methodology was developed such that at each cell parameter point and for each equilibrium point of the uncoupled CNN cell one can compute three coefficients T, Δ, and a_{22}. According to corollary 4.4.1 in [Chua, 1998], for any reaction-diffusion system with two state variables and one diffusion coefficient these three scalars suffice to classify the emergent dynamics into one of the following three categories:

- ***Locally active and stable*** (usually points in the cell parameter space are labeled with color **red** when belonging to this class);
- ***Locally active and unstable*** (usually points in the cell parameter space are labeled with color **green** when belonging to this class);
- ***Locally passive*** (usually points in the cell parameter space are labeled with color **blue** when belonging to this class);

A similar classification can be done using the four coefficients $a_{11}, a_{12}, a_{21}, a_{22}$ obtained in Step 3, Equation (13') above, for the case of two-diffusion coefficients. As shown next, the *local activity domain* for the case of *two diffusion coefficients* includes *the local activity domain* for the case of *a single diffusion coefficient*. In what follows, a detailed

[1] See [Chua, 1998] for the associated *circuit model*

[2] For the one-diffusion coefficient case, the second resistive grid is absent.

treatment of the theoretical issues and test procedures used to determine the membership of a cell parameter point to one of these three categories will be given.

Let $v_1(t) \stackrel{\Delta}{=} V_1(t) - \overline{V}_1(Q_i)$, $v_2(t) \stackrel{\Delta}{=} V_2(t) - \overline{V}_2(Q_i)$, $i_1(t) \stackrel{\Delta}{=} I_1(t) - \overline{I}_1(Q_i)$ and $i_2(t) \stackrel{\Delta}{=} I_2(t) - \overline{I}_2(Q_i)$ denote respectively an *infinitesimal perturbation* in the voltages and currents about their equilibrium values $\overline{V}_1, \overline{V}_2, \overline{I}_1$, and $\overline{I}_2$ at an equilibrium point Q_i. Now, given the *local power flow* $p(t) = v_1(t)i_1(t) + v_2(t)i_2(t)$ entering the CNN cell *at a cell equilibrium point* Q_i, the cell is said to be *locally active at the equilibrium point* Q_i, if, and only if, there exists infinitesimal (small amplitude) coupling currents i_1 and i_2 such that the cell will "inject" a net amount of energy (the cell acts like an energy source) into an external *passive* coupling system over *some* time interval $T > 0$. This passive coupling system is formed of resistive grids. The infinitesimal voltages (v_1, v_2) represent the response of the cell to the infinitesimal excitation currents (i_1, i_2) about the equilibrium point Q_i; namely,

$$\delta E(Q) \stackrel{\Delta}{=} \int_0^T p(t)\,dt < 0 \tag{17}$$

The test for *local activity at a given equilibrium point* depends on the number of diffusion coefficients:

Local activity test for case 1: One diffusion coefficient

A one-diffusion cell is locally active at an equilibrium point Q_i if at least one of the following conditions holds:

$$\begin{array}{l} \text{C1. } T > 0 \quad \text{OR} \quad \Delta < 0 \\ \text{C2. } T > a_{22} \quad \text{OR} \quad (T \le a_{22} \ \text{ AND } \ a_{22}\Delta > 0) \\ \text{C3. } T = 0 \quad \text{AND} \quad \Delta > 0 \quad \text{AND} \quad a_{22} \ne 0 \\ \text{C4. } T = 0 \quad \text{AND} \quad \Delta = 0 \quad \text{AND} \quad a_{22} \ne 0 \end{array} \tag{18}$$

where $T = a_{11} + a_{22}$ and $\Delta = a_{11}a_{22} - a_{12}a_{21}$ are the *trace* and the *determinant* of the Jacobian matrix $A = \begin{bmatrix} a_{11} & a_{12} \\ a_{21} & a_{22} \end{bmatrix}$ defined earlier in Eqs. (13), (15) and (16), *evaluated at* Q_i, and where the local cell coefficients a_{ij} are defined by (13'). The first two parameters T and Δ are closely related to the *stability* of the equilibrium points of the *decoupled* cell obtained by setting $I_1 = I_2 = 0$: it can be shown [Chua, et al., 1985] that $T < 0$ AND $\Delta > 0$ is the only region which corresponds to *locally asymptotically stable* [Hartman, 1982] equilibrium points. It follows that the only region where the cell can be locally passive is the stable region defined by $T < 0$ AND $\Delta > 0$. Observe that if $a_{22} > 0$, then (18) implies that condition C2 is always satisfied in this quadrant. It follows that *the cell can be locally passive only if* $a_{22} \le 0$. In fact, C2 is the only condition which allow us to "draw" a boundary, situated in the stable region of the $T-\Delta$ plane, between the *local passivity* domain and a sub-region of the *local activity* domain where the equilibrium point Q_i is both *stable* and *locally active*, henceforth coded by $S(Q_i)\ A(Q_i)$. This sub-region has a special physical significance since it corresponds to the operating mode of all active electronic devices used in information processing (e.g. transistors, logic gates, etc.). We will henceforth define the *edge of chaos* as a subset of the locally active parameter domain where *at least* one equilibrium point lies in the $S(Q_i)\ A(Q_i)$ region; namely,

Stable and locally active region $S(Q_i)\ A(Q_i)$ at Q_i for Reaction-Diffusion Equation (1 diffusion coefficient) :

$$(a_{22}<0 \text{ AND } a_{22}<T<0) \text{ OR } (a_{22}>0 \text{ AND } T<0) \\ \text{AND} \\ \Delta>0 \tag{19}$$

Substituting the parameters from the FitzHugh-Nagumo Equation into (19) we obtain the following:

$S(Q_i)A(Q_i)$ region at Q_i for FitzHugh-Nagumo Equation (one diffusion coefficient):

$$\varepsilon\left(bc-b\overline{V}_1^2-1\right)>0 \\ \text{AND} \\ (b\varepsilon<0 \text{ AND } c+b\varepsilon<\overline{V}_1^2<c) \text{ OR } (b\varepsilon>0 \text{ AND } \overline{V}_1^2>c+b\varepsilon) \tag{20}$$

Since $a_{22}(Q_i)\leq 0$ is a necessary condition for local passivity, and since the *locally passive region* $P(Q_i)$ (at a given equilibrium point Q_i) can be located only in the $T<0$ AND $\Delta>0$ quadrant, it follows from (18) that it is completely determined by imposing the additional restriction $T\leq a_{22}(Q_i)$. Hence, we have:

Local passivity region $P(Q_i)$ at Q_i:

$$a_{22}(Q_i)\leq 0, \text{ AND } T(Q_i)<0, \text{ AND } \Delta(Q_i)>0, \text{ AND } T(Q_i)\leq a_{22}(Q_i) \tag{21}$$

For the particular case of a FitzHugh-Nagumo cell:

Local passivity region $P(Q_i)$ at Q_i for FitzHugh-Nagumo Equation (one diffusion coefficient):

$$b\varepsilon\leq 0 \text{ AND } \overline{V}_1^2\geq c \text{ AND } \varepsilon\left(bc-b\overline{V}_1^2-1\right)>0 \tag{21'}$$

If, for a given set of cell parameters, a particular equilibrium point Q_i is neither in $P(Q_i)$ nor in $S(Q_i)A(Q_i)$, then it must lie in the remaining domain $A(Q_i)U(Q_i)$ (Locally Active and Unstable). Thus, given a set of cell parameters, and a specified equilibrium point Q_i , one can determine first the value of $\overline{V}_1^2$ and then test (20) and (21) to determine whether Q_i belongs to either the *stable and locally active* region $S(Q_i)A(Q_i)$, or to the *locally passive* $P(Q_i)$ region. If it doesn't lies in either region, then it must lies in the *locally active and unstable* region $A(Q_i)U(Q_i)$. Observe that this region is equivalent to the fulfillment of at least one of the conditions C1,C3, or C4 in the general test (18) for local activity. It can be easily proven that for any Reaction-Diffusion CNN with a diffusion coefficient and two state variables, the following logical relationships holds:

$$A(Q_i)U(Q_i) = \text{C1} \cup \text{C3} \cup \text{C4}$$
$$S(Q_i)A(Q_i) = \text{C2} \setminus A(Q_i)U(Q_i)$$
$$P(Q_i) = \overline{A(Q_i)U(Q_i) \cup S(Q_i)A(Q_i)}$$

where the bar denotes the set complement operation.
We will use the notation $A(Q_i)$ to denote the region $A(Q_i)U(Q_i) \cup S(Q_i)A(Q_i)$ of *local activity at an equilibrium point* Q_i, *regardless* of stability.

Figure 3 gives a graphical interpretation of the above domains, where each of the above three domains in the $a_{22} - T - \Delta$ space are assigned a specific color as follows: *blue* for the locally passive domain (); *green* for the locally active and unstable domain, and *red* for the locally active and stable domain. This distribution stands for any arbitrary cell with 2 state variables. Note that within the 16 sectors of the space, only one (labeled 14 in the figure) corresponds to local passivity. Most of the sectors (twelve) correspond to *locally active and unstable* and three of them (13,15, and 16) to the most interesting regime from the point of view of emergence; namely, the *locally active and stable*.

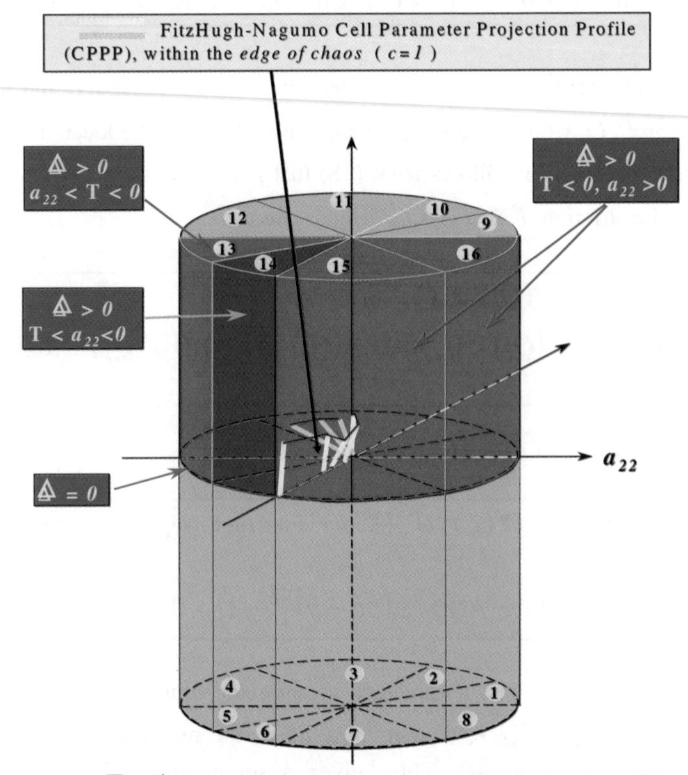

Fig. 3. The $a_{22} - T - \Delta$ space where different colors identify the three classes of locally active cells

A cross section trough the upper part of the space ($\Delta > 0$) is presented in Fig.4. Here, the *cell parameter point* positions (in the case of a FitzHugh-Nagumo cell) are specified using numerical labels. A detailed analysis of each *cell parameter point* is given in the next sections. Note that although the above plots are identical for any reaction-diffusion system with two state variables, each cell type is defined trough its own signature, a compact sub-domain within the $a_{22} - T - \Delta$ space. The exact shape of this sub-domain, called a *cell parameter projection profile* (CPPP) depends on the nonlinear functions defining the cell and on its specific set of tunable parameters. In particular, a CPPP is defined as the projection of any prescribed subset Π of the cell parameter domain of interest into the (a_{22}, T, Δ)-space. In Figs. 3 and 4 the CPPP corresponding to the FitzHugh-Nagumo cell is emphasized using yellow lines.

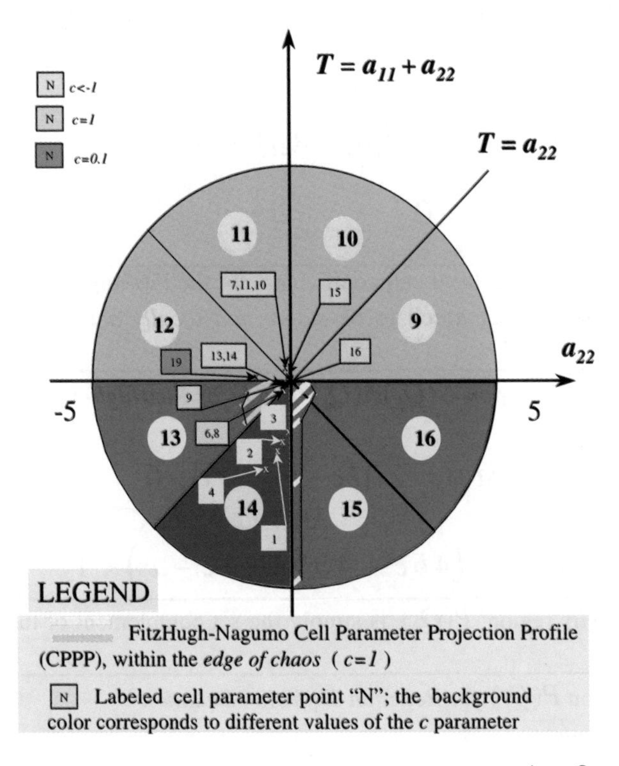

Fig. 4. The T- a_{22} parameter plane, cross section for $\Delta > 0$

The plot of the CPPP within the $a_{22} - T - \Delta$ space gives an indication on whether the underlying cell will give rise to emergent phenomena or not. For reasons that will be explained lately, for emergent phenomena to occur it is necessary that the CPPP intersect the sector 13 of the $a_{22} - T - \Delta$ space. As we will see for several examples consisting in different cell types, for certain ranges of parameters their associated CPPPs intersect sector 13, thereby leading to emergent properties. However, in order to locate better the *domains of cell parameter points* for which emergence occurs, additional *bifurcation diagrams* have to be determined for each cell. A bifurcation diagram is essentially a structure similar to the $a_{22} - T - \Delta$ space but where the color labels for each point are plotted directly within the particular cell parameter space or within projections of this space.

Local Activity Test for two diffusion coefficients

Following corollary 4.4.2. in [Chua, 1998], a 2-diffusion cell is *locally active* at an equilibrium point Q_i if, and only if, the matrix:

$\mathbf{G}_{Q_i} = \begin{bmatrix} -2a_{11} & -(a_{12}+a_{21}) \\ -(a_{12}+a_{21}) & -2a_{22} \end{bmatrix}$ is *not* positive semi-definite at the equilibrium point Q_i. By applying the standard test for positive semi-definite matrices, it follows that the above condition is equivalent to the following system of inequalities:

Stable and locally active region $S(Q_i)A(Q_i)$:

$$a_{22} > 0 \quad \text{OR} \quad 4a_{11}a_{22} < (a_{12}+a_{21})^2$$
$$\text{AND} \tag{22}$$
$$T < 0 \quad \text{AND} \quad \Delta > 0$$

Substituting the particular local cell coefficients for the FitzHugh-Nagumo Equation, we can express the condition for a cell to be *stable and locally active* ($S(Q_i)A(Q_i)$) as follows:

Stable and locally active region $S(Q_i)A(Q_i)$ *at* Q_i *for FitzHugh-Nagumo Equation (2 diffusion coefficients)*:

$$\overline{V}_1^2 > c + b\varepsilon \quad \text{AND} \quad \varepsilon \cdot (bc - b\overline{V}_1^2 - 1) > 0$$
$$\text{AND} \tag{23}$$
$$b\varepsilon > 0 \quad \text{OR} \quad \varepsilon(4bc - 4b\overline{V}_1^2 - 2 - \varepsilon) < 1$$

The *local passivity* region $P(Q_i)$ is simply the set complement of the region $A(Q_i)$; namely,

Local passivity region $P(Q_i)$ *for Reaction-Diffusion Equation (2 diffusion coefficients)*:

$$a_{22} \le 0 \quad \text{AND} \quad 4a_{11}a_{22} \ge (a_{12}+a_{21})^2 \tag{24}$$

Substituting the parameters for the FitzHugh-Nagumo Equation into (24) we obtain:

Local passivity region $P(Q_i)$ at Q_i for the FitzHugh-Nagumo Equation (two diffusion coefficients):

$$b\varepsilon \le 0$$
$$\text{AND } \varepsilon(4bc - 4b\overline{V}_1^2 - 2 - \varepsilon) \ge 1 \tag{25}$$

As in the one-diffusion coefficient case, if a cell is found to be neither *stable and locally active* ($S(Q_i)A(Q_i)$), nor *locally passive* ($P(Q_i)$) at the equilibrium point Q_i, then it is *locally active and unstable* ($A(Q_i)U(Q_i)$) at Q_i.

4.3.4. Unrestricted versus restricted local activity, the edge of chaos

In the above sections we defined three general classes of local activity and explained why the necessary condition for emergent computation implies that a cell should be locally active and stable. In the framework of the local activity theory, a infinity of equilibrium points corresponding to coupled cells should be considered in the above definition. This is the consequence of choosing $\bar{I}_1 \in (-\infty,\infty)$ and $\bar{I}_2 \in (-\infty,\infty)$. Although this choice is correct from a pure theoretical point of view, as we will see in this section, it leads to a situation where most of the cell parameter points are *locally active* and consequently likely to produce emergent behaviors. However, the simulations show that the locally active domains defined in this way are still too large and in fact many cell parameter points within such local activity domains do not lead to emergent behaviors. Partially motivated by the fact that in practice both $\bar{I}_1$, and $\bar{I}_2$ actually would vary within a limited domain (otherwise the power within the coupling resistors in a physical realization would become infinity), in this section we will narrow even more the domains where emergence is likely to occur. First, *restricted local activity domains* will be defined by taking the assumption $\bar{I}_1 = 0$ and $\bar{I}_2 = 0$, then additional constrains will be imposed leading to the definition of the "edge of chaos" domain and a methodology for identifying it.

Unrestricted local activity and passivity

The *local activity domain* $\mathbf{A}$ of a Reaction-Diffusion CNN is defined in [Chua, 1998] as $\mathbf{A} = \bigcup_{Q_i \in \Omega} A(Q_i)$, where Ω is the set of *all* cell equilibrium points $Q_i(\bar{I}_1, \bar{I}_2)$, for all $\bar{I}_1 \in (-\infty,\infty)$ and $\bar{I}_2 \in (-\infty,\infty)$, and where $A(Q_i)$ is a subset of *the cell parameter space* in which the cell is *locally active* ($A(Q_i)$) at the equilibrium point Q_i. Consequently, *the local passivity domain* $\boldsymbol{P}$ can be defined as the intersection $\mathbf{P} = \bigcap_{Q_i \in \Omega} P(Q_i)$ of the subsets $P(Q_i)$ in the cell parameter space where the cell is *locally passive* ($P(Q_i)$) at Q_i. The tests for local activity or local passivity at Q_i are made by checking condition (22) for the case of one diffusion coefficient, or condition (26) for the case of two diffusion coefficients.

For the FitzHugh-Nagumo Equation (1), the cell parameter space is 4-dimensional, corresponding to the 4 parameters, a, b c and ε. Following the above definitions, it can be shown that the unrestricted *local passivity domain* $\boldsymbol{P}$ for one and two diffusion coefficients is defined as:

Local passivity domain P for FitzHugh-Nagumo Equation

(a) One-diffusion coefficient $(D_1 > 0\ ,\ D_2 = 0)$

$$b\varepsilon \ \leq \ 0 \quad \text{AND} \quad c < 0 \quad \text{AND} \quad c < \frac{1}{b} \tag{26}$$

(b) Two-diffusion coefficients $(D_1 > 0\ ,\ D_2 > 0)$

$$b\varepsilon \ \leq \ 0 \qquad \text{AND} \qquad c < \frac{(1+\varepsilon)^2}{4b\varepsilon} \tag{27}$$

The unrestricted *local activity domain* is simply defined as the complement of **P**. As it can be easily seen the parameter a is not influencing **A** and **P**. Therefore it suffices to consider only the 3-dimensional (b, c, ε) parameter space. Moreover, for practical purpose we will confine our analysis to the *cell parameter cube* bounded by the cube $-2 \leq c \leq 2$, $-2 \leq b \leq 2$, and $-2 \leq \varepsilon \leq 2$, as shown in Figs. 5(a) and 5(b) for the one and two diffusion case, respectively.

The *red* regions in Fig 5 correspond to the set of cell parameters where the FitzHugh-Nagumo cell is *locally active,* i.e. *locally active* with respect to *at least one equilibrium point* Q_i corresponding to some $(\bar{I}_1, \bar{I}_2)$. The remaining *blue* regions depict the local passivity domain.

Observations:

(i) For either 1 or 2 diffusion coefficients, the local activity domain does not depend on the parameter a of the cell. This is because the coefficient $a_0 = a/b - I_2/(\varepsilon b) - I_1$ in (12) spans the entire range $a_0 \in (-\infty, \infty)$ when $\bar{I}_1$ varies from $-\infty$ to $+\infty$, regardless of the value of a;

(ii) The local passivity domain in the case of two non-zero diffusion coefficients is a *subset* of the local passivity domain for the case of only one diffusion coefficient for *any* reaction-diffusion equation [Chua, 1998]. This fact can be confirmed by examining the *blue* areas in the two cross sections corresponding to $c = -5$ in Fig. 5(c), and $c = -1$ in Fig. 5(d), as well as in the three-dimensional plots in Fig. 5(a,b). It can be easily proved that the parameter domain defined by (27) is a subset of that of (26) for all parameters, i.e. **P**(two-diffusion case) $\subset$ **P**(one-diffusion case), thereby confirming the preceding observation;

(iii) Determining the unrestricted (theoretical) local passivity and local activity domains by numerical methods can sometimes lead to errors in the general case because it is impractical to conduct the test for all $\bar{I}_1 \in (-\infty, \infty)$. Fortunately, for cells with two state variables, analytical techniques can be applied so that the local passivity and local activity domains can be determined analytically.

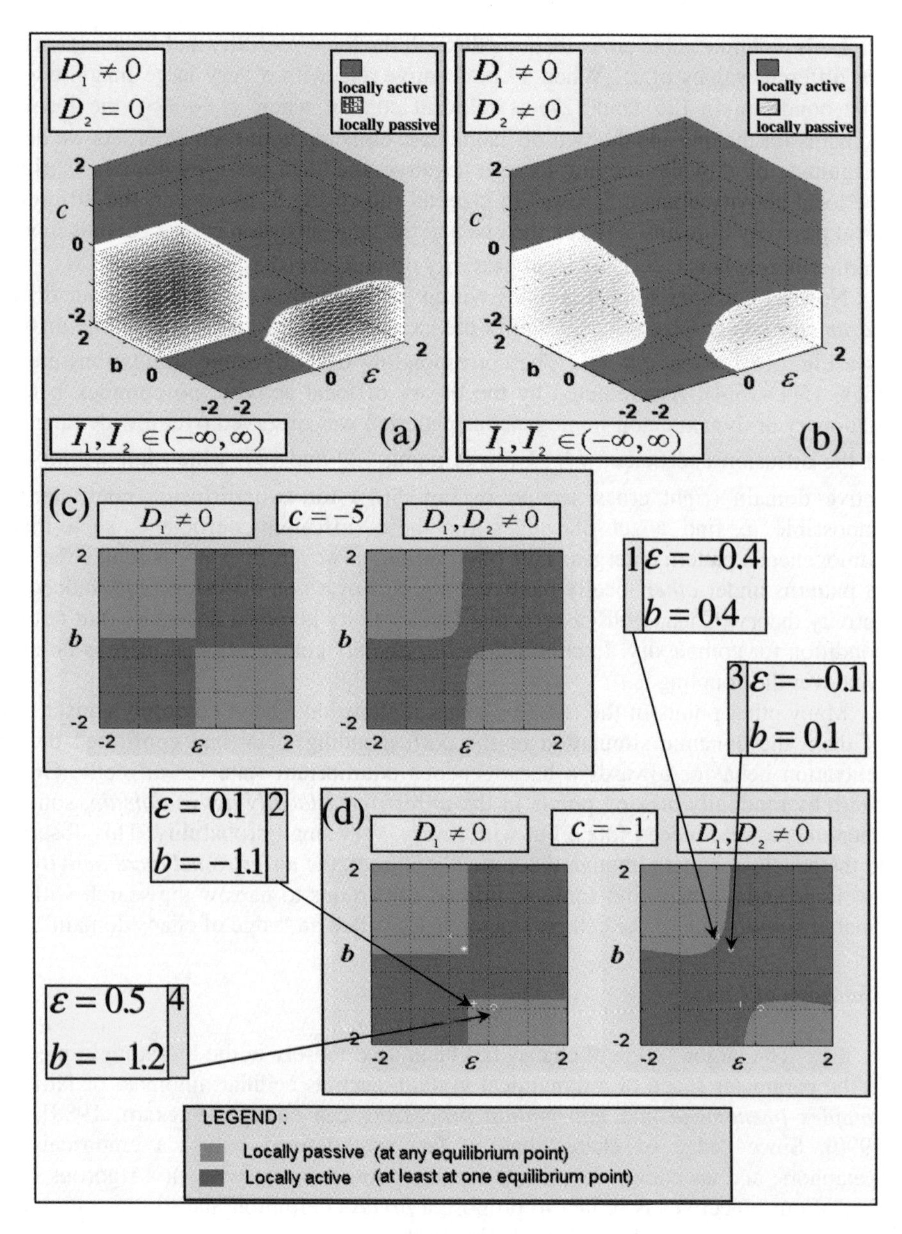

Fig. 5. Several cross sections identifying the local activity domain (coded in red) and local passivity domain (coded in blue), of the FitzHugh-Nagumo Equation. The *locally passive* regions in (a) and (b) are printed in non-uniformly faded blue in order to enhance their 3-dimensional perspective. FitzHugh-Nagumo cells from these bluish regions can *not* exhibit any form of complexity regardless of the coupling coefficients: (a) Three-dimensional domain: one diffusion coefficient case; (b) Three-dimensional domain: two diffusion coefficients case; (c)Two-dimensional cross-sections in b and ε for fixed $c = -5$; The left side corresponds to the one diffusion case and the right side to the two diffusion case, respectively; (d) Two-dimensional cross-sections in b and ε for fixed $c = -1$. Each point marked on these cross-sections is associated with a set of cell parameters shown in the yellow "caption box". Dynamic simulations of the corresponding CNNs are shown in Figs. 15(1) - 15(4) where the number in parenthesis is the same as the label shown in the light blue "caption box" attached to each parameter point at the tip of an arrowhead.

Let's examine some cross sections through the local passivity and local activity domains, for different values of c. When c is negative and with a very large magnitude, only the first condition in (26) and (27) is relevant so that when $c \to -\infty$ the local passivity domains for the one and the two-diffusion case coincide with each other. As we decrease the magnitude of c while keeping its sign negative, the local passivity domain "shrinks" while the local activity domain "grows" in size, as shown Fig.5. Moreover, the difference of the local passivity domains between the one and the two- diffusion cases becomes more evident. Furthermore when $c > 0$ the local passivity domain vanishes.

Now let consider some examples within the local passivity domain. Four different *cell parameter points* were selected within the local passivity domain for the one diffusion case (see Fig. 5(d)) when $c = -1$. The corresponding CNN dynamic simulations are shown in Figs. 15(1)-15(4). As predicted by the theory of local activity, no complex behavior (i.e. stationary or dynamic non-homogeneous patterns) was observed over a wide range of values of the diffusion coefficient D_1. Even at points "2" and "3", which fall within the locally active domain (right cross section in Fig. 5(d)) for two diffusion coefficients, it was impossible to find a set of values for these diffusion coefficients such that a non-homogeneous pattern emerges. This observation, however, does not preclude the emergence of patterns under *other* locally passive couplings of a non-diffusion type. Indeed, the local activity theory [Chua, 1998] asserts that local activity is *only* a necessary, but not sufficient, condition for complexity. In contrast, *local passivity* guarantees that there is no emergence, whatever the coupling is.

Many other points in the *local passivity* regions have been randomly chosen and for all of them the dynamic simulation of the corresponding CNN had confirmed the predicted relaxation behavior towards a homogeneous equilibrium state for all cells. On the other hand, by randomly picking points in the *unrestricted locally active domain*, some complex phenomena were indeed found but with a very, very small probability. This observations let to the conclusion that although theoretically correct, the *unrestricted local activity* domain is too large and we may find some additional constraints to narrow the search within a much smaller sub-domain of the cell parameter space called an "edge of chaos domain".

The Edge of Chaos

The jargon "edge of chaos" has been used loosely in the literature to mean a region in the parameter space of a dynamical system, such as cellular automata or lattices, where *complex phenomena* and *information processing* can emerge [Packard, 1988], [Langton, 1990]. Since "edge of chaos" has so far been defined only via empirical examples, metaphors, and anecdotes, it remains an ambiguous concept without a rigorous foundation. One of our objectives is in fact to propose a *precise* definition and to illustrate its scientific significance as a "complexity" *predicting tool.*

Definition: Edge of chaos $\boldsymbol{C}$

A Reaction-Diffusion CNN with one diffusion coefficient D_1 (resp., two diffusion coefficients D_1 and D_2) is said to be operating on the *edge of chaos* $\boldsymbol{C}$ if, and only if, there is at least one equilibrium point Q_i which is both *locally active* and *stable* at Q_i when $\bar{I}_1 = 0$ (resp., $\bar{I}_1 = 0$ and $\bar{I}_2 = 0$).

The algorithm for finding the edge of chaos C is given below:

Edge of chaos algorithm for the one and two diffusion case

1. Set $\bar{I}_1 = 0$ and $\bar{I}_2 = 0$ in the equilibrium equations:

$$\dot{V}_1 = f_1(V_1, V_2) + \bar{I}_1 = 0 \tag{28}$$

$$\dot{V}_2 = f_2(V_1, V_2) + \bar{I}_2 = 0 \tag{29}$$

and solve for

$$V_2 = h(V_1) \tag{30}$$

from (29).

2. Substitute $V_2 = h(V_1)$ for V_2 in (28) and solve

$$f_1(V_1, h(V_1)) = 0 \tag{31}$$

for $V_1(Q_i)$.

For the example of the FitzHugh-Nagumo Equation, (30) can have one, two or three real solutions $V_1(Q_1)$, $V_1(Q_2)$ and $V_1(Q_3)$.

3. For each equilibrium point Q_i, use Eq. (29) to calculate

$$V_2(Q_i) = h(V_1(Q_i)) \tag{32}$$

4. Calculate the local cell coefficients a_{11}, a_{12}, a_{21} and a_{22} from Eq. (13') about each equilibrium point Q_i,($i = 1, 2, 3$) for the FitzHugh-Nagumo example).
5. For each point in the cell parameter space, use the tests in Section 4.3.3. to determine the nature (*stable and locally active* ($S(Q_i)A(Q_i)$), *locally active and unstable* ($A(Q_i)U(Q_i)$), or *locally passive* $P(Q_i)$) of each equilibrium point determined from step 2. Here Q_i is defined to be *stable* if, and only if, the *uncoupled cell* (i.e. $\bar{I}_1 = \bar{I}_2 = 0$) is *locally asymptotically stable* [Hartman, 1982] at Q_i. These tests are different for the cases of one and two diffusion coefficients.
6. Identify the *edge of chaos* domain C in the cell parameter space according to the above definition.

Let us now examine the edge of chaos domains for the FitzHugh-Nagumo Equation[3] with $c = 1$, as well as several cross sections of the local activity regions for several values of $(\bar{I}_1, \bar{I}_2)$ as shown in Figs. 6-8.

Recall from Fig. 5 that the FitzHugh-Nagumo cell for $c = 1$ is *always* (unrestrictedly) locally active. This means that there is *at least* one $(\bar{I}_1, \bar{I}_2)$ where the cell is locally active. The *local passivity* (blue) region in Figs. 6-8 is therefore valid *only* for the specified values of $(\bar{I}_1, \bar{I}_2)$. We will next refer to this conditional passivity property as *restricted local passivity*. Similarly, the term "*restricted local activity*" implies that the cell

---

[3] In the literature, the FitzHugh-Nagumo Equation usually assumes $c = 1$ [FitzHugh, 1969], [Cronin, 1987].

is locally active at an equilibrium point when the point $(\bar{I}_1, \bar{I}_2)$ has some specified values. The *restricted* local activity domain $A(\bar{I}_1, \bar{I}_2)$ and the *restricted* local passivity domain $P(\bar{I}_1, \bar{I}_2)$ are depicted in *red* and *blue* respectively in the three-dimensional (a, b, ε) parameter cube in Fig 6(a) for the one-diffusion coefficient case, and in Fig. 6(b) for the two-diffusion coefficients case, when the cell parameter cube is bounded by $-0.5 \le \varepsilon \le 0.5$, $-20 \le a \le 20$, and $-20 \le b \le 20$. We emphasize here that our definition of *edge of chaos* stipulates that $\bar{I}_1 = 0$ for the one-diffusion case, and $\bar{I}_1 = \bar{I}_2 = 0$ for the two-diffusion case[4]. This is why we introduced the adjective "*restricted*" when speaking about local activity and local passivity domains in the context of edge of chaos. However, to avoid clutter, in the following the word "restricted" will be removed since it is implicitly understood that $\bar{I}_1 = \bar{I}_2 = 0$.

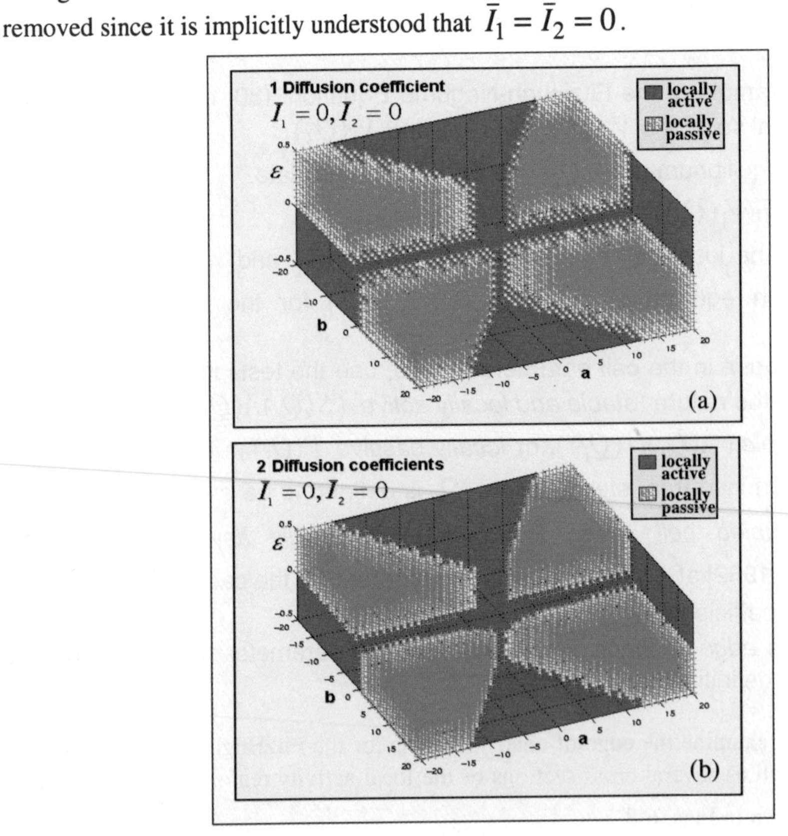

Fig 6. Restricted local activity (red) and local passivity (blue) domains in the (a, b, ε) parameter space for $c = 1$. The coupling inputs $\bar{I}_1$, and $\bar{I}_2$ have been restricted to $\bar{I}_1 = \bar{I}_2 = 0$. (a) The one diffusion coefficient case; (b) The two diffusion coefficients case. Since the red region here is larger (especially clear near the origin) than the red region from above, the local activity domain in this case includes the local activity domain shown in (a) as a proper subset.

---

[4] The restriction $\bar{I}_1 = \bar{I}_2 = 0$ (which corresponds to an uncoupled cell) is imposed in order to define the *edge of chaos* precisely as a *cell property*, which expresses its potential for emergent computation without requiring any knowledge about the passive coupling.

When $\bar{I}_1 = \bar{I}_2 = 0$, the *local activity* domain will depend on the cell parameter a, and is symmetrical with respect to the $a = 0$ plane. Moreover, observe from Fig. 6(a) and Fig. 6(b) that the local activity domain of the one-diffusion case coincides with that of the two-diffusion case for large magnitudes of a, b and ε. However, a careful examination near $(a,b,\varepsilon) = (0,0,0)$ in Figs. 6(a) and 6(b) reveals that the local activity domain (red area) in the one-diffusion case is a *subset* of the corresponding domain in the two-diffusion case.

Let us now conduct a detailed analysis of the local activity domain in Figs. 7 and 8, where the *edge of chaos* domain is shown in *red* and the remaining *locally active and unstable* domain is coded in *green.* The *yellow* regions correspond to locally active domains in the cell parameter space where there are more equilibrium points, but none of them is simultaneously locally active and stable. The blue regions denote the *restricted locally passive domains.* They are in fact *unrestrictedly locally active,* when $\bar{I}_1, \bar{I}_2 \in (-\infty,+\infty)$.

The edge of chaos domain of the FitzHugh-Nagumo Equation with *one* diffusion coefficient $D_1 \neq 0$ is shown in Fig. 7(a), and in Fig. 7(b) on a larger scale in *a* and *b*. Several ε - cross sections of Fig. 7(b) are shown in Fig. 7(c). Observe that the profile of the "edge of chaos" domain depends on ε and changes dramatically in the neighborhood of $\varepsilon = 0$. For negative values of ε, Figure 7(c) shows that the *edge of chaos* consists of two narrow strips intersecting each other near the point $(a = 0,\ b = 1.4)$, and two swallow-tail areas located to the left of $b = 1$ on each ε - cross section. Observe that the two narrow pointed tips of each swallow tail become shorter as ε becomes more negative. For positive values of ε, the domain of edge of chaos vanishes completely.

The edge of chaos domain (coded in red) in Fig. 7(c) is *symmetrical* with respect to the $a = 0$ plane. This symmetry is broken as $\bar{I}_1$ increases from $\bar{I}_1 = 0$ to $\bar{I}_1 = 1$ in Fig. 7(d). Here, the red area can be interpreted as an elastic (topological) deformation of the edge of chaos domain, henceforth called a *perturbed edge of chaos domain.*

The edge of chaos of the FitzHugh-Nagumo Equation with *two* diffusion coefficients ($D_1 \neq 0$ and $D_2 \neq 0$) is coded in red in Fig. 8(a). It is symmetrical with respect to $a = 0$. Several ε - cross sections are shown in Fig. 8(b), where the *red* areas denote the edge of chaos for different ε values. Comparing these cross sections with those shown in Fig. 7(c), we find that the former includes the latter as a proper subset. In other words, the edge of chaos domain for the two-diffusion coefficient case is larger and includes the corresponding region for the one diffusion coefficient case as a subset. When at least one of $\bar{I}_1, \bar{I}_2$ changes, the edge of chaos domain in Figs 8(a) and 8(b) undergoes an "elastic" deformation, resulting in a corresponding *perturbed edge of chaos domain,* shown in Fig. 8(c) for $\bar{I}_1 = 0.5, \bar{I}_2 = 0$, and in Fig. 8(d) for $\bar{I}_1 = 0, \bar{I}_2 = 0.5$.

In the following we will examine several cross sections in the parameter space for fixed values of the cell parameters c and ε. A detailed structure of several *bifurcation diagrams* derived from the definitions of *local activity* and *edge of chaos* will be next presented. In Section 4.3.6, for several representative cross-sections, we will randomly pick some points lying on or nearby the edge of chaos domain and demonstrate that they indeed give rise to complex *non-homogeneous* static and dynamic patterns, which can be interpreted as examples of emergent computation.

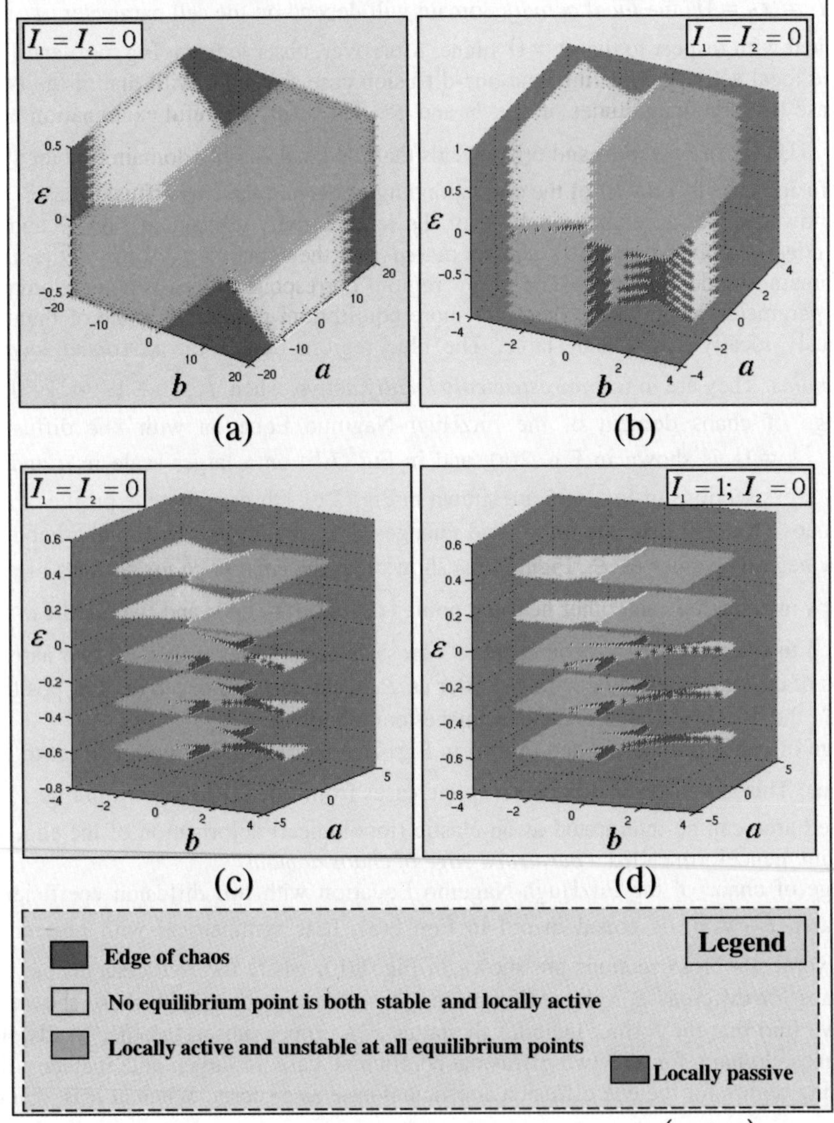

Fig. 7. Cross sections containing the *edge of chaos* domain (red) in the (a, b, ε) parameter space for $c = 1$ (one diffusion coefficient case).

(a) $a \in [-20,20]$, $b \in [-20,20]$, $\varepsilon \in [-0.5,0.5]$. Observe the rather small parameter domain near the origin where the *edge of chaos* domain (red) has a significant magnitude.

(b) A detailed cross-section for $a \in [-4,4]$, $b \in [-4,4]$, $\varepsilon \in [-1,1]$.

(c) Five ε - cross sections of the *edge of chaos* domain within the same parameter domain as in (b)

(d) Five ε - cross sections of the *perturbed edge of chaos* domain in (c) when $\bar{I}_1 = 1$; Observe the "elastic" cantilever-like deformation to the right of the "fixed" base located near $b = 0$.

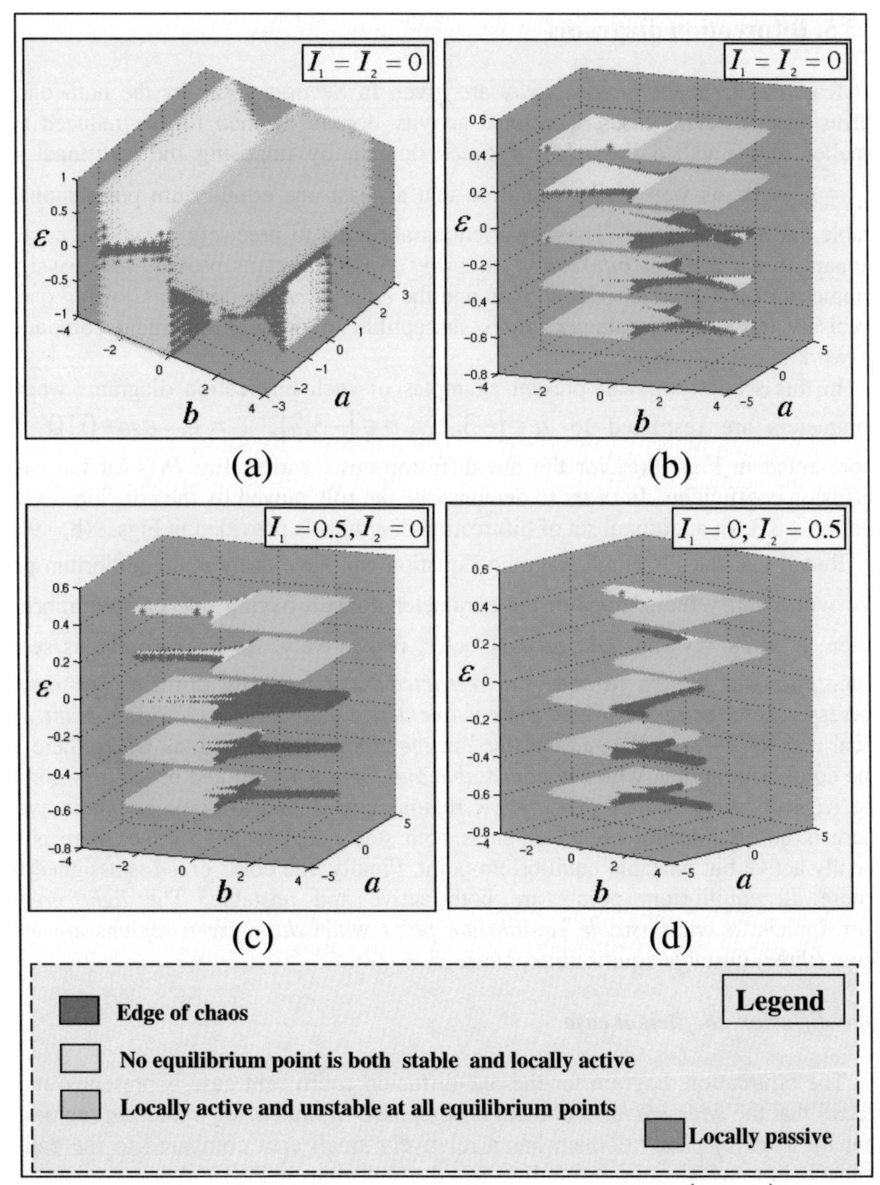

Fig. 8. Cross sections within the *edge of chaos* domain (red) in the (a, b, ε) parameter space for $c = 1$ (two diffusion coefficient case). Observe that the *edge of chaos* domain is now larger and includes the one diffusion coefficient case as a subset: (a) A section through the same (a, b, ε) parameter domain as in Fig. 7(b); (b) Five ε - cross sections of the *edge of chaos* domain within the same parameter domain as in (a). Also, compare with the one diffusion case in Fig. 7(c). (c) Five ε - cross sections of the perturbed *edge of chaos* domain in (b), when $\bar{I}_1 = 0.5$; As in Fig. 7(d), observe the "elastic" cantilever-like deformation to the right of the "fixed" base located near $b = 0$; (d) Five ε - cross sections of the perturbed *edge of chaos* domain in (b) when $\bar{I}_2 = 0.5$. Again, observe the "elastic" deformation, but now perceived as a slightly compressed translation to the left (with respect to the *a*-axis near *b*=0). The magnitude of the translation is proportional to $1 / \varepsilon$.

4.3.5. Bifurcation diagrams

Rigorous tests for *local activity* are given in Section 4.3.3 for the both one and two diffusion coefficient cases. The local activity domain is then further reduced to a much smaller region called the *edge of chaos* domain by imposing the additional restriction $\bar{I}_1 = \bar{I}_2 = 0$, as well as the condition that at least one equilibrium point should be both stable and locally active. These restrictions allow us to precisely locate the edge of chaos domain in the form of *bifurcation diagrams*. Such diagrams provide much insight on the geometry of the *local activity* domain and the *edge of chaos* domain allowing one to locate precisely those cell parameter points susceptible to produce emergent computation in a network of coupled cells.

In this section we will present examples of such bifurcation diagrams when the cell parameters are restricted to $a \in [-3,3]$, $b \in [-3,3]$, $c = 1$, $\varepsilon = -0.10$. They are represented in Figs. 9(a) for the one-diffusion case, and in Fig. 9(f) for the case of two diffusion coefficients. In order to demonstrate the role played by the conditions C1 to C4 in section 4.3.3, an additional set of bifurcation diagrams is provided in Figs. 9(b) - 9(e).

Recall that the FitzHugh-Nagumo Equation can have at most 3 equilibrium points Q_i. We will identify the regions in the parameter domain having one or 3 equilibrium points when $\bar{I}_1 = \bar{I}_2 = 0$ by *light* or *dark* color, respectively. As in the previous sections, the color *blue* was chosen to code the *restricted* local passivity domain. The *orange color* corresponds to the region where there is *one active and stable equilibrium point.* Within the local activity domain, the *red* and the *orange* colors denote regions where there is at least one equilibrium point, which is *both stable and locally active*, and hence they correspond to the *edge of chaos* domain. The *yellow* regions correspond to domains where even though there is no equilibrium point, which is both stable and locally active, there is at least a locally active but unstable equilibrium point. Finally, the color *green* is assigned to regions where all equilibrium points are both active and unstable. The *light green region* corresponds to *one unstable equilibrium point* while *dark green* regions are assigned to more (three) unstable equilibrium points.

One-diffusion coefficient case

The bifurcation diagram for the one-diffusion coefficient case is presented in Fig. 9(a). Recall that the *edge of chaos* domain is composed of both *orange and red* regions. Observe that the *edge of chaos* domain has a relatively small area compared to the entire locally active region, which consists of all except the blue areas in Fig. 9(a).

What is the role played by the *local activity conditions* (see section 4.3.3.) in defining the "edge of chaos"?. Let us first consider the conditions C3 and C4 (section 2.2.1). They are represented by dark blue lines[5] (condition C3) intersecting each other at a point which corresponds to the fulfillment of C4. It is interesting to observe that these 2 lines form a sort of skeleton for the entire *edge of chaos* domain. The intersection point $(a = 0, b \approx 1.4)$ defined by condition C4 has several interesting topological properties. Indeed, not only does it represent a discontinuity in the shape of the *edge of chaos domain* but in its vicinity, points from all possible *local activity* and *local passivity* regimes can be found. We have

[5] Condition C3 ($T = 0,\ \Delta > 0$) is artificially "relaxed" in our computation to the condition $(|T| < 0.03,\ \Delta > 0)$ in order to produce visible bold lines on the bifurcation diagrams.

found that cell parameter points situated in this neighborhood give rise to a large variety of complex dynamic behaviors.

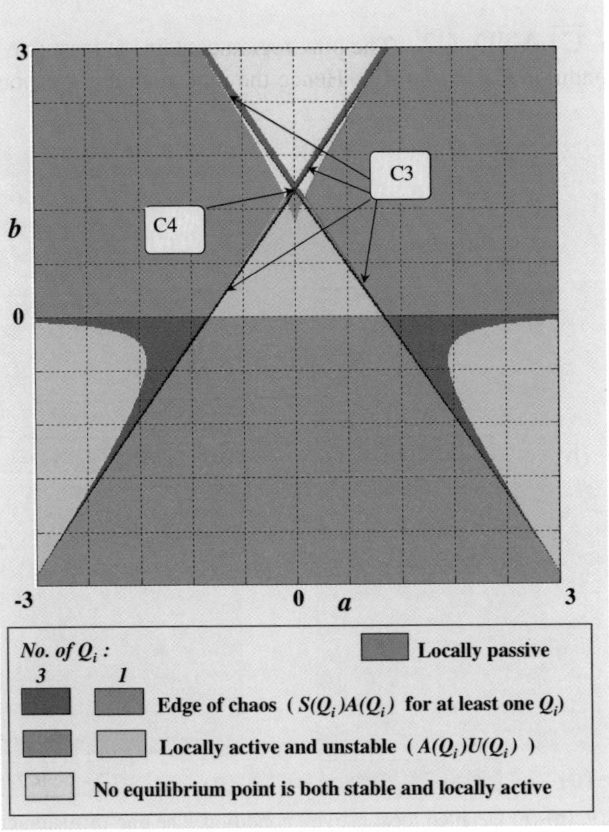

Fig. 9.(a) Bifurcation diagrams in the local activity domain at cross-section $a \in [-3,3]$, $b \in [-3,3]$, $c = 1$, and $\varepsilon = -0.10$; (a) The one-diffusion case.

Although conditions C1 and C2 in equation (19) are disjoint from conditions C3 and C4, they are not disjoint among themselves. Thus, in order to have a better understanding of how these conditions are related to the *local activity* areas shown in Fig. 9(a), an additional set of 4 bifurcation diagrams, shown in Figs. 9(b)-(e), is presented. The pink color in Figs 9(b)-(e) codes the domain where one of the following combination is satisfied, while the remaining domains keep the same color code as in Fig. 9(a).

1) Figure 9(b): C1 AND C2 : In the pink region both C1 and C2 are simultaneously satisfied by at least an equilibrium point.

2) Figure 9(c): $\overline{\text{C1}}\text{ AND }\overline{\text{C2}}$: In the pink region, there is at least one equilibrium point, which satisfies neither C1 nor C2 and is in fact locally passive at $\bar{I}_1 = \bar{I}_2 = 0$. Since the pink region is located within the locally active domain, it follows that there should be another equilibrium point (with the same cell parameters) which is locally active. Moreover, the pink region includes a part of the *edge of chaos* domain, since there is another equilibrium point (having the same cell parameter) which is both stable and locally active, as shown in Fig. 9(e).

3) Figure 9(d): $\text{C1 AND }\overline{\text{C2}}$: The pink region contains at least one equilibrium point, which satisfies condition C1 but *not* C2. This region includes the two narrow tails

from the *edge of chaos* domain, as well as most of the locally active but *unstable* domain. Observe that in the light green region, there is only an equilibrium point and it satisfies both C1 and C2.

4) Figure 9(e): $\overline{C1}$ AND C2 : The pink region contains at least one equilibrium point, which satisfies condition C2 but not C1. Hence the pink region correspond precisely to the *edge of chaos.*

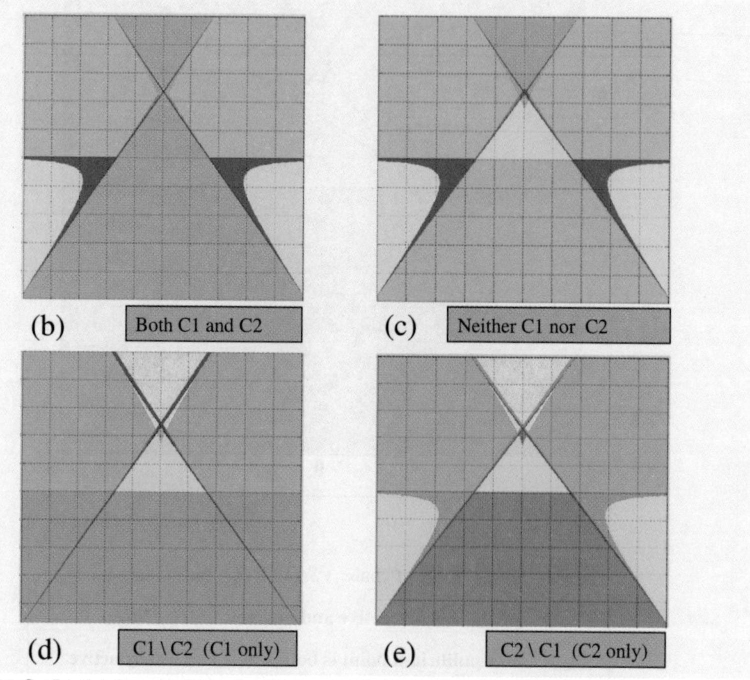

Fig. 9. (b)-(e) Detailed local activity condition. The one-diffusion case.

Observe that the 4 bifurcation diagrams displayed in Figs. 9(b)-(e) together with the bifurcation diagram shown in Fig. 9(a) contain all information needed to characterize uniquely a specific cell parameter point with respect to: (i) the number of equilibrium points, (ii) stability, and (iii) local activity.

Moreover, in the case of 3 equilibrium points one can infer the qualitative nature of each equilibrium point by examining the entire set of 4 bifurcation diagrams in Figs. 9(b)-(e). Indeed, if a cell parameter point is colored pink in more than one of the four figures, then necessarily there are equilibrium points corresponding to different regimes. The most interesting examples are given by cell parameter points located in the two upper thin strips of the edge of chaos domain. Since this domain is colored in pink in Figs. 9(b), (c) and (e) it follows that each of the 3 distinct equilibrium points is in a different qualitative regime: The first is *stable* and *locally passive* (Fig. 9(c)), the second is *unstable* and *locally active* (Fig. 9(b)), and the last is both *stable* and *locally active* (Fig. 9(e)). Observe that inside the *edge of chaos* domain several failure boundaries *separating* the various *"edge of chaos" sub-domains* where the equilibrium points exhibit different qualitative properties can be identified.

Finally it is important to observe that the edge of chaos domain cannot be identified using the stability criterion alone. Indeed, the boundary of the *edge of chaos* domain is determined by cell parameters, satisfying the condition "only C2, *not* C1".

The two-diffusion case

In the case of two diffusion coefficients, since there is a single test for local activity, given by equation (22), there are no corresponding bifurcation diagrams of the type presented in Figs. 9(b)-(e). Based on this test, the resulting bifurcation diagram is presented in Fig. 9(f), where the same color code as in Fig. 9(a) is used.

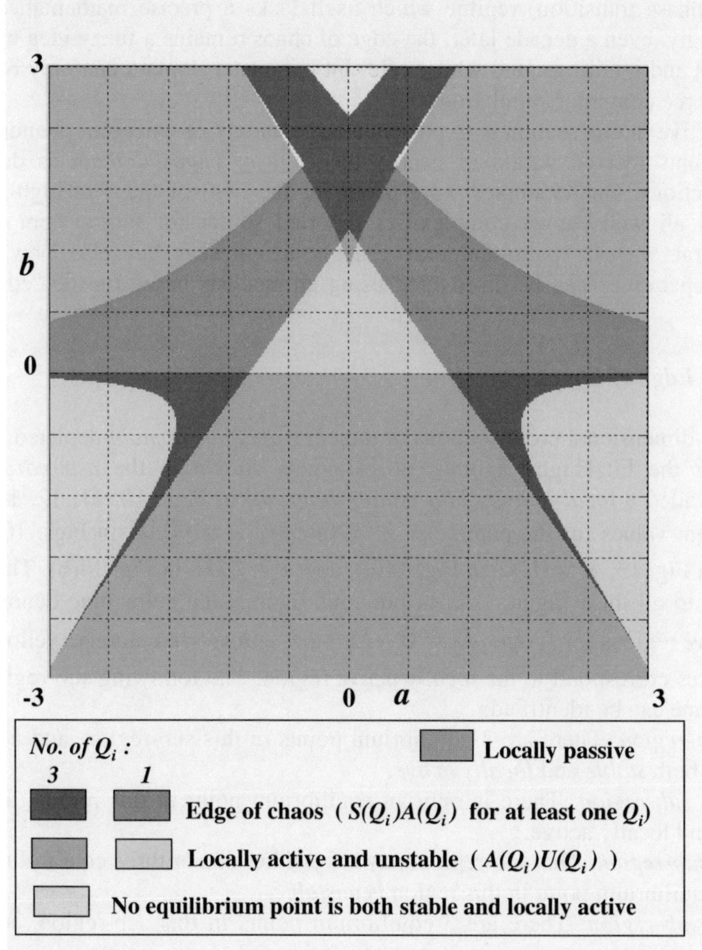

Fig.9(f) The two-diffusion case. The edge of chaos domain is identified by the red *and* orange areas.

Observe that the *edge of chaos* domain in this case includes the *edge of chaos* domain of the one diffusion case as a proper subset. In particular, observe that the two "swallow-tail" areas on each side of the green "pyramid" in both the one and two-diffusion cases are identical. Finally observe that the *unstable* region (light and dark green areas) in both the one-diffusion (see Fig.9(a)) and two-diffusion (see Fig.9(f)) case always coincide with each other because, by definition, stability is defined here only for isolated uncoupled cells (i.e. $\bar{I}_1 = \bar{I}_2 = 0$).

4.3.6. Emergent behaviors near and within the "edge of chaos"

Systems made of a large number of simple components interacting with each other in accordance with some *coupling laws* tend to operate in one of three possible regimes;

namely, an *ordered* regime such as crystals, a *disordered* regime such as gases, and a *phase transition* regime which separates them. The intriguing concept of the *edge of chaos* [Packard, 1988], [Langton, 1990] argues that only systems operating in the phase transition regime are capable of *information processing, emergence,* and *complexity*, and the domain of parameters which gives rise to such a regime is called the *edge of chaos*. The problem with this definition is that it is somewhat circular since it is defined in terms of an ambiguous (phase transition) regime which itself lacks a precise mathematical definition. Not surprisingly, even a decade later, the edge of chaos remains a fuzzy idea best described by metaphors and intuitions, and whose relevant parameter domain can only be determined using brute force computer simulation.

Our objective in this section is to present a large variety of emergent phenomena for cell parameter points located within or nearby the *edge of chaos domain* as derived in the preceding sections. Our examples were taken for the case of the FitzHugh-Nagumo cell showing that all well-known complexities reported so far for this system (e.g., Turing patterns, spiral waves, spatio-temporal chaos), are present but also new information processing capabilities can be discovered using an analysis based on the "edge of chaos" concept.

Mapping the Edge of Chaos

Several 2-dimensional cross-sections of the *edge of chaos domain* depicted previously in Figs. 7-8 for the FitzHugh-Nagumo cell, are now shown in the *a-b parameter plane* (henceforth called a *local activity bifurcation diagram*) in Figs. 10, 11, 12, 13 and 14, by fixing different values for the parameter ε. Namely, $\varepsilon = -0.10$ in Figs. 10, 11 and 12, $\varepsilon = -2.0$ in Fig. 13, $\varepsilon = 0.10$ in Fig. 14(a), and $\varepsilon = 0.01$ in Fig. 14(b). The same *color code* applies to all these figures. As in previous figures, the color *blue* denotes *restricted locally passive* regions for $\bar{I}_1$ and $\bar{I}_2 = 0$. All other colors (red, orange, yellow and green) in these figures correspond to the *locally active* region. The following sub-region in the *a-b* parameter plane can be identified:

1. *Red sub-region*: There are 3 equilibrium points in this sub-region, and at least one of them is both *stable* and *locally active*.
2. *Orange sub-region*: There is only an equilibrium point in this region, and it is both stable and locally active.
3. *Green sub-region*: This sub-region may have either one or three equilibrium points and every equilibrium point in this region is *unstable*.
4. *Yellow sub-region*: There are 3 equilibrium points in this sub-region, at least one *is locally active*, but none is simultaneously *stable* and *locally active*.

By definition, the *red* and *the orange* sub-regions in Figs. 10, 12, 13, and 14 *together* constitute the *edge of chaos*. We do not include the case of *3 stable equilibrium* points because this case can not exist in a planar system in view of *index theory*.

Each bifurcation diagram in Figs. 10, 11, 12, 13, and 14 is composed of 4 sections. The two upper sections are associated to the one diffusion coefficient case, whereas the two lower sections correspond to the case of two diffusion coefficients. In each case, the *rightmost* diagram gives a finer subdivision of the *"red" local activity* region on the *leftmost* diagram into the above 4 locally active sub-regions.

Let us pick a few representative points (a,b) on, or nearby the *edge of chaos* domain in the bifurcation diagrams in Figs. 10, 12, 13, and 14.

For ease of comparison, the *same set* of points is identified using solid black dots in all 4 parts of these figures. In addition, the coordinates (a,b) of each of these points are specified, along with an identification number (from 5 to 20), and a one or two letter code

(C, P, S, T, IC) indicating the nature of the associated patterns depicted in Fig. 15, where "C" denotes a *chaotic dynamic pattern*, "P" denotes a *periodic dynamic* pattern, "S" denotes a *spiral wave* pattern, "T" denotes a *Turing pattern*, and IC denotes an *information computation* pattern.

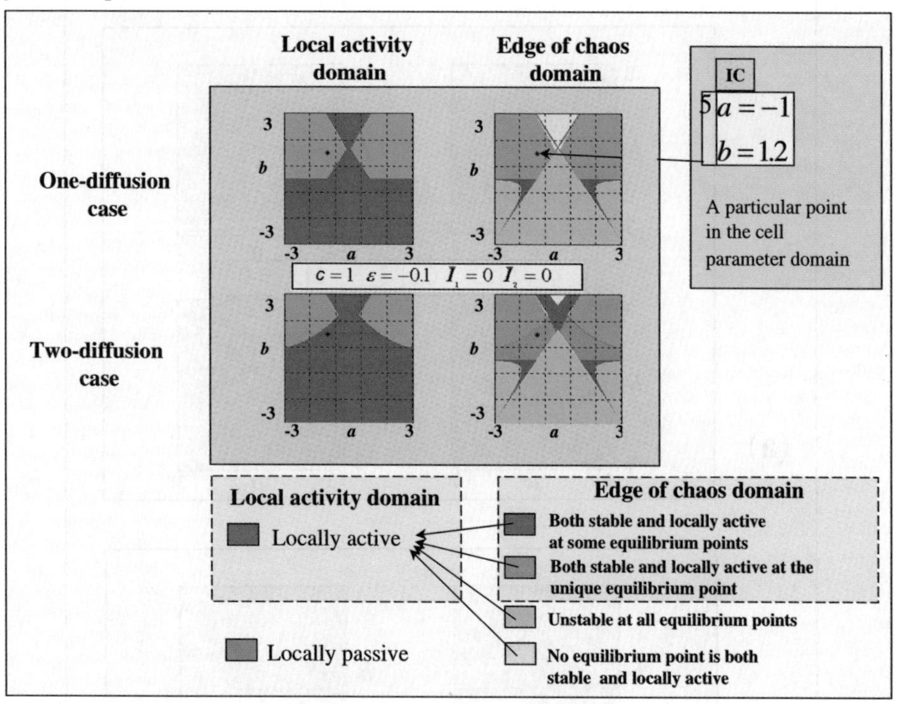

Fig. 10. Color legend for two dimensional local activity and edge of chaos bifurcation diagrams in the (a,b) parameter plane for Figs. 11-14: Particular points of interest in the parameter domain are assigned a numerical label (highlighted with a thin light-blue box next to a light-yellow "caption box" specifying the parameters (a,b)) so that they can be easily identified from the corresponding CNN dynamic simulation shown in Fig. 15. For example, the parameter point labeled "5" in this figure has its associated CNN dynamics presented in Fig. 15(5a) and Fig. 15(5b) corresponding to two different choices of diffusion coefficients.

Since the *edge of chaos* domain in the 2-diffusion case is in general *larger* and includes 1-diffusion edge of chaos domain, among the 16 representative points in Figs. 10,12, 13, and 14, several points are located *outside* of the edge of chaos domain in the 1-diffusion case. For example, point "5" in Fig. 10 and point "11" in Fig. 12 both are located inside the *restricted locally passive* (blue) region (when $\bar{I}_1 = \bar{I}_2 = 0$) for the 1-diffusion case, but *on* the *edge of chaos* domain for the 2-diffusion case.

For readers not familiar with the concept of *local passivity*, it is important to remember that the blue regions in Figs. 10,12, 13, and 14 are *locally passive* only in a restricted sense; namely, when there is only one coupling input $I_1(j,k)$, i.e. $D_2 = 0$. Such parameter points may change into locally active points when a second coupling input $I_2(j,k)$ is added ($D_2 \neq 0$), provided that they are *not* locally passive in the *absolute* sense depicted within the blue regions in Fig. 5(b). From the perspective of emergence and complexity, the blue regions in Fig. 5(b) are of no interest because non-homogeneous patterns can *not* occur when the FitzHugh-Nagumo cell parameters (b,c,ε) are located in these regions, no matter what diffusion coefficients $\left(D_1 \geq 0,\ D_2 \geq 0\right)$ are chosen [Chua, 1998]. Such FitzHugh-Nagumo cells are truly *lifeless*!

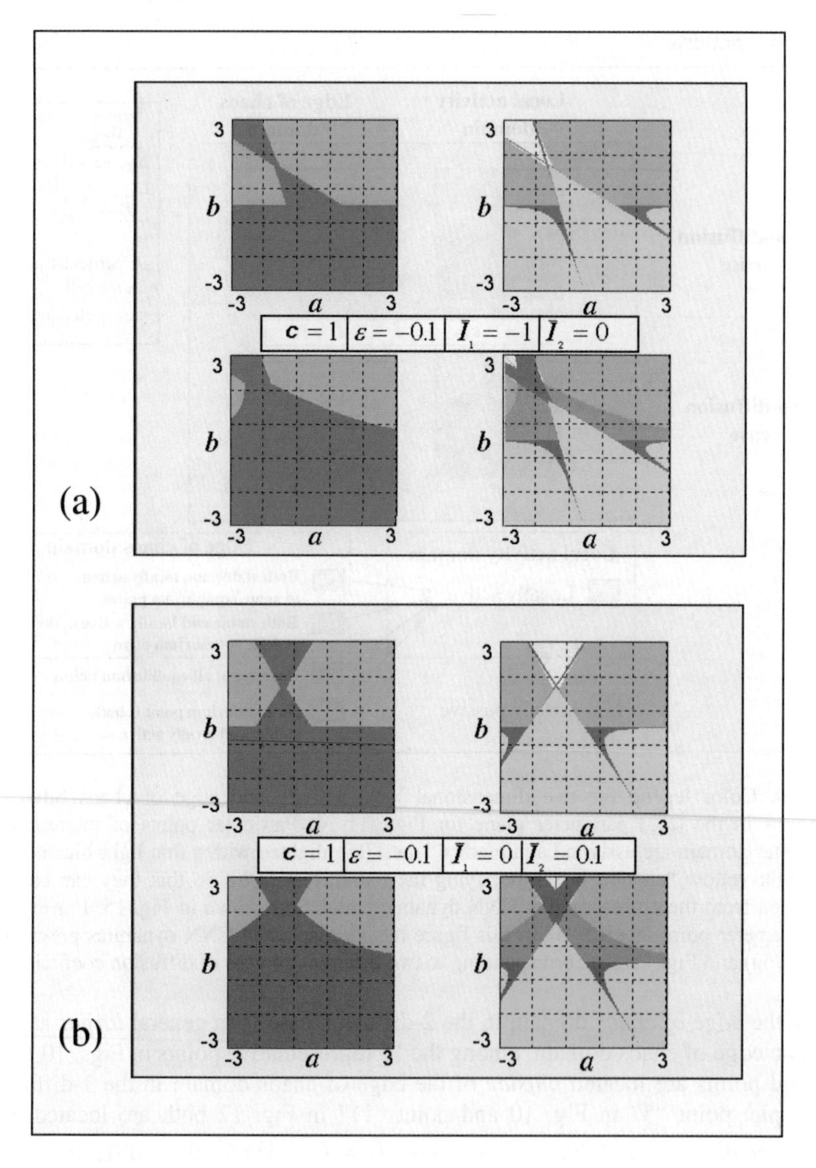

Fig. 11. Two *perturbed local activity* and *edge of chaos* bifurcation diagrams when $a \in [-3,3]$, $b \in [-3,3]$, $c = 1$, and $\varepsilon = -0.1$ For comparison, see the corresponding unperturbed ($\bar{I}_1 = \bar{I}_2 = 0$) domains in Fig. 10. (a)The effect of changing $\bar{I}_1$ from $\bar{I}_1 = 0$ to $\bar{I}_1 = -1$; (b) The effect of changing $\bar{I}_2$ from $\bar{I}_2 = 0$ to $\bar{I}_2 = 0.10$.

Since the restricted locally-passive (blue) regions (when $\bar{I}_1 = \bar{I}_2 = 0$) in Figs. 10,12,13, and 14 are located *outside* of the above "lifeless" region, *adding* a second coupling (i.e., $D_1 \neq 0$, $D_2 \neq 0$), may transform portions of the blue region into a *locally active* (red) region not only at $\bar{I}_1 = \bar{I}_2 = 0$. The same may happen if choosing a coupling different from D_1 (i.e., $D_1 = 0$, $D_2 \neq 0$). For example, the 4 bifurcation diagrams in Fig. 10 deform into the 4 corresponding diagrams shown in Fig. 11(a) when $\bar{I}_1 = -1$ and $\bar{I}_2 = 0$, or in Fig. 11(b) when $\bar{I}_1 = 0$ and $\bar{I}_2 = 0.10$. Observe that the blue region in fig. 11(a) has "skewed" towards the left in Fig. 11(a), and "translated" to the left in Fig. 11(b). By changing the sign of $\bar{I}_1$ and $\bar{I}_2$, similar deformations but in the opposite direction (right) are observed. In other words, portions of the blue regions, which are locally passive for $\bar{I}_1 = \bar{I}_2 = 0$ have the "potential" of being converted into local activity. In fact, it can be proved that a *locally passive* coupling, possibly more complicated than "diffusion", can always be found to convert a *restricted* locally passive cell parameter into a locally active one. In sharp contrast, if the blue regions in Fig. 10 had belong to the unrestricted locally passive blue regions in Fig. 5(b), then these regions will remain *invariant* no matter what values of $\bar{I}_1$ and $\bar{I}_2$ are chosen.

Static and Dynamic Patterns on the Edge of Chaos

For each cell parameter (a,b,c,ε) of the FitzHugh-Nagumo Equation identified in Figs. 5, 10, 12, 13, and 14, Fig. 15 displays 5 snapshots from the time evolution of the FitzHugh-Nagumo Reaction-Diffusion CNN (i.e., space-discretized version of Eqs. (1) for the 1-diffusion coefficient case, and Eq. (2) for the 2-diffusion coefficient case) for various choices of the diffusion coefficients D_1 and/or D_2. Each set of dynamical evolutions in Fig. 15 is identified by an integer (N=1,2...20) on the left (blue margin) which coincides with the integer label of the corresponding cell parameter point identified in Figs. 5, 10, 12, 13, and 14. In those cases where different diffusion coefficients (D_1, D_2) are chosen for the *same* cell parameter point, an additional code "a", "b", "c", etc., is attached to the above integer label. For example, three different values of (D_1, D_2) are associated with the cell parameter point "6" identified in Fig. 12. Their corresponding dynamic evolution is identified using the labels (6a), (6b), and (6c), respectively in Fig. 15.

The time evolution of the FitzHugh-Nagumo Reaction-Diffusion CNN is uniquely determined by specifying the 6 relevant parameters $(a,b,c,\varepsilon,D_1,D_2)$, the *initial condition* $(V_1(j,k),V_2(j,k))$, *j,k*=1,2,.. M, (where M^2 is the number of cells in the array) at *t=0,* and the *boundary condition,* which is assumed to be toroidal (i.e. periodic) throughout this book. Extensive numerical simulations [Chua, 1998] have shown that except for a few contrived cases, numerical solutions of a reaction-diffusion PDE are virtually identical to those calculated from the space-discretized version; namely, the associated Reaction-Diffusion CNN. We will infer therefore that whatever *emergence, complexity,* and *information computation* exhibited in Fig. 15 must also apply to the FitzHugh-Nagumo Reaction-Diffusion PDE. Let us now examine the various patterns generated by the FitzHugh-Nagumo Reaction-Diffusion CNN.

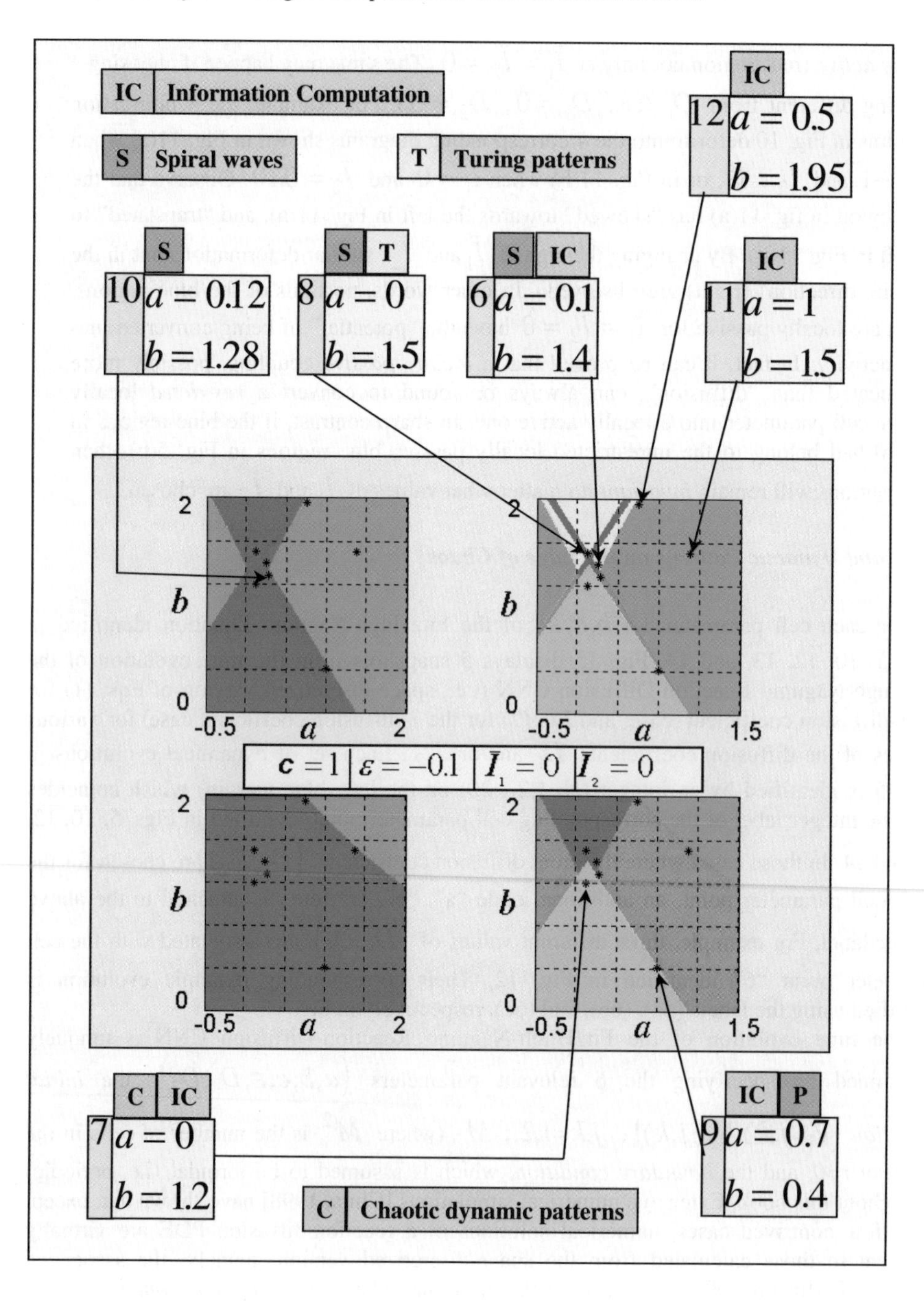

Fig. 12. Details of the *local activity* and *edge of chaos* domains from Fig. 10, for $a \in [-5,.5]$, $b \in [0,8]$, $c = 1$, and $\varepsilon = -0.1$. Seven parameter cell points (7 to 12) are indicated, each one is assigned an additional literal label ("S","T","IC","P" or "C") identifying the type of dynamical behavior exhibited by the corresponding CNN. The corresponding results of the CNN dynamic simulations are exhibited in Figs. 15(7) to 15(12).

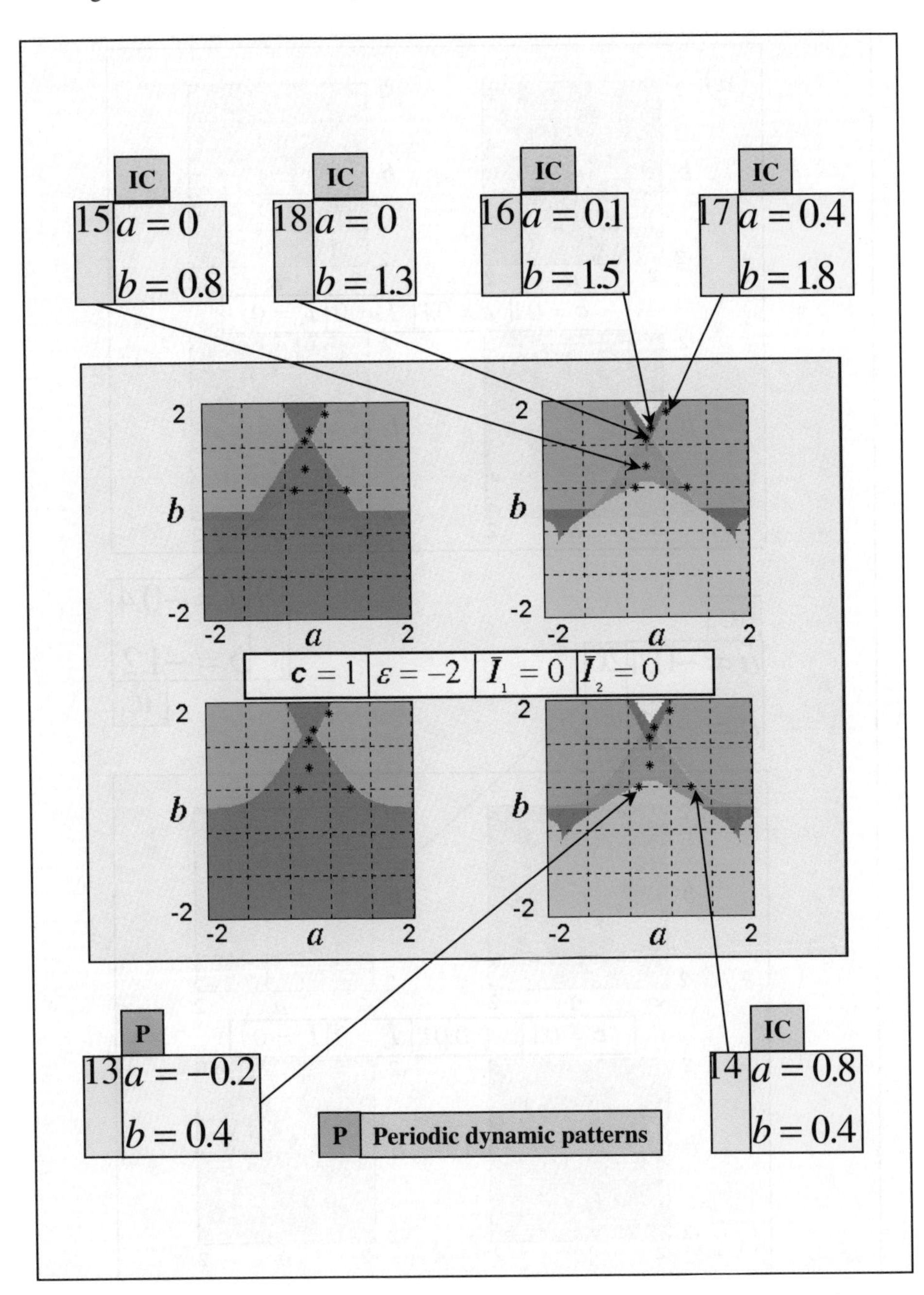

Fig. 13. Local activity and edge of chaos bifurcation diagrams for $\varepsilon=-2$, $c=1$, $a\in[-2,2]$, $b\in[-2,2]$. Six different cell parameter points are indicated using the same notations as in Figs. 10 and 12.

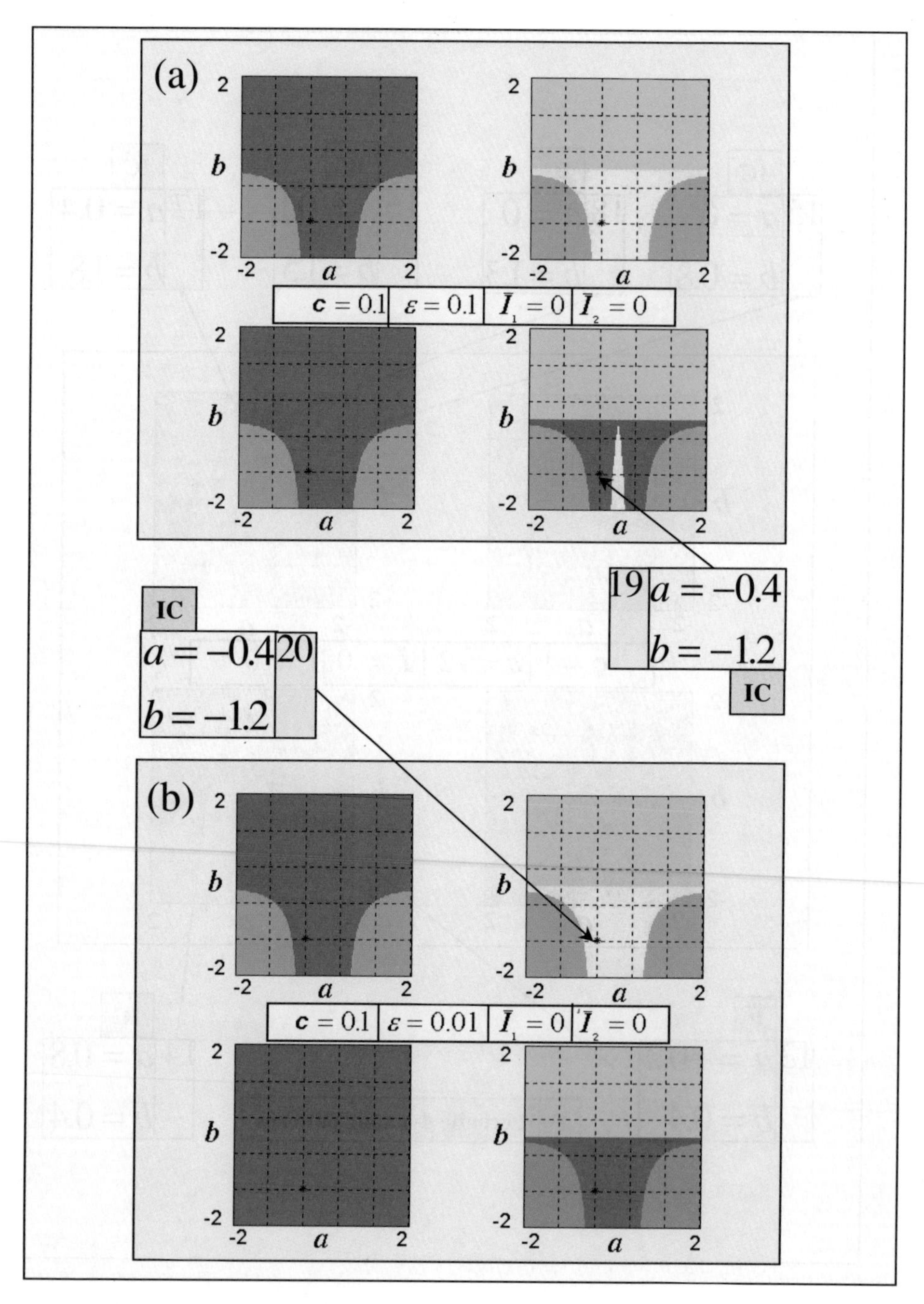

Fig.14. Local activity and edge of chaos bifurcation diagrams for $c = 0.1$, $a \in [-2,2]$, $b \in [-2,2]$. Two cell parameter points (19,20) are indicated. (a) $\varepsilon = 0.10$; (b) $\varepsilon = 0.010$.

Homogeneous static patterns

Consider first the two cell parameter points labeled "1" and "4" in Fig. 5(d), which lie near the border but nevertheless are inside the *locally passive* (blue) regions in Fig. 5(b) for the 2 diffusion case. As expected, the computer simulations confirmed that all cells converge to a homogenous steady state $V_1(j,k) \to 0$ and $V_2(j,k) \to 0$ for all $j,k = 1,2,..M$. Using the pseudo color code defined in the bottom of Fig. 15(3), the color code for the corresponding steady state distribution for $V_1(j,k)$ is shown in a *light green* color in Figs. 15(1)-15(4). The same steady state distribution is also observed for $V_2(j,k)$ [6]. Indeed, *all* cell parameters lying within the locally-passive (blue) regions in Fig. 5(b) are found to converge to the same homogeneous light-green static pattern, *regardless of the initial states.* These results can of course be predicted from the concept of local passivity, without carrying out any computer simulations!

Consider next the two cell parameter points labeled "2" and "3" in Fig. 5(d). Although they are locally passive for the 1-diffusion case (blue regions on the left of Fig. 5(d)), they are *locally active* in the 2-diffusion case (red region) on the right of Fig. 5(d). For the parameters shown in (2a), (2b), and (3) in Fig. 15, the corresponding steady state is seen to converge also to the same trivial homogeneous light-green pattern. These two examples illustrate that local activity is only a *necessary but not a sufficient* condition for complexity. In particular, extensive simulations have confirmed our conjecture that the farther a *locally active cell* parameter point is from the edge of chaos domain (of which "2" and "3" is a case in point), the smaller is the probability of observing complexity.

Turing-like patterns

Consider the cell parameter point "5" located at $(a,b) = (-1,\ 1.2)$ in Fig. 10. Observe that although this point lies in the locally passive region in the one-diffusion case, it lies on the *edge of chaos* domain in the two-diffusion case. For any *random* initial state and for parameters indicated in Fig. 15(5a), the array dynamics converges to a static *Turing-like pattern* [7].

Now let us have a look at the cell parameter point "8" located at $(a,b) = (0,\ 1.5)$ in Fig. 12. Observe that this point lies on the *edge of chaos* domain in both the 1-diffusion and 2-diffusion case. For the choice of parameters and initial states shown in Fig. 15(8c) the array dynamics also converges to a *static Turing-like pattern.* Indeed, extensive computer simulations had revealed that such patterns are quite abundant for cell parameters lying on the edge of chaos domain, when $D_2 >> D_1$, e.g., $D_2 = 4D_1$.

Two steady state Turing-like patterns for $V_1(j,k)$ and $V_2(j,k)$, respectively, are shown in Fig. 16(a). They correspond to the cell parameter point "6" in Fig. 12 when $D_1 = 0.3$, $D_2 = 1.40$ and for a random initial state. The pair of waveforms $V_1(t)$ and $V_2(t)$ at cell location $(j,k) = (50,50)$ is shown in Fig. 16(b).

---

[6] To conserve space, only the evolution of $V_1(j,k)$ is shown for all snapshots in Fig. 15.

[7] We call this a "Turing-like" pattern because $D_1 = 0$ in Fig. 15(5a), whereas in the literature on "Turing" patterns [Murray, 1989], both D_1 and D_2 are assumed to be non-zero.

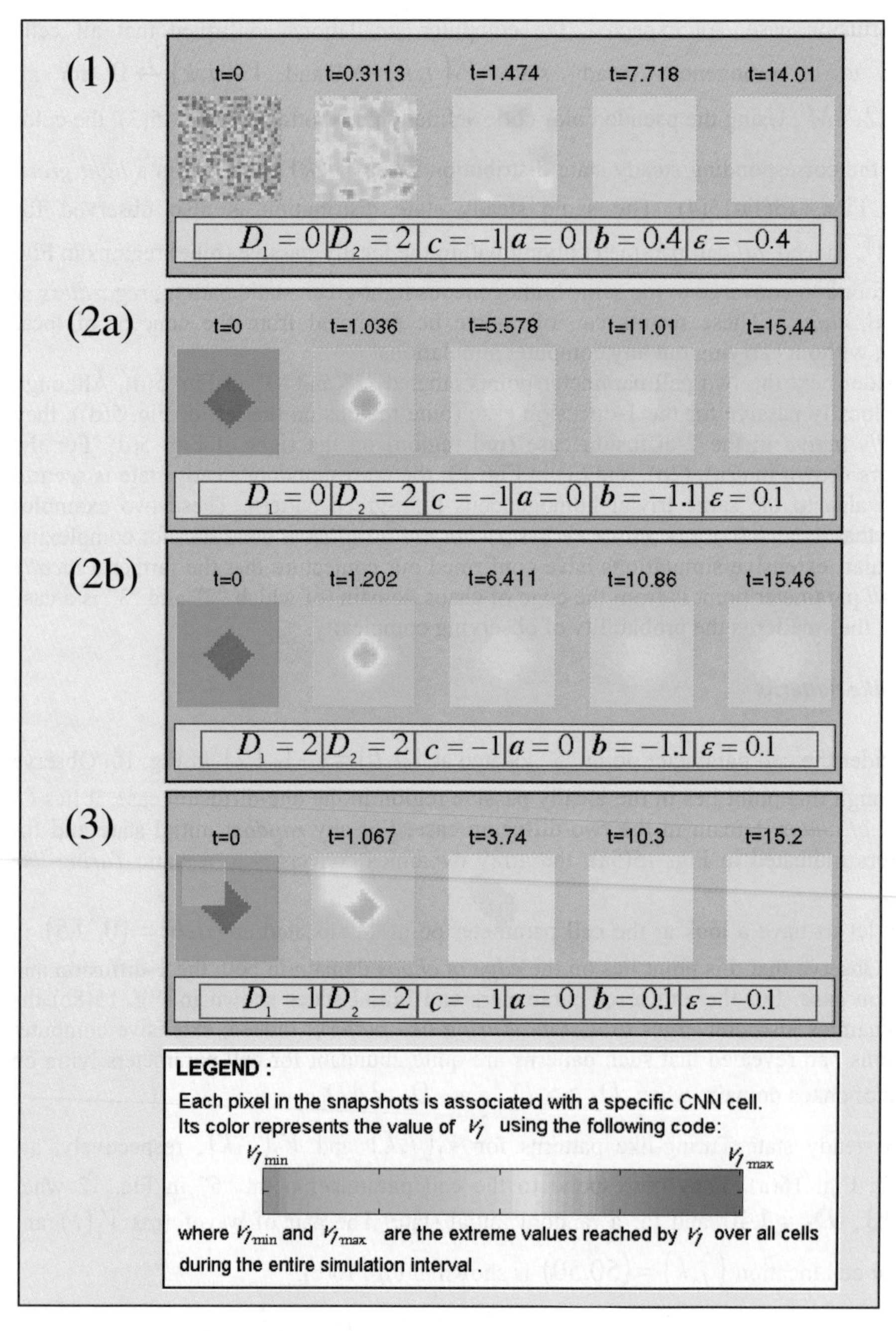

Fig. 15. Examples of dynamic simulations at different cell parameter points indicated in Figs. 2,10,11,12,13 and 14. Except for the two gray scale figures (where black corresponds to minimum and white to maximum), the color code is defined in the legend. The label assigned in the upper left (light-blue margin) of each sequence of 5 snapshots consists of a number indicating the cell parameter point in one of the previous figures, and followed by a letter which differentiates among different choices of diffusion coefficients, for *the same* cell parameter point.

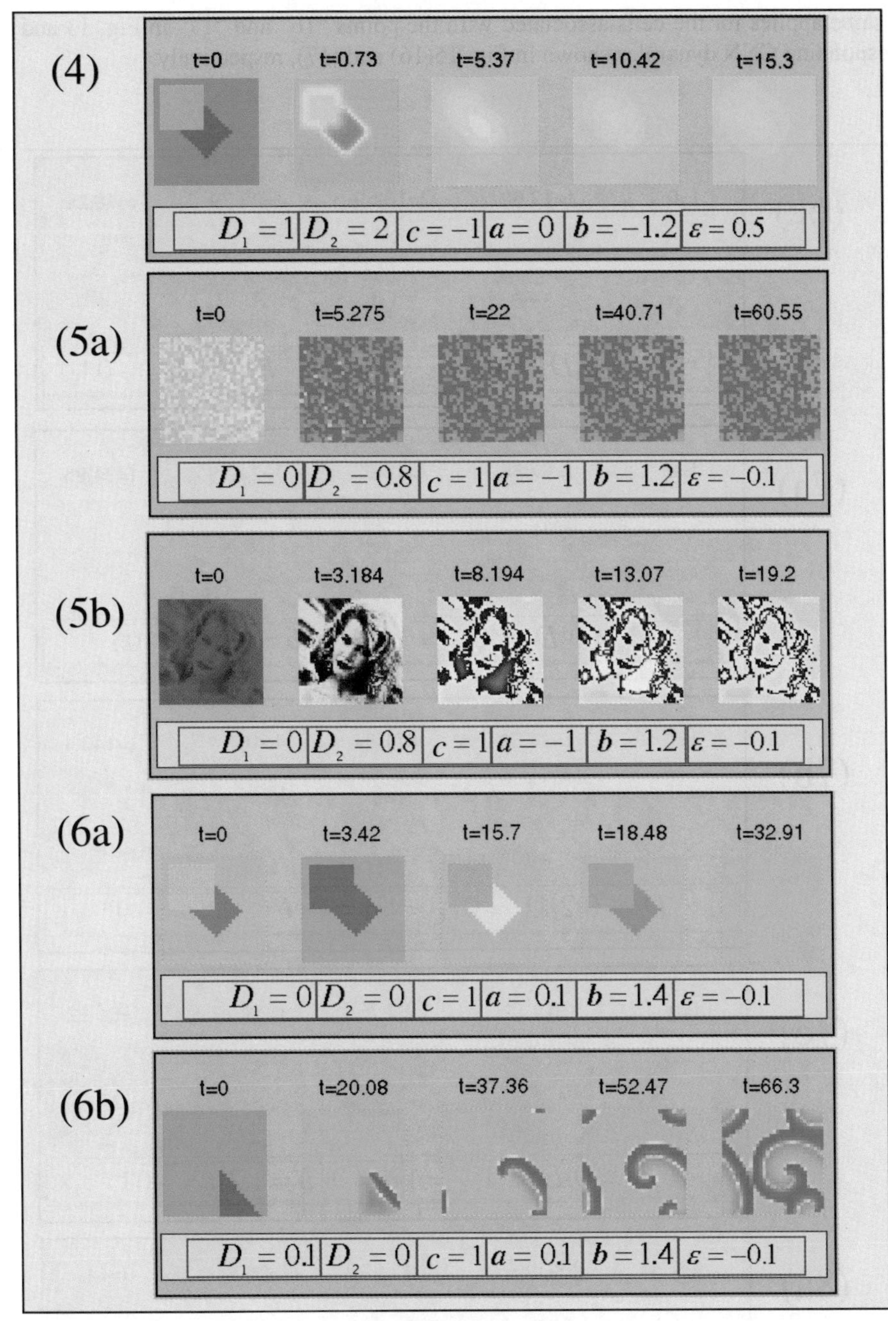

Fig. 15 *(Continued)*

Other examples of Turing-like patterns were observed (when a random initial state was chosen) at points "11" and "12" in Fig. 12 which lie at the border of the *edge of chaos* domain in the 2-diffusion coefficient case. The dynamic evolutions presented in Fig. 15(11) and Fig. 15(12) correspond to an initial condition chosen to be a pattern having some meaningful information content. Since the resulting steady-state pattern in all these cases represents a meaningful transformation (computation) of the initial pattern, the associated dynamics will be described in detail in the information computation patterns sub-section.

The same applies for the cells associated with the points "16" and "17" in Fig. 13 and their corresponding CNN dynamics shown in Fig. 15(16) and (17), respectively.

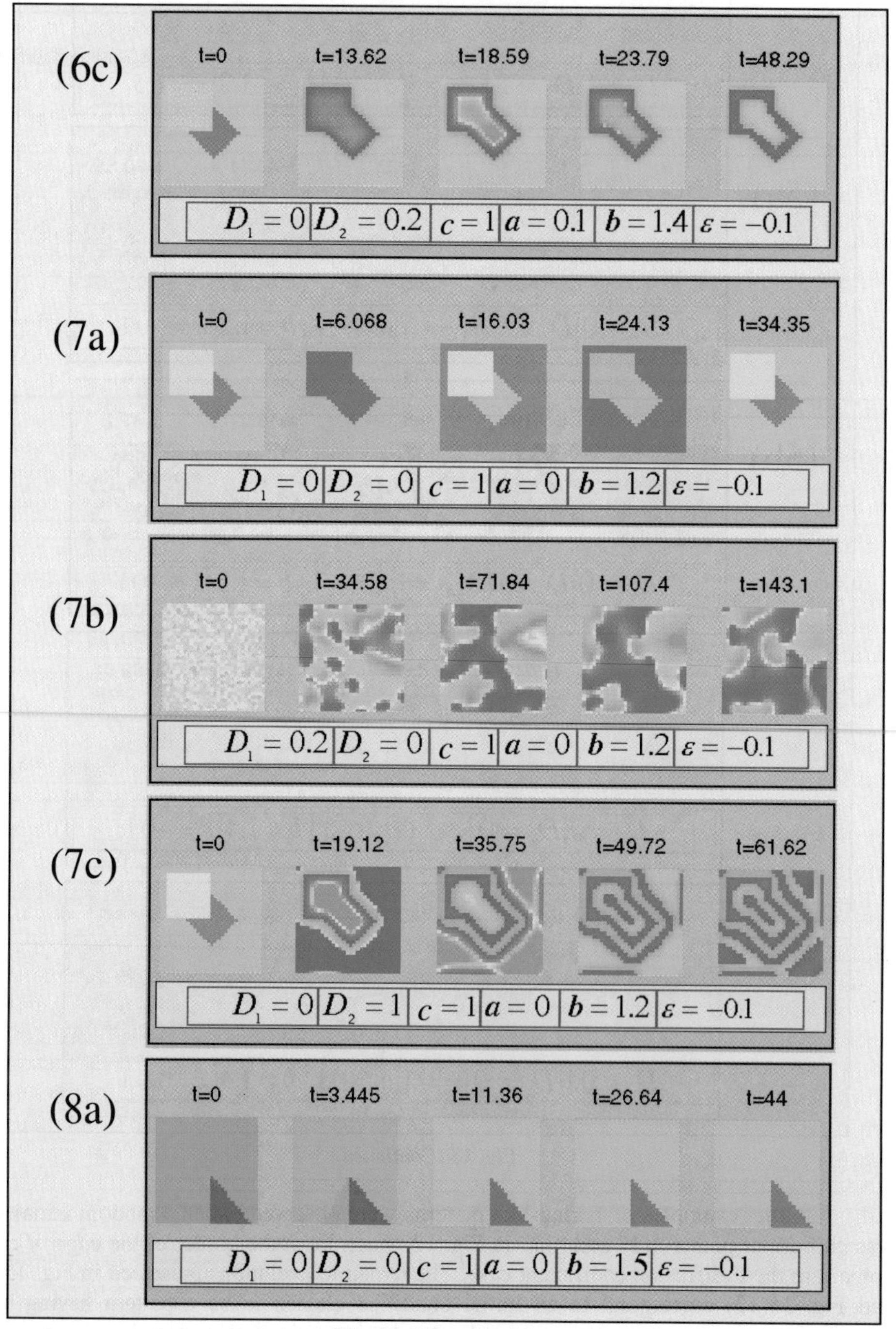

Fig. 15. *(Continued)*

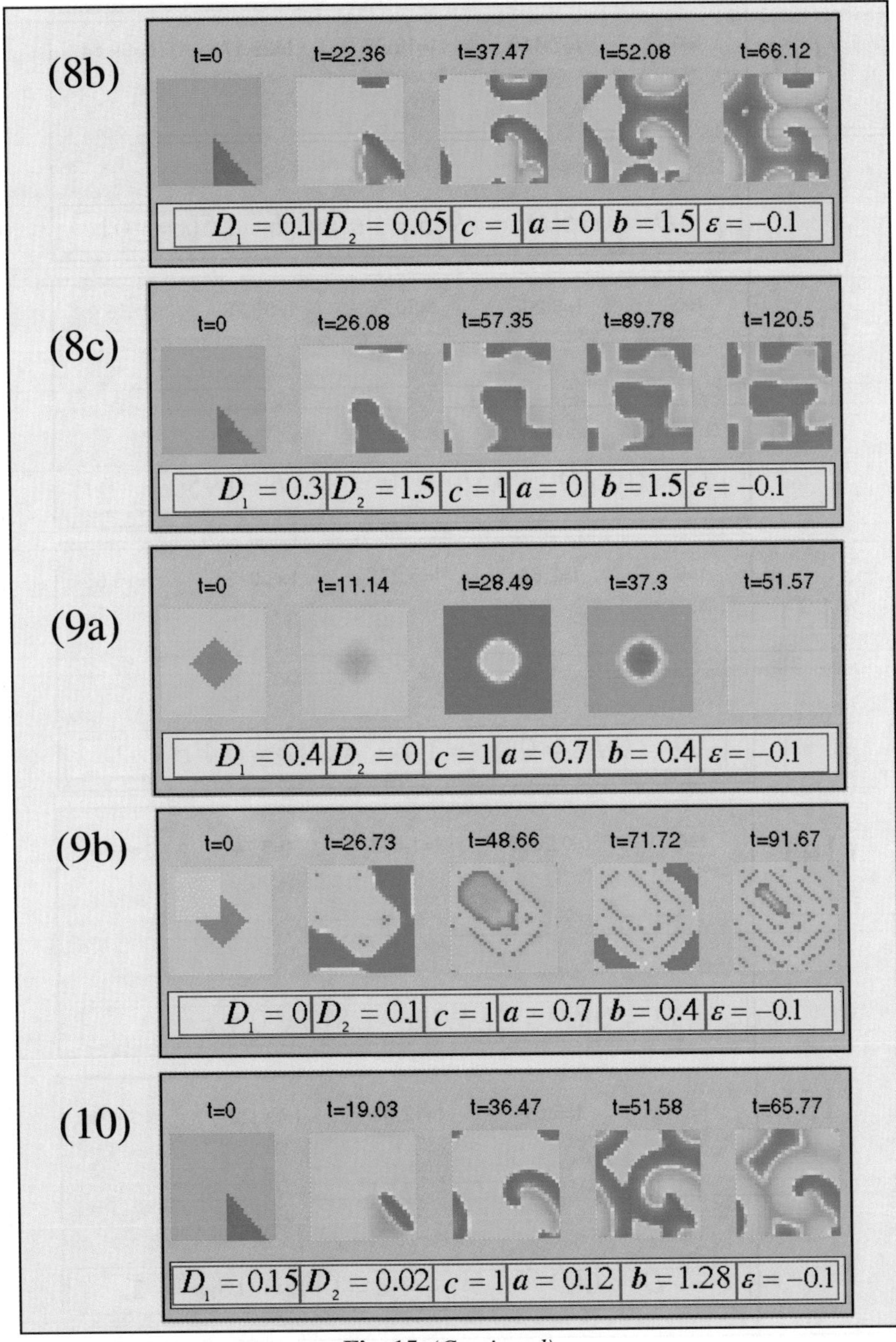

Fig. 15. *(Continued)*

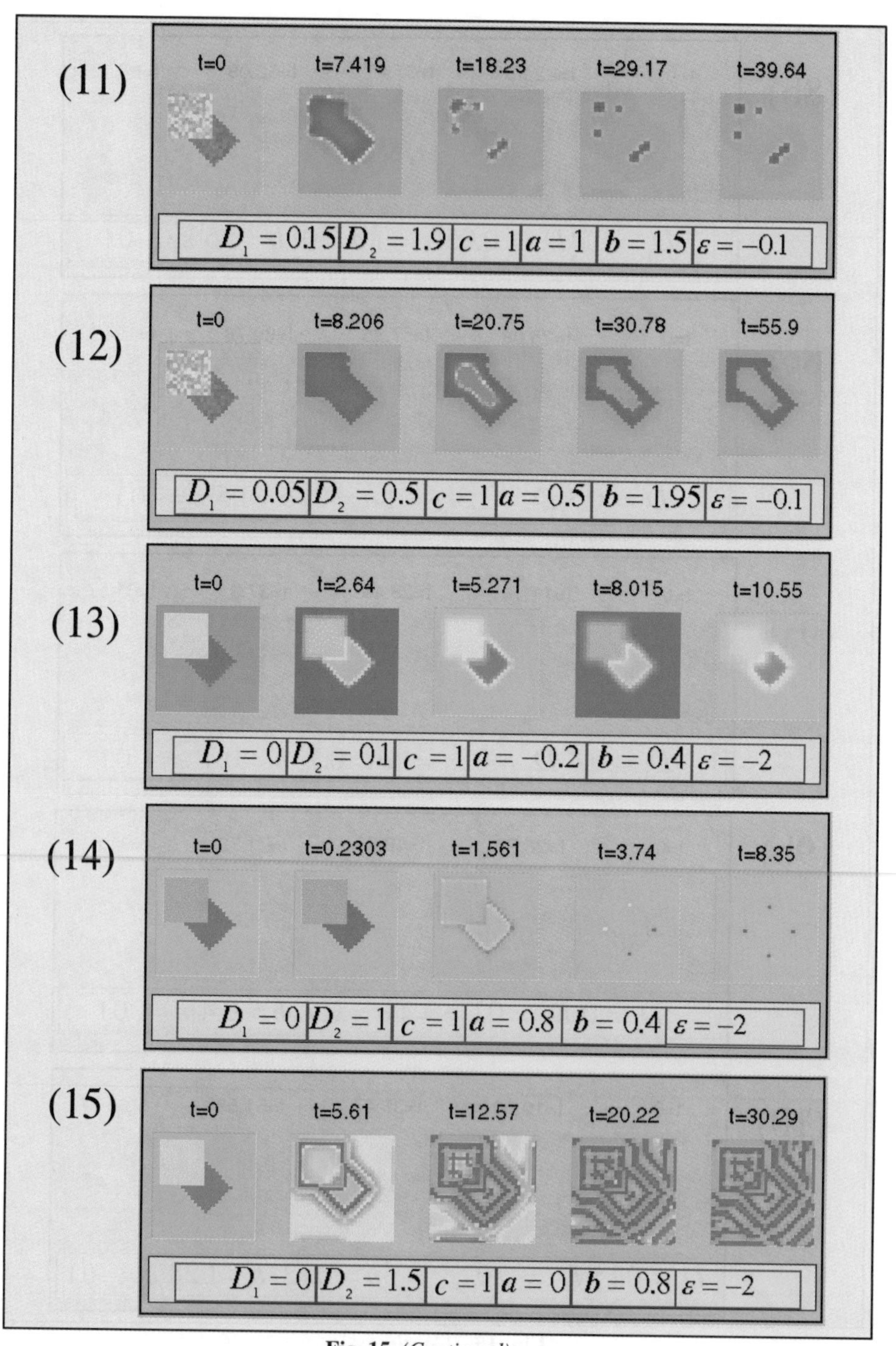

Fig. 15. *(Continued)*

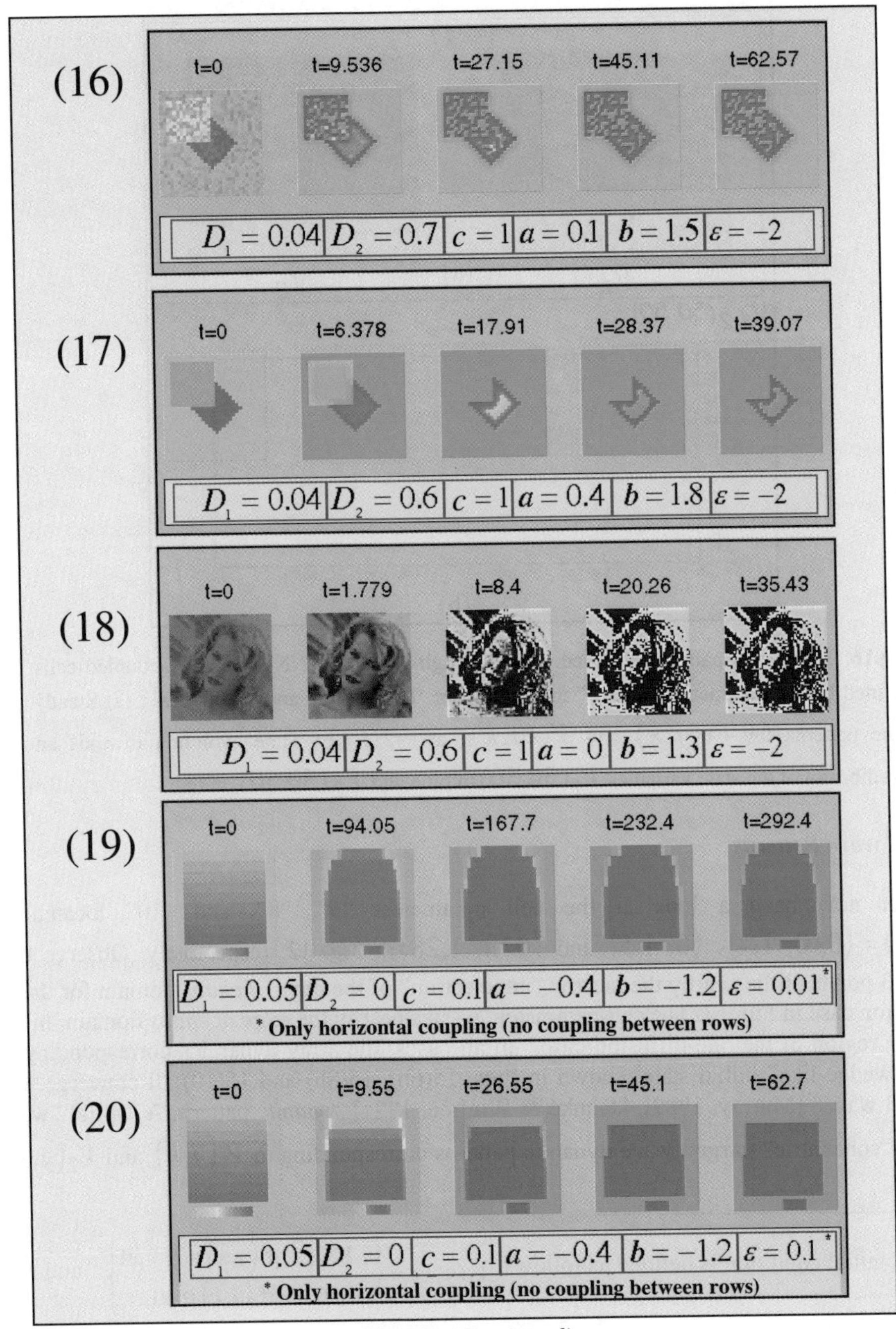

Fig. 15. *(Continued)*

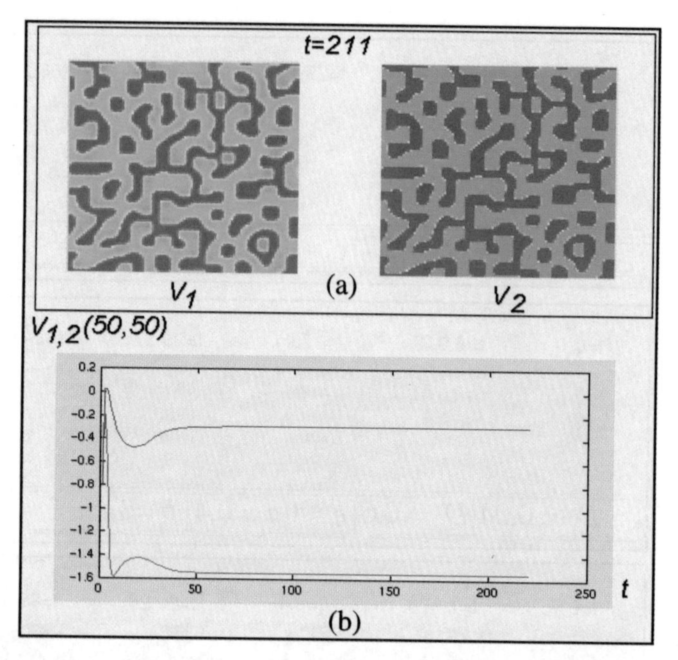

Fig.16. Turing-like patterns obtained in a FitzHugh-Nagumo CNN formed by coupled cells defined by the parameter point "6" in Fig. 12, for $D_1 = 0.3$, and $D_2 = 1.4$; (a) Steady state patterns for $V_1(j,k)$ and $V_2(j,k)$, at *t=211*; (b) Time evolution towards an equilibrium of the state variables $V_1(50,50)$ in blue and $V_2(50,50)$ in red.

Spiral wave Patterns

Let now have a look at the cell parameters "6", "8", and "10" located at $(a,b) = (0.10,\ 1.4)$, $(0,\ 1.5)$ and $(0.12,\ 1.28)$ in Fig. 12, respectively. Observe that these 3 points all lie nearby the narrow "intersection" of the *edge of chaos* domain for the 1-diffusion case in Fig. 12. The cell parameter "6" lies nearby the *edge of chaos* domain, in the *yellow* region of the one-diffusion case. In all cases, the array dynamics corresponding to the "wedge-like" initial state[8] shown in Figs. 15(6b), 15(8b) and 15(10) all converge to a "spiral wave" [Murray, 1989], [Jahnke & Winfree, 1991] *dynamic* pattern. A "spiral" wave and a "concentric" (target) wave dynamic patterns corresponding to $V_1(j,k)$ and $V_2(j,k)$

---

[8] The initial condition is defined as follows: $V_1(j,k) = \begin{Bmatrix} -1.67, & \text{if } (j,k) \in G \setminus A1 \\ 1.67, & \text{if } (j,k) \in A1 \end{Bmatrix}$ and $V_2(j,k) = \begin{Bmatrix} -0.2, & \text{if } (j,k) \in G \setminus A2 \\ 0.3, & \text{if } (j,k) \in A2 \end{Bmatrix}$, where G is the entire set of cell locations within the two-dimensional grid, and $A1$, $A2$ are two sets of cell locations defined by an angular sector having its origin in the middle of the CNN grid and extended up to the border of the grid. The two corresponding angular domains are: $\theta_1 \in (270^0, 312^0)$ for $A1$, and $\theta_2 \in (235^0, 270^0)$ for $A2$, respectively.

are shown in Fig. 17(a), and 18(a), respectively. The parameter cell point in this case is "6", and the coupling is given by $D_1 = 0.10$ and $D_2 = 0$. The "concentric" wave in Fig. 18(a) resulted from two colliding spiral waves obtained by using a different initial condition, while the cell and coupling parameters were kept unchanged. The associated waveforms $V_1(t)$ and $V_2(t)$ at cell location $(j,k) = (50,50)$ are shown in Fig. 17(b) and 18(b), respectively.

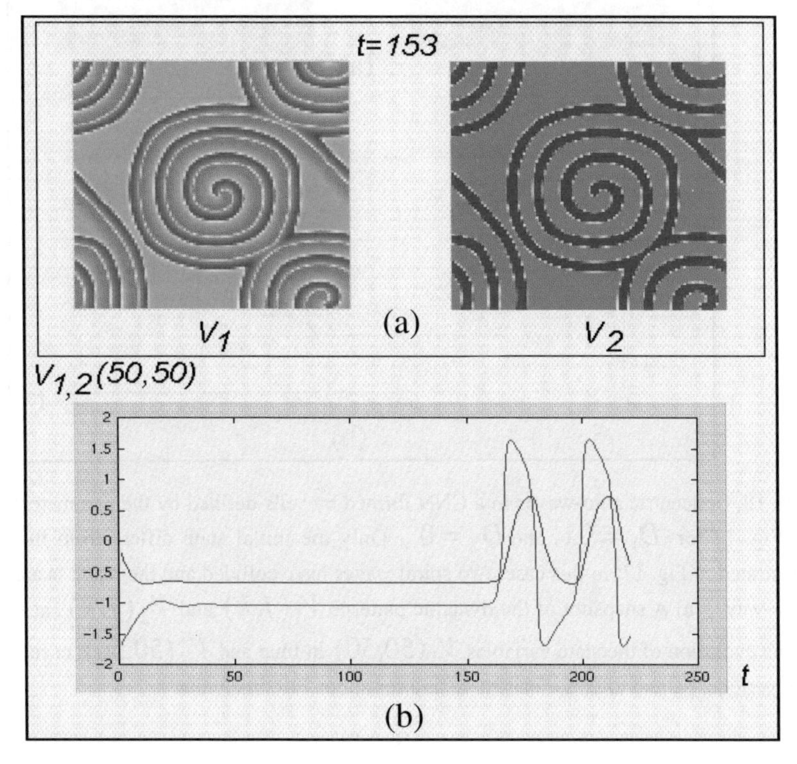

Fig. 17. Spiral waves in a CNN formed by coupled cells defined by the parameter point "6" in Fig. 12 for $D_1 = 0.10$, and $D_2 = 0$. (a) A snapshot of the dynamic patterns $V_1(j,k)$ and $V_2(j,k)$, at *t=153*; (b) Time evolution of the state variables $V_1(13,39)$ in blue and $V_2(13,39)$ in red. Observe the periodic pattern after a transient lasting for about 170 time units.

In all these examples, relatively complex dynamics was observed as an effect of coupling, when the cell parameters were on or nearby the *edge of chaos* domain. Uncoupled cells with the same parameters, however, exhibit only trivial relaxation dynamics as shown in Fig. 15(6a) and Fig. 15(8a) respectively.

Information Computation Patterns

Consider the cell parameters "5", "6", "7", "9", "11", "12", "14", "15", "16", "17","18", "19" and "20" in Figs. 12, 13, and 14. Observe that all of these points lie either on the *edge of chaos* domain, or nearby its boundary, in either the one-diffusion or the two-diffusion case. In each of these examples, the cell dynamics converges to *static* patterns, representing certain *features* of the initial state, which in these cases consists of a meaningful *image*.

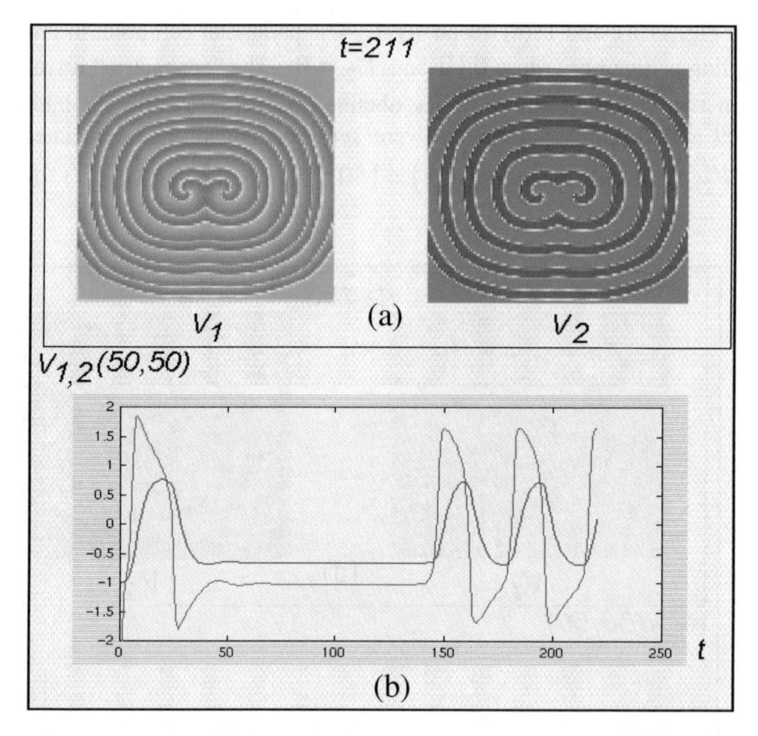

Fig. 18. Concentric auto-waves in a CNN formed by cells defined by the parameter point "6" in Fig. 12 for $D_1 = 0.1$, and $D_2 = 0$. Only the initial state differs from the situation illustrated in Fig. 17; in this case, two spiral waves have collided and the result is a concentric auto-wave; (a) A snapshot of the dynamic patterns $V_1(j,k)$ and $V_2(j,k)$, at *t=211*; (b) Time evolution of the state variables $V_1(50,50)$ in blue and $V_2(50,50)$ in red. Observe the periodic pattern after a transient lasting for about 140 time units.

For example, the steady state patterns in Figs. 15(5b), 6(c), and 12 can be interpreted as "edges" of the associated initial image. Another interesting example of selective edge detection is the static pattern shown in Fig. 15(17). In this case, the "orange" square in the initial state had been completely removed while in the remaining part only the edges were detected. Several examples of selective corner detection are presented in Figs. 15(11) and 15(14) respectively. The static pattern in Fig. 15(18) can be interpreted as a *halftoning* operation which converts a gray scale image into a binary image while preserving as much as possible the information content of the original image. The steady state pattern in Fig. 15(16) can be interpreted with respect to the initial state as its enhanced version where the background noise had been removed. The patterns produced in Figs. 15(7b), 15(9b) and 15(15) can be interpreted as textures having the initial state pattern as a seed. Observe that many of the above information processing tasks are similar to the early vision image processing in the human retina.

The final examples in Figs. 15(19) and 15(20) on information computation are related to a form of *analog-to-digital* conversion. Observe that the cell points "19" and "20" lie within the edge of chaos domain only for the two-diffusion case. Our examples, however, have only one diffusion coefficient, which in this case pertain to the *yellow* region in Figs. 15(19) and 15(20) and therefore corresponds to at least one locally active equilibrium point but not stable at the same time. These two examples demonstrate that even the *yellow* region may give rise to emergent behaviors, albeit at a considerably lower probability than when the cell parameters where chosen from within the edge of chaos.

Since there is no coupling between the rows in figures 15(19) and 15(20), each row in the CNN grid acts like an independent one-dimensional CNN with a ring topology. Depending on the initial state (chosen to be the same for a majority of neighboring cells in the same row), and except for several cells near the margin, the final steady state displays a number of "on" cells (shown in *red*) proportional to the analog value of the initial state. Observe that saturation occurs when the initial state is below some critical threshold, or is above a higher critical threshold. In the first case, all cells in the steady state pattern are in the "off" state (coded blue) while in the second case, except for some marginal cells, all cells remain in the "on" state (coded red). However, when the initial state is allowed to vary within the domain determined by the two critical thresholds, an emergent information processing similar to an analog to digital (binary state) information conversion is observed. Since such a phenomenon is *not* possible with uncoupled cells, these examples provide a demonstration of emergent computation resulting from coupling elementary cells defined by parameters lying within the *local activity* domain.

Since computation is in general understood as some meaningful transformation of information, it is reasonable to group all of these examples in a class called "information computation". Observe that the same information computation task (e.g. edge detection) will be implemented as long as the cell and the coupling parameters (a,b,c,ε) and (D_1,D_2) lie within some compact domain (i.e. *basin of attraction*) bounded by a *failure boundary* [Chua, 1998]. Programming such an analog computational device is equivalent to the identification of various information computation functions and their corresponding failure boundaries in the parameter space. This task may be very tedious even if the dimension of the parameter space is relatively small (6 in our case). However, since our local activity theory had already dramatically shrunk the potentially useful domains to a relatively small edge of chaos, the task is now much easier since the failure boundary had been reduced to a small subset of the two-dimensional parameter space (D_1,D_2) of coupling coefficients.

Periodic Dynamic Patterns

Let us now have a look at the cell parameter "13" located at $(a,b)=(-0.2,\ 0.4)$ in Fig. 13. Observe that it lies on the boundary of the *edge of chaos* domain. The corresponding array dynamics shown in Fig. 15(13) converges to a periodically oscillating dynamic pattern, as shown in the 42 snapshot sequence in Fig. 19. Observe that, as an effect of coupling, the shape of the initial pattern changes progressively from a rhombic to a circular pattern.

Chaotic Dynamic Patterns

Consider the cell parameter "7" located at $(a,b)=(0,\ 1.2)$ in Fig. 12. Observe that this point lies on the *unstable* region near the border of the *edge of chaos*. In fact, the uncoupled cells display trivial, periodic, oscillatory dynamics as shown in Fig. 15(7a). For the same cell parameters, but a non-zero diffusion coefficient $D_1=0.2$, the corresponding CNN dynamics shown in Fig. 15(7b) consists of an *irregular* spatio-temporal pattern. Figure 20 displays 42 snapshots of $V_1(j,k)$ illustrating the "chaotic" nature of the dynamics. A magnified view of the dynamic pattern for $V_1(j,k)$ and $V_2(j,k)$ taken at $t=211$ is shown in Fig. 21(a).

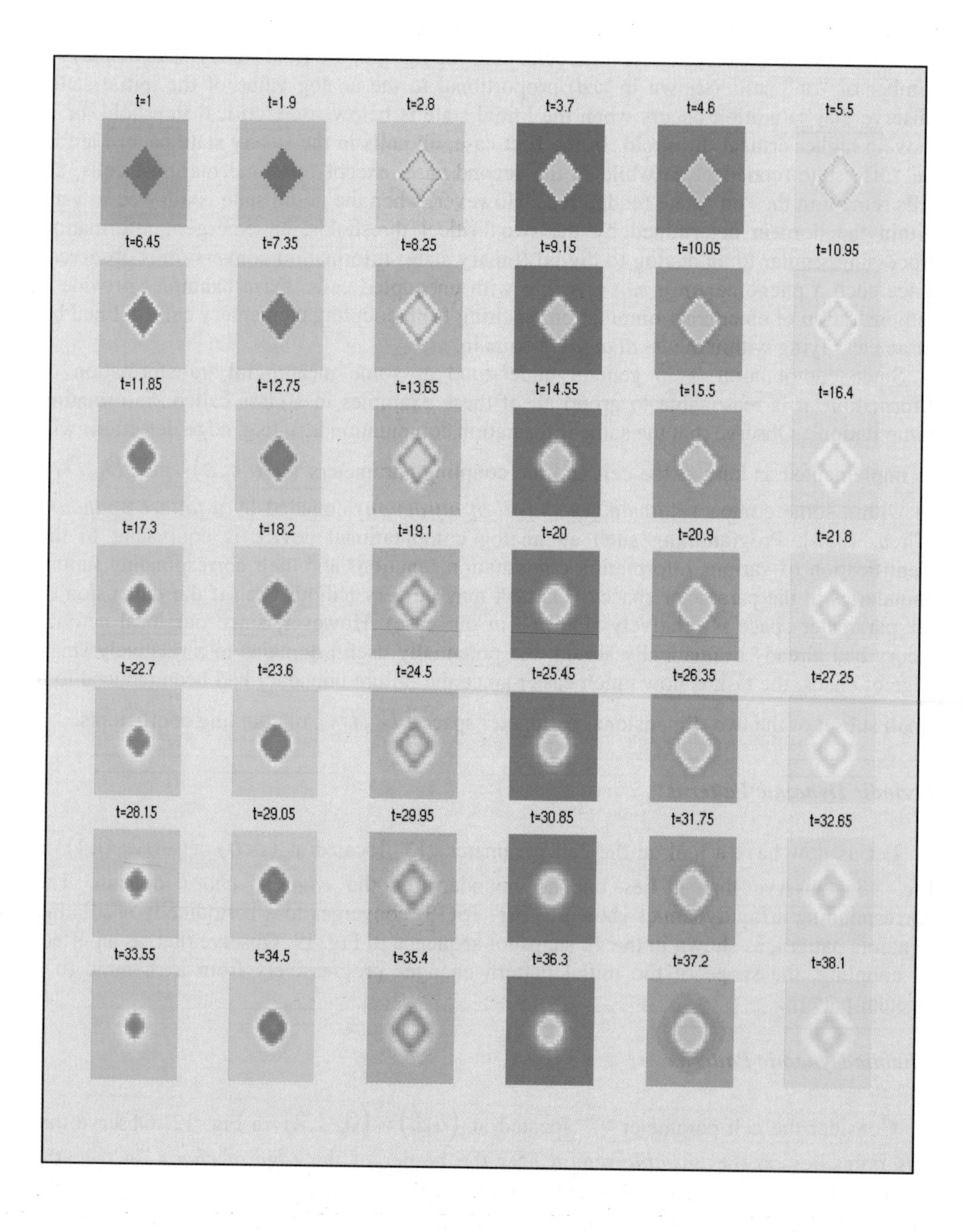

Fig. 19. Periodic dynamic patterns in a CNN formed by cells defined by the parameter point "13" in Fig. 13 for $D_1 = 0$, and $D_2 = 0.1$. A sequence of 42 time evolution snapshots of the state variable $V_1(j,k)$.

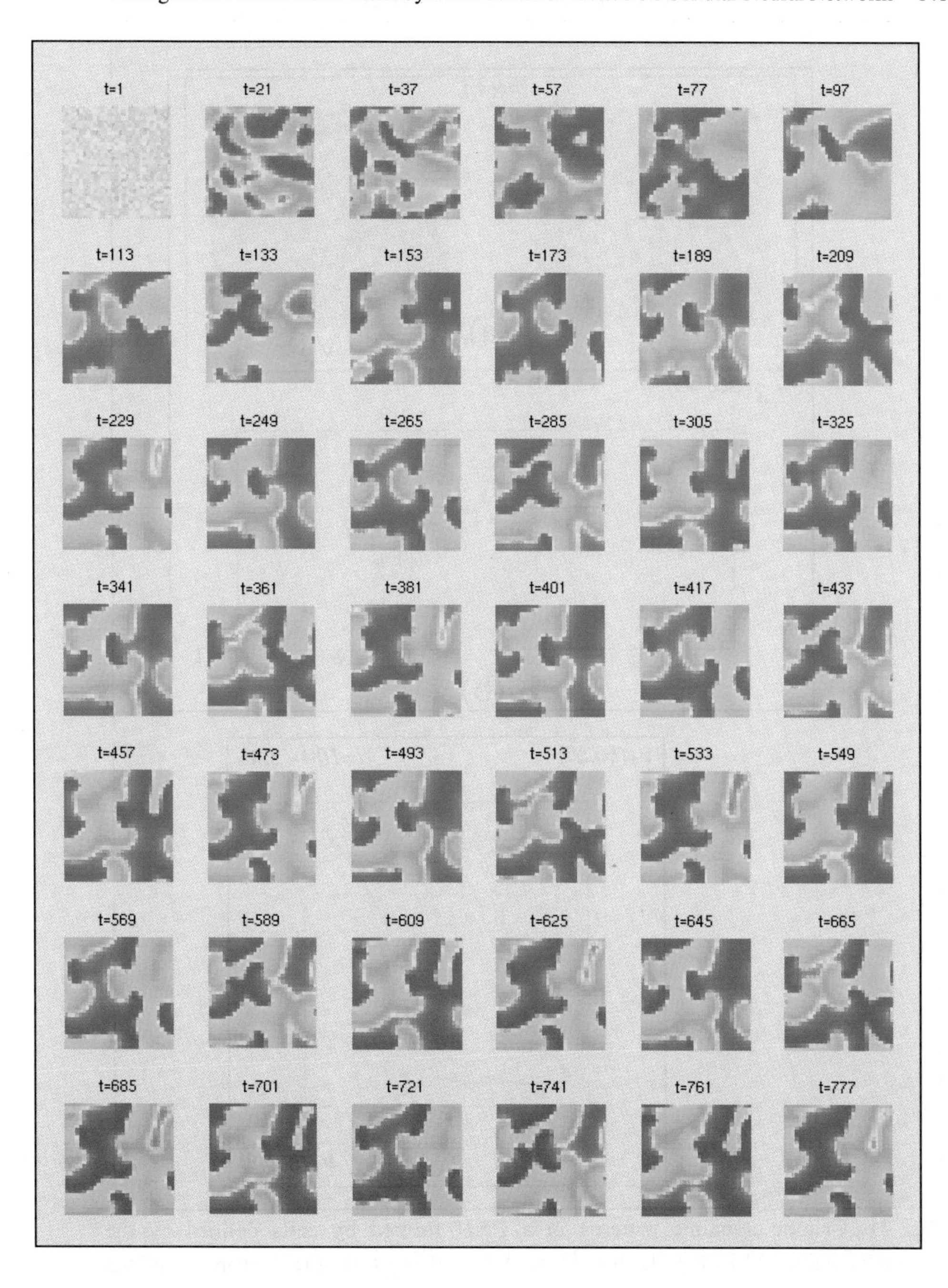

Fig. 20. Chaotic dynamic patterns in a CNN formed by cells defined by the parameter point "7" in Fig. 12 for $D_1 = 0.2$, and $D_2 = 0$. A sequence of 42 time evolution snapshots of the state variable $V_1(j,k)$.

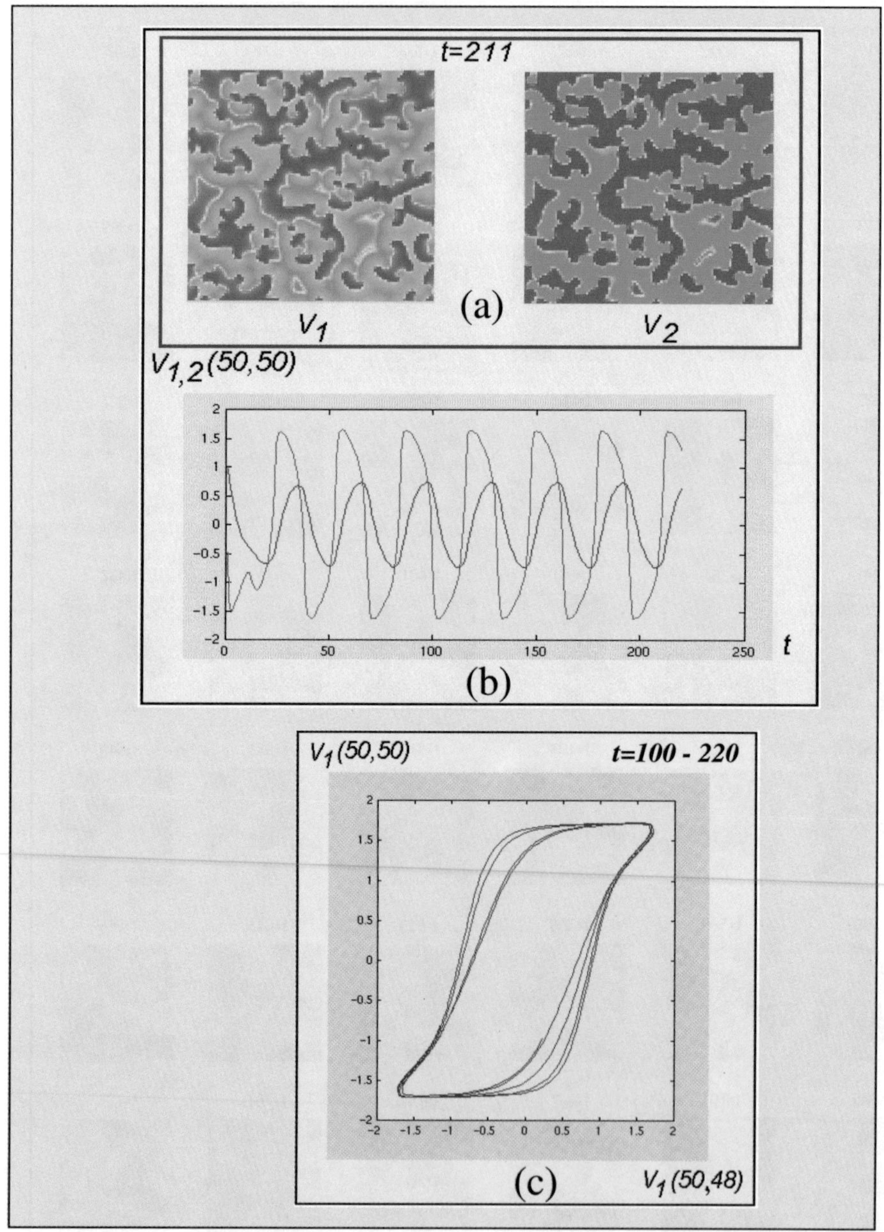

Fig. 21. Chaotic dynamic patterns in a CNN formed by cells defined by the parameter point "7" in Fig. 12 for $D_1 = 0.2$, and $D_2 = 0$; (a) A snapshot of the dynamic patterns $V_1(j,k)$, and $V_2(j,k)$, at *t=211*; (b) The time evolution of the state variables $V_1(50{,}50)$ in blue and $V_2(50{,}50)$ in red. Although the oscillation appears to be periodic, after careful examinations slight differences can be discerned in the time-domain signals. (c) A better illustration of the non-periodic nature of the oscillation can be visualized by projecting the high-dimensional attractor (in this case 101x101) into a 2-dimensional plane associated with the voltage $V_1(j,k)$ at the two cells $(50{,}48)$ and $(50{,}50)$ respectively. For a periodic oscillation, this projection should appear as a one-dimensional simple closed curve.

4.4. Emergent behaviors within the "edge of chaos" domains of the Brusselator CNN

In this section, the local activity and the "edge of chaos" methods are applied to a different system; namely, the *Brusselator* CNN. The Brusselator model [Prigogine, 80] is chosen here for its historical significance: It is the first mathematical model used to explain the self-organization phenomena observed in chemical reactions of the reaction-diffusion type. Based on this model, a theory *of dissipative structures operating far from thermodynamic equilibrium* had been developed by Prigogine. Using stability theory techniques such as the normal mode analysis, Prigogine and his colleagues had derived a *critical bifurcation boundary* for the *uncoupled* cell [Prigogine, 1980]. They have *postulated* and verified, by numerical simulations, that when cell parameters are located in a *neighborhood of the critical bifurcation boundary* two types of emergent behaviors can occur. They correspond to either dynamic or stationary patterns (in the latter case, they are called Turing patterns). However, except for the stability boundaries which the Prigogine school has derived for several particular cases, the *far from thermodynamic equilibrium* theory is too coarse to predict the precise domain in the cell parameter space responsible for emergent behaviors.

As shown next, the local activity theory offers a rigorous and effective tool for sharpening existing results on *systems operating far from thermodynamic equilibrium* in the sense that it can identify more precisely those regions in the cell parameter space which are capable of emergent behaviors. It is also shown that using the methodology previously exposed in sections 4.1 to 4.3 it is possible to finely tune such regions into a relatively small subset called *the edge of chaos,* where the emergence of complex phenomena is most likely. In particular it is confirmed that all results previously reported by the Prigogine school using the *far from thermodynamic equilibrium theory* can be derived *directly* from the *local activity theory*. Indeed such a result is expected since the local activity is a much more general and mathematically rigorous theory derived from basic physical principles, namely from the principle of conservation of energy.

4.4.1. Local Activity and Edge of Chaos Domains

The mathematical model of the unstirred Brusselator [Prigogine, 1980] is described by the following system of partial differential equations (PDE):

Brusselator PDE with two diffusion coefficients D_1 and D_2:

$$\begin{aligned} \frac{\partial V_1(x,y)}{\partial t} &= a-(b+1)\,V_1(x,y)+V_2(x,y)\big(V_1(x,y)\big)^2+D_1\nabla^2V_1(x,y) \\ \frac{\partial V_2(x,y)}{\partial t} &= bV_1(x,y)-V_2(x,y)\big(V_1(x,y)\big)^2+D_2\nabla^2V_2(x,y) \end{aligned} \tag{33}$$

where V_1 and V_2 are the two state variables characterizing the cell dynamics. The cell parameters are denoted by a, and b, and the spatial coordinates are denoted by x and y. In keeping with the physical meaning of diffusion we will assume the coupling coefficients $D_1 \geq 0$ and $D_2 \geq 0$ throughout this paper.

Step ❶
To apply the concept of *local activity*, the *first step* (see Section 4.3.1) is to map the Brusselator PDE into the following *discrete-space version*, called a Brusselator CNN:

Brusselator CNN with two diffusion coefficients:

$$\frac{d\,V_1(j,k)}{d\,t} = a-(b+1)\,V_1(j,k)+V_2(j,k)\big(V_1(j,k)\big)^2+I_1(j,k)$$
$$\frac{d\,V_2(j,k)}{d\,t} = bV_1(j,k)-V_2(j,k)\big(V_1(j,k)\big)^2+I_2(j,k) \tag{33'}$$

The meaning of the variables used above was explained in section 4.3.1.

Next we will perform two separate local activity and edge of chaos tests: one for the case *of one diffusion coefficient* ($D_1 \neq 0, D_2 = 0$), and the other for the case of *two diffusion coefficients* ($D_1 \neq 0, D_2 \neq 0$). Throughout this section we consider only the *restricted local activity domain* (see Section 4.3.4).

Step ❷
The *second step* in applying the local activity theory is to determine the equilibrium points Q_i when $I_1 = I_2 = 0$[9]. In the case of the Brusselator cell there is a *unique equilibrium point* Q_1 defined as follow:

Equilibrium point Q_1 *of the Brusselator CNN:*

$$V_1(Q_1)=\frac{b}{a},\quad V_2(Q_1)=a \tag{34}$$

Step ❸
The *third step* consists of determining the four coefficients of the Jacobian matrix $A=\begin{bmatrix} a_{11}(Q_1) & a_{12}(Q_1) \\ a_{21}(Q_1) & a_{22}(Q_1) \end{bmatrix}$ of (1)' at the equilibrium point Q_1. For the Brusselator cell they are given by:

$$\begin{aligned} a_{11}(Q_1) &= b-1 \\ a_{12}(Q_1) &= a^2 \\ a_{21}(Q_1) &= -b \\ a_{22}(Q_1) &= -a^2 \end{aligned} \tag{35}$$

---

[9] If the reader is interested in finding the *local passivity* region for the Brusselator CNN, i.e., the subset of the *a-b* parameter plane where *no complexity* can emerge for *any* $D_1 \geq 0$ and $D_2 \geq 0$, then the equilibrium points must be found for *all* $(I_1, I_2) \in \mathbb{R}^2$. In the case of the Brusselator CNN it can be easily shown that the *unrestricted locally passive* region is an *empty set*. In real systems however, the couplings I_1 and I_2 usually vary only within a small bounded set $[-\varepsilon, \varepsilon]$ and it is sufficient in most cases to identify only the *restricted* local passivity domain where $I_1 = I_2 = 0$.

As shown in Section 4.3.1, it is also important to determine the trace $T(Q_1)$ and the determinant $\Delta(Q_1)$ of the Jacobian matrix A :

$$\boxed{\begin{aligned} T(Q_1) &= a_{11}(Q_1) + a_{22}(Q_1) = -a^2 + b - 1 \\ \Delta(Q_1) &= a_{11}(Q_1)a_{22}(Q_1) - a_{12}(Q_1)a_{21}(Q_1) = a^2 \end{aligned}} \tag{36}$$

The information contained in (35) and (36) is the input of a test algorithm (*step four*) which classifies each cell parameter point (a,b) at the equilibrium point Q_1 into one of the following three disjoint categories:

(a) *Locally Active and Stable* $S(Q_1)A(Q_1)$. In the case of the Brusselator CNN, because there is only one equilibrium point, this region coincides with the *edge of chaos* domain. Based on the color code introduced in Section 4.3, this region will be coded in *red* when referred from the perspective of "local activity", and in *orange* when referred from the perspective of "edge of chaos" (see also Fig. 23).

(b) *Locally Active and Unstable* $A(Q_1)U(Q_1)$. This region corresponds to the oscillatory or unstable region of an isolated CNN cell, and for the case of the Brusselator it is considered in [Prigogine, 1980] as a region where the cell presents "a real instability". It is coded in *red* when *local activity* plots are considered regardless of the stability issues, and in *green* otherwise (see also the legend in Fig.23).

(c) *Locally Passive* $P(Q_1)$. This region (in fact a restricted locally passive region) is coded in *blue* and is the region in the cell parameter space where complex phenomena are unlikely to occur. The restricted local passivity region can be partially or totally included in the *local activity region* (defined for all possible equilibrium points when $I_1, I_2 \in (-\infty, \infty)$. However, for *reaction-diffusion* equations, the assumption $I_1 = I_2 = 0$ is usually strong enough to ensure that cells belonging to the $P(Q_1)$ will not lead to emergent dynamic behaviors.

The test for classifying in which one of the above 3 categories does a cell parameter point belongs, depends on the number of diffusion coefficients:

Step ❹
Local activity and edge of chaos domain, one diffusion coefficient

The general local activity tests presented in section 4.3, when applied to the Brusselator cell lead to the following domains:

$$S(Q_1)A(Q_1): \quad b > 1 \text{ AND } b < a^2 + 1 \tag{37}$$
$$A(Q_1)U(Q_1): \quad b > 1 \text{ AND } b \geq a^2 + 1 \tag{38}$$
$$P(Q_i): \quad b \leq 1 \tag{39}$$

For the Brusselator CNN, the subset of parameters (a,b) which satisfies either C3 or C4 from the general local activity test (see Section 4.3.3) coincides with the *critical bifurcation boundary* $a^2 = b - 1$ derived in [Prigogine, 1980]. From the perspective of the local activity theory it corresponds to the condition $T = 0, \Delta \geq 0$. This *bifurcation boundary set* has already been found to play an important role in the FitzHugh-Nagumo

CNN (see section 4.3.5). Indeed, it was found that many interesting complex behaviors are obtained when cell parameter points are located not only within the *edge of chaos domain* but further restricted to a neighborhood of this boundary. In this case, the *new* information provided by the local activity theory is to identify a part of the neighborhood of this boundary, which need not be checked for complexity because it is locally passive when $I_1 = I_2 = 0$. As long as a cell parameter point is located near the bifurcation boundary $T = 0, \Delta \geq 0$ but in the *locally passive region*, its associated CNN *cannot* produce any emergent, non-homogeneous pattern for any initial condition near the equilibrium point. The condition "C2 AND NOT C1" in the general test condition for local activity defines not only the edge of chaos domain but it also reveals an important boundary; namely, the one separating the *local activity* domain from the *local passivity* domain. This boundary *cannot* be derived using the stability theory alone. In fact, this is precisely where the *local activity* theory becomes indispensable.

The cell parameter projection profile (CPPP)

A cell parameter projection profile (CPPP) is defined as the projection of any prescribed subset Π of the cell parameter domain of interest into the (a_{22}, T, Δ)-space. In the case of the Brusselator CNN this projection is determined by Eqs. (34) and (35). For $\Pi = \{(a,b) | a \in (-5,5), b \in (-5,5)\}$ the corresponding Brusselator CPPP is displayed in Fig. 22(a) in the 3-dimensional (a_{22}, T, Δ)-space. Its projection onto the $\Delta = 0$ plane is shown in Fig. 22(b). For each of the two parameters, a_{22} and T, the range of variation was quantified into 100 points so that the CPPP visualized in Fig.22 is a discretized version containing 10000 points of the continuous $CPPP(\Pi)$. The colors in Fig.22 indicate the membership of each cell parameter point (a,b) in one of the three classes defined above: *locally active and stable (edge of chaos)* in *red, locally active and unstable* in *green,* and *locally passive* in *blue.* The location of 16 cell parameter points in the (a_{22}, T, Δ)-space is specified in Fig. 22(b). All of these points share a common property: They are all located within the *locally active* domain and thus it follows from the local activity theory that CNNs made of such cells can exhibit emergent behaviors. The dynamic simulations to be presented next will confirm this prediction for the locally active Brusselator cells.

Bifurcation diagrams

The local activity domain and the edge of chaos domains are shown in Fig. 23 (the upper row) for the case $a \in [-5,5]$, and $b \in [-4,6]$. Most of the cell parameter points in [Prigogine, 1980] which were found capable of "self-organization" are located nearby the *bifurcation boundary* separating the stable region from the unstable region.

Our local activity analysis provided not only this information by a *unified* approach, versus a problem-dependent "normal mode" approach used in [Prigogine, 1980], but additional information on what region in the parameters space should be ignored, by identifying the *local passivity* domain $b \leq 1$. The boundary $b = 1$ (or $T = a_{22}$ AND $a_{22} < 0$ in the general case for any cell model - see Fig 22.(b)) between the *local passivity* domain and the *local activity* domain *cannot* be determined via a local stability analysis, since cells defined by parameter points located in both $S(Q_1)A(Q_1)$ and

$P(Q_i)$ regions share the same stability property, i.e. they are always *stable*. To illustrate the lack of complexity in such a region, we have randomly picked several cell parameter points in the locally passive domain $(b \leq 1)$ and have found no emergent behavior in the associated CNN when $D_2 = 0$ (condition for which the result of this test applies).

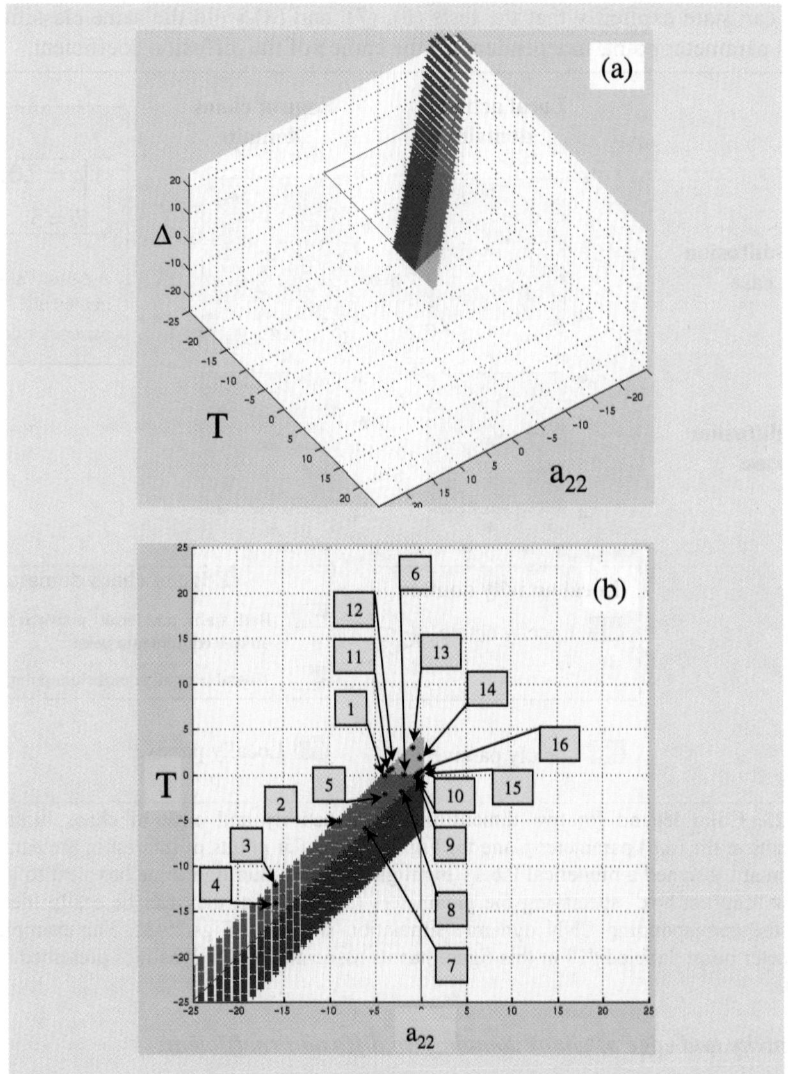

Fig.22. Cell Parameter Projection Profile $CPPP$ for the Brusselator cell with one diffusion coefficient. The cell parameter domain is represented by a set of 10000 points: $\{(a,b) | \ a = -5 + 0.1 \cdot i ; b = -5 + 0.1 \cdot j ; i = 1,..100; \ j = 1,..100\}$. Each cell parameter point (a,b) maps to a $CPPP$ point in the (a_{22}, T, Δ)-space. The color at each point indicates its classification into one of the following classes: *locally active and stable (or edge of chaos)* - color *red; locally active and unstable* - color *green*; *locally passive* - color blue; (a) Spatial view; (b) Projection onto the $\Delta = 0$ plane. The location of 16 cell parameter points arbitrarily chosen within the *locally active* domain for the one diffusion coefficient is indicated. Each point is assigned a blue box containing its identification label. The same number is used to identify the cell parameter point in the (a,b) parameter plane (Figs. 23, and 24) and in Figs.25-32.

The case of one diffusion coefficient presented above assumes the coupling is made via the first cell state variable. However, we may also use the second state variable for the coupling. In [Dogaru & Chua, 1998b] it is proved that for any cell having two state variables, the *edge of chaos domain,* the *local passivity domain,* and the *unstable local activity* domain do not depend on whether $D_1 \neq 0$ or $D_2 \neq 0$. Thus, for the *Brusselator* case, we can state explicitly that the tests (6), (7), and (8) yield the same classification for each cell parameter point, independent of the choice of the diffusion coefficient.

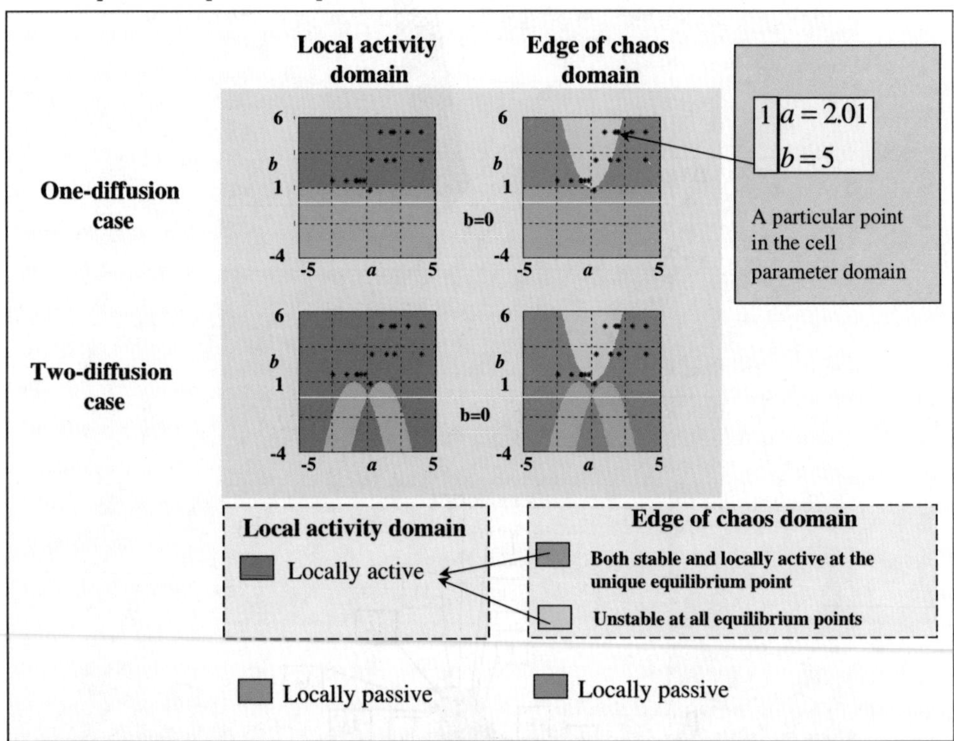

Fig. 23. Color legend for two dimensional local activity and edge of chaos bifurcation diagrams in the (a,b) parameter plane for Fig.24: Particular points of interest in the parameter domain are assigned a numerical label (highlighted with a thin light-blue box next to a light-yellow "caption box" specifying the parameters (a,b)) so that they can be easily identified from the corresponding CNN dynamic simulation shown in Figs.25-32. For example, the parameter point labeled "1" in this figure has its associated CNN dynamics presented in Fig. 25(1).

Local activity and edge of chaos domain, two diffusion coefficients

For the Brusselator CNN if the general test in section 4.3.3 is applied, it leads to the following domains:

$$S(Q_1)A(Q_1): \quad -4(b-1)a^2 < (a^2-b)^2 \text{ AND } b < a^2+1 \tag{40}$$

$$A(Q_1)U(Q_1): \quad -4(b-1)a^2 < (a^2-b)^2 \text{ AND } b \geq a^2+1 \tag{41}$$

$$P(Q_i): \quad -4(b-1)a^2 \geq (a^2-b)^2 \tag{42}$$

Consequently, the local activity domain and edge of chaos domain for the Brusselator CNN with two positive diffusion coefficients are displayed in the lower row of Fig. 23. Observe that now, comparing to the case of one diffusion coefficient ($D_2 = 0$) the local passivity (blue) region shrinks to a much smaller subset. As in the case of the FitzHugh-Nagumo CNN, this situation indicates that there may be cell parameter points situated in the *restricted* locally passive region (when $D_2 = 0$) for which complex patterns may emerge when $D_2 > 0$. Indeed, the cell parameter point (CPP) "17" in Fig. 24 is such a case.

Let us next pick 16 cell parameter points located within the *local activity domain* which are common to both the one-diffusion coefficient case and the two-diffusion coefficients case, and demonstrate by computer simulations that complexity did emerge at these points. Among these, 10 points are located within the *edge of chaos* domain. Another point (CPP "17" in Fig. 24) where complexity also emerges, was chosen in the *edge of chaos domain* for two diffusion coefficients, but *outside* of the edge of chaos domain for one diffusion coefficient. These examples clearly illustrate the "fine-tuning" ability of the local activity theory.

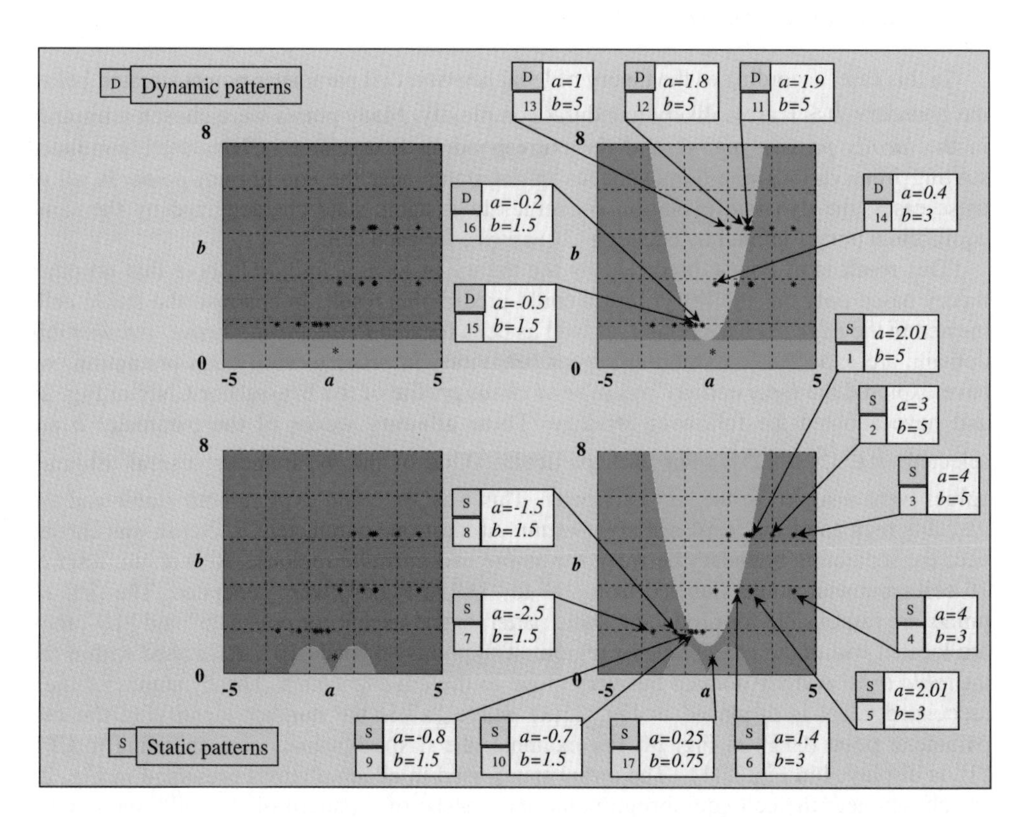

Fig.24. Details of the *local activity* and *edge of chaos* domains from Fig. 2, for $a \in [-5, .5]$, and $b \in [0,8]$. Seventeen cell parameter points (1 to 17) are indicated, each one is assigned an additional literal label ("S" or "D") identifying the type of dynamical behavior exhibited by the corresponding CNN. The corresponding results of the CNN dynamic simulations are shown in Fig. 25(1-10,17) for the first 10 and the seventeenth cell parameter points, and in Figs. 26-32 for the remaining cell parameter points.

4.4.2. Emergence at the edge of chaos

As shown in the previous section, the cell parameters associated with each CNN cell (in this case the Brusselator cell) can be mapped into a subset called the *cell parameter projection profile* (CPPP) in the (a_{22}, T, Δ)-space, as shown in Fig. 22. This profile provides some useful information concerning the potential of the cell to exhibit complexity. For example, a cell model may have a CPPP such that all of its points lie inside the "blue" sector in Fig.22 defined by $T \leq a_{22}$ and $a_{22} \leq 0$ and $\Delta \geq 0$. Such a cell is useless since, according to the local activity theory, it cannot produce complexity when coupled with other identical cells in a reaction-diffusion CNN. In the Brusselator, both *red* and *green* regions are found in the CPPP. The location of these cell parameter points (CPP) and their membership in either the *locally active (stable and unstable) domain,* or the *locally passive* domain, is shown in Fig. 24 following the legend in Fig.23. Let us now examine the various patterns generated by the Brusselator CNN. The boundary condition is assumed to be toroidal (i.e. periodic) for all examples.

The one-diffusion coefficient case

In this case, according to the theory of local activity, cell parameter points situated below the boundary $b \leq 1$ are unlikely to exhibit complexity. Many points were chosen arbitrarily in the *locally passive* region, and the corresponding Brusselator CNNs were simulated starting from various non-homogeneous initial states near the equilibrium point. In all of these cases, the dynamic evolution converges to a stable state characterized by the same equilibrium point (defined by equation (34) above) for each cell.

This result is of course predicted by the theory of local activity. Observe that no other theory based only on stability criterion could predict this result. In contrast, the stable cells located in the *edge of chaos* region, as well as cells located in the *locally active and unstable* domain, are expected to exhibit emergent behaviors. In order to verify this prediction, we have explored the *local activity* and *edge of chaos* profile of the Brusselator CNN in Fig. 24 and have adopted the following strategy: Three arbitrary values of the parameter *b* are selected: $b \in \{5, 3, 1.5\}$. For each particular value of the *b* parameter several arbitrary values were assigned to the "*a*" parameter. This way we could explore both *stable* and the *unstable* regions of the *local activity domain.* At least one parameter "*a*" value was chosen near the separating boundary between the *stable* and *unstable* regions. As a result, a set of 16 cell parameter points (labeled from "1" to "16" in Fig.24) were generated. The first 10 points are responsible for producing static patterns, and except for points "6" and "10", they are located within the *edge of chaos* region. The points "6" and "10" are located within the *unstable local activity* domain but very close to the *edge of chaos.* The dynamics of their associated CNN is displayed in Fig. 25(x) where "x" is the number identifying the cell parameter point (CPP) in Fig. 24. For example, the CNN dynamics corresponding to CPP "1" is displayed in Fig.25(1). The initial states used in all simulations presented in Fig. 25 are chosen near the cell equilibrium point and consist of a pattern of "small" magnitudes whose "color codes" are given in the legend at the bottom of Fig. 25.

It is interesting to observe that for all points "1" to "10", the emergent patterns consists of *stationary patterns* when $D_2 \neq 0$. When $D_1 \neq 0$, no emergent patterns were observed at points "1" to "5" and "7" to "9". For points located in the *green* domain (points 11-16) associated with the *unstable* sub-region of the *locally active* domain, *dynamic patterns* were usually observed when $D_1 \neq 0$. In this case ($D_1 \neq 0$), static patterns were also observed for cell parameter points located in the unstable region but very close to the boundary

$a^2 = b - 1$ between the *stable* and the *unstable* local activity domains, and for large enough values of D_2. This is the case for points "6" and "10". These observations coincide not only with the results reported in the literature on the Brusselator, but also with the previous results based on local activity analysis for the FitzHugh-Nagumo cell reported in Section 4.3.6. The dynamic patterns associated with points "11" to "16" are shown in Figs. 26 to 30.

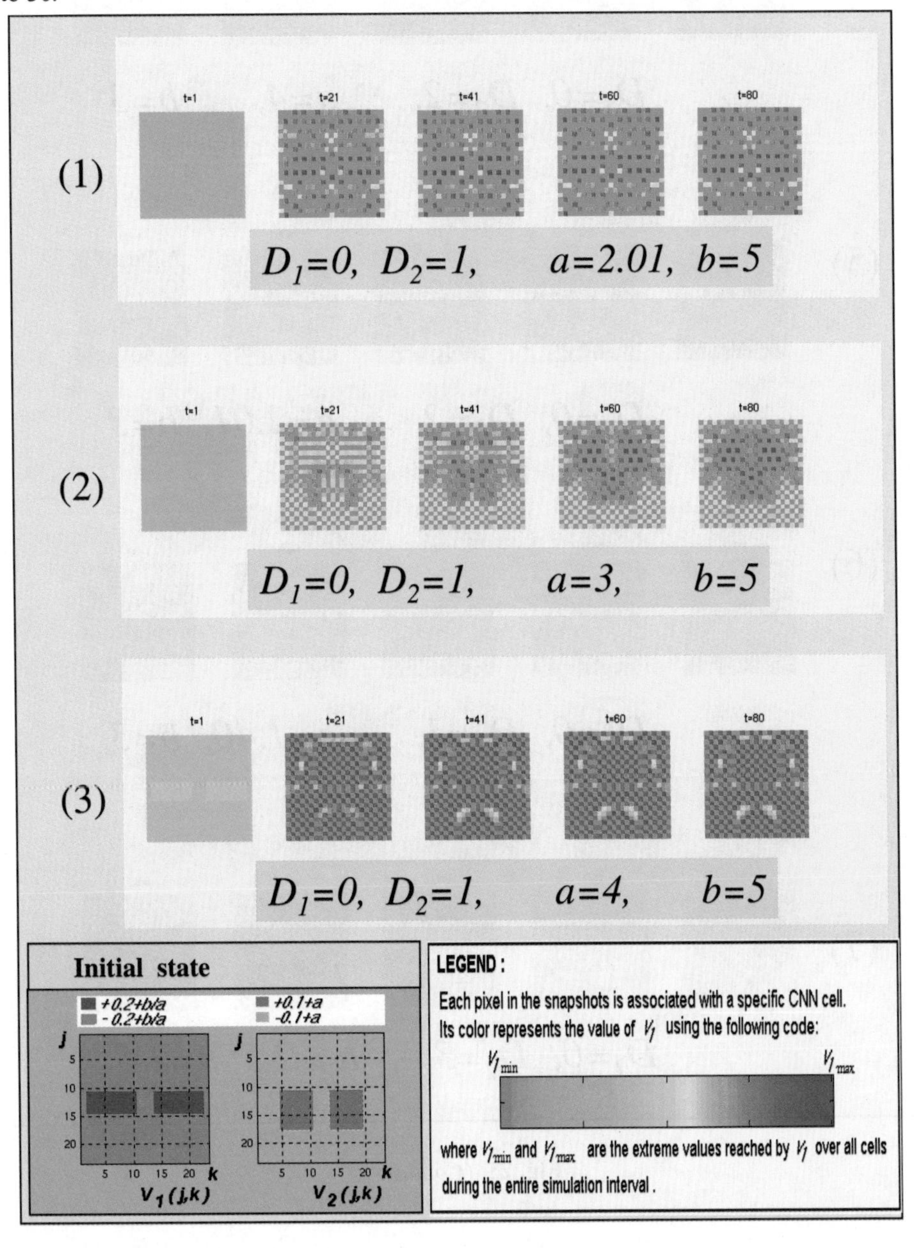

Fig. 25. Examples of dynamic simulations at different cell parameter points responsible for producing static patterns. The color code and the initial state are defined in the legend. The label assigned in the upper left (light-blue margin) of each sequence of 5 snapshots consists of a number indicating the cell parameter point in one of the previous figures.

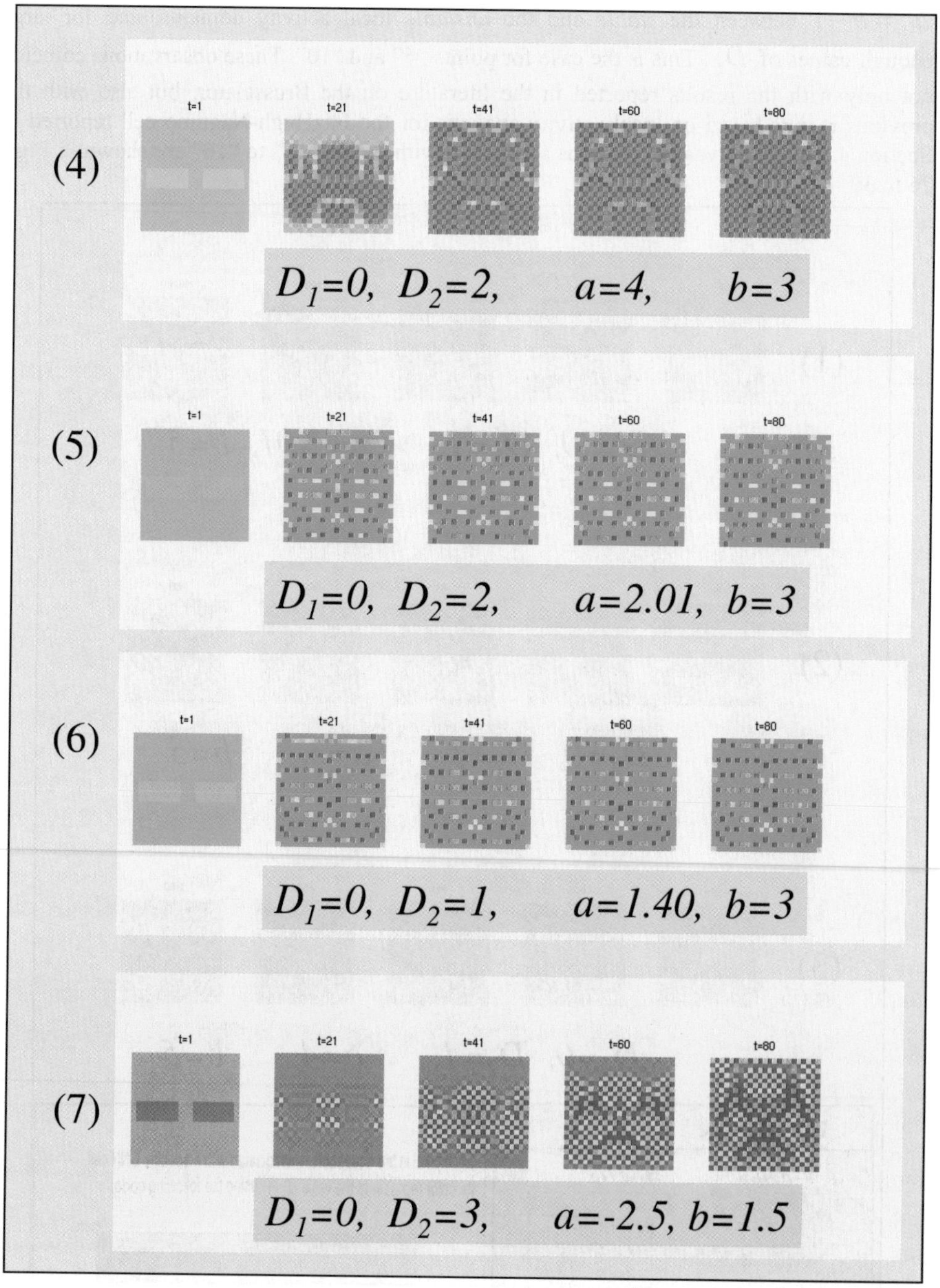

(4) t=1 t=21 t=41 t=60 t=80

$D_1=0$, $D_2=2$, $a=4$, $b=3$

(5) t=1 t=21 t=41 t=60 t=80

$D_1=0$, $D_2=2$, $a=2.01$, $b=3$

(6) t=1 t=21 t=41 t=60 t=80

$D_1=0$, $D_2=1$, $a=1.40$, $b=3$

(7) t=1 t=21 t=41 t=60 t=80

$D_1=0$, $D_2=3$, $a=-2.5$, $b=1.5$

Fig. 25. *(Continued)*

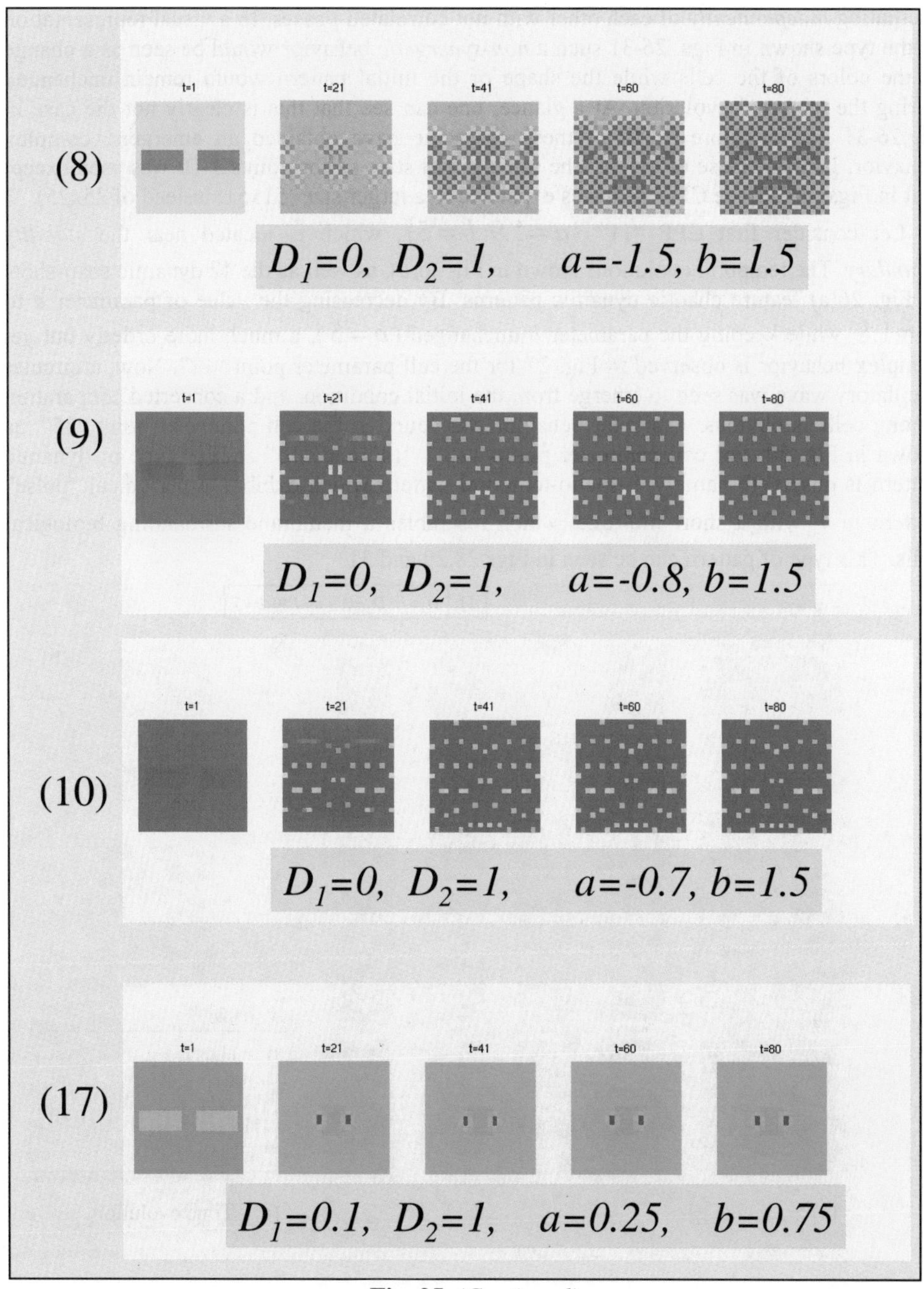

Fig. 25. *(Continued)*

Dynamic patterns

The dynamic patterns presented in this paper are associated with cell parameter points located within the *green* region in Fig. 24, which corresponds to the *locally active and unstable* domain. A "non-emergent" behavior in this case will correspond to cells oscillating *independently* of each other with not correlated phases. In a visual representation of the type shown in Figs. 26-31 such a *non-synergetic* behavior would be seen as a change in the colors of the cells while the shape of the initial pattern would remain unchanged during the temporal evolution. At a glance, one can see that this is clearly not the case in Fig.26-31 and therefore in all of these cases we have obtained an emergent, complex behavior. In all of these examples, the same initial state as for points 1-10 was used except that in Figs. 26-32, the CNN grid was expanded to a larger size (51x51, instead of 25x25).

Let consider first CPP "11" $(a=1.9, b=5)$, which is located near the *stability boundary*. The temporal evolution, shown in Fig.26(c), as well as the 42 dynamic snap-shots in Fig. 26(a) feature chaotic dynamic patterns. By decreasing the value of parameter *a* to $a=1.8$ while keeping the parameter *b* unchanged ($b=5$), a much more orderly but yet complex behavior is observed in Fig. 27 for the cell parameter point "12". Now, a circular oscillatory wave was seen to emerge from the initial condition, and a concerted cooperation among cells is obvious. A similar behavior was found at the cell parameter point "15", as shown in Fig. 30. For cell parameter points "13", "14" and "16" another type of dynamic pattern is observed; namely, a spatio-temporal pattern which exhibits a periodical "pulse" pattern in V_1 with a short life-time, which resembles a membrane surrounding biological cells. This type of pattern can be seen in Figs 28,29 and 31.

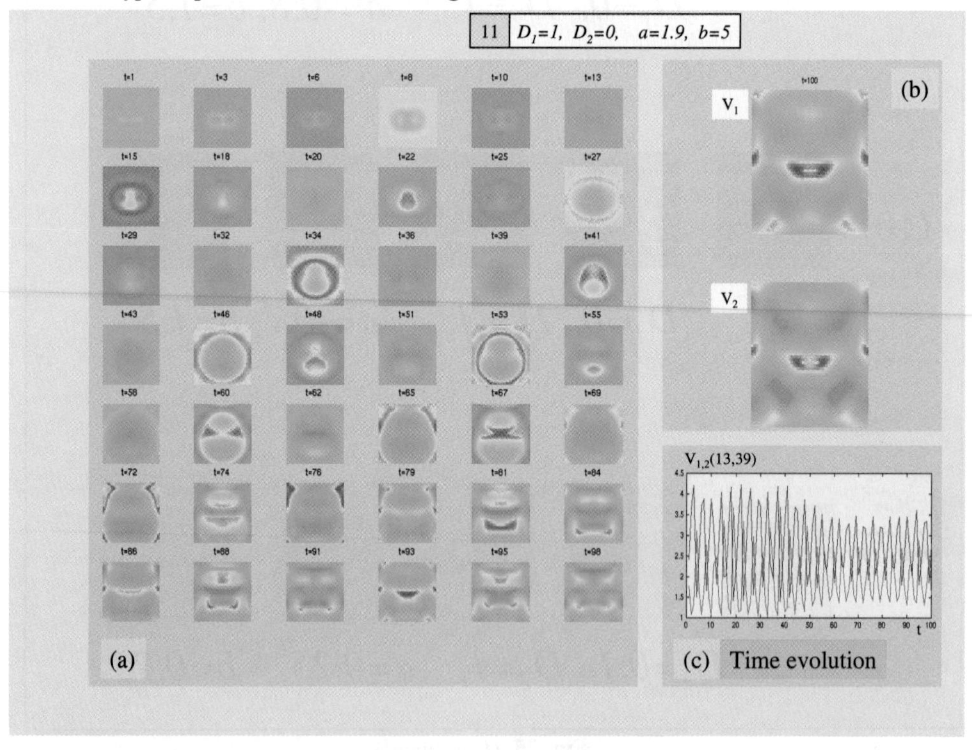

Fig. 26. Dynamic patterns of the Brusselator CNN defined by cell parameter point "11" in Fig. 24; (a) A sequence of 42 time evolution snapshots of the state variable $V_1(j,k)$; (b) A snapshot of the dynamic patterns $V_1(j,k)$ and $V_2(j,k)$, at *t=100*; (c) Time evolution of the state variables $V_1(13,39)$ in red and $V_2(13,39)$ in blue.

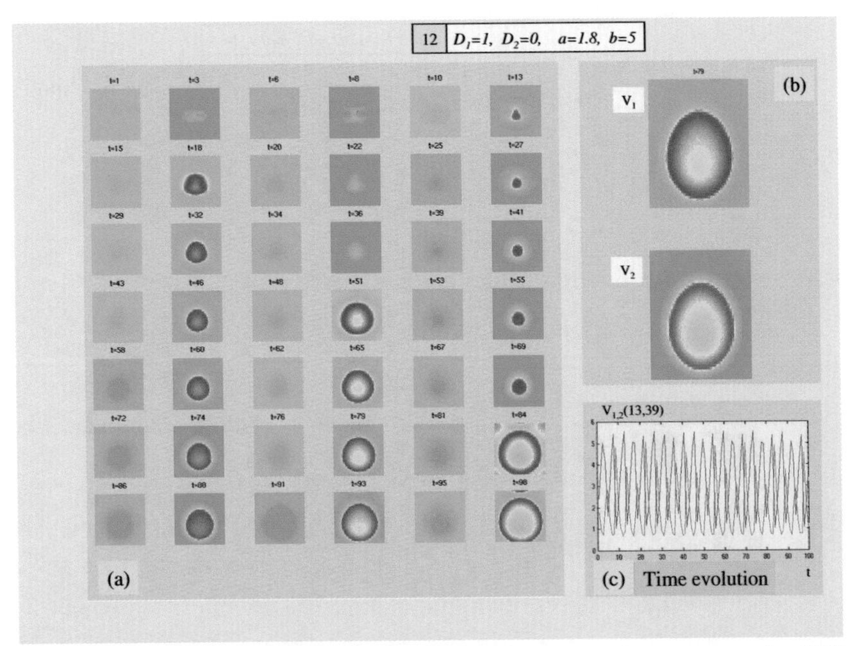

Fig. 27. Dynamic patterns of the Brusselator CNN defined by cell parameter point "12" in Fig. 24; (a) A sequence of 42 time evolution snapshots of the state variable $V_1(j,k)$; (b) A snapshot of the dynamic patterns $V_1(j,k)$ and $V_2(j,k)$, at *t=79*; (c) Time evolution of the state variables $V_1(13,39)$ in red and $V_2(13,39)$ in blue.

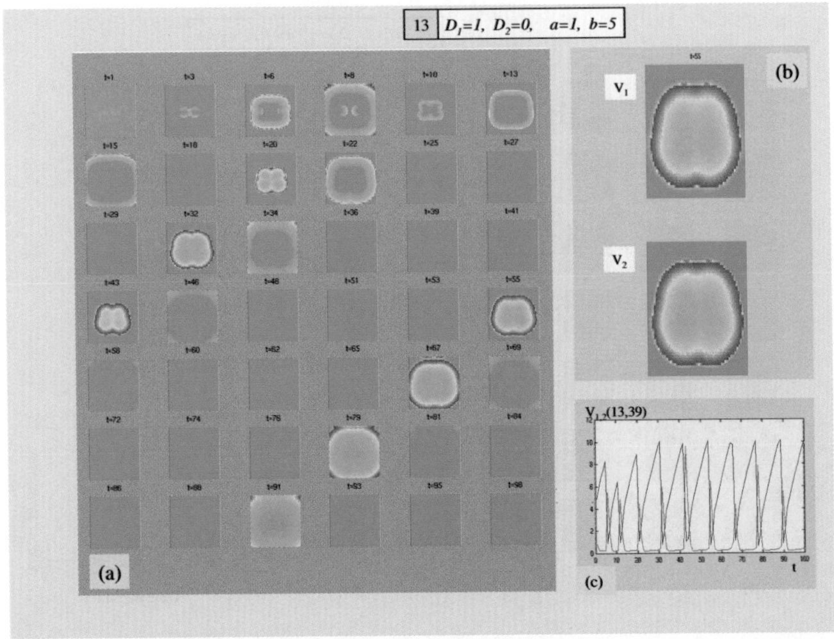

Fig. 28. Dynamic patterns of the Brusselator CNN defined by cell parameter point "13" in Fig. 24; (a) A sequence of 42 time evolution snapshots of the state variable $V_1(j,k)$; (b) A snapshot of the dynamic patterns $V_1(j,k)$ and $V_2(j,k)$, at *t=55*; (c) Time evolution of the state variables $V_1(13,39)$ in red and $V_2(13,39)$ in blue.

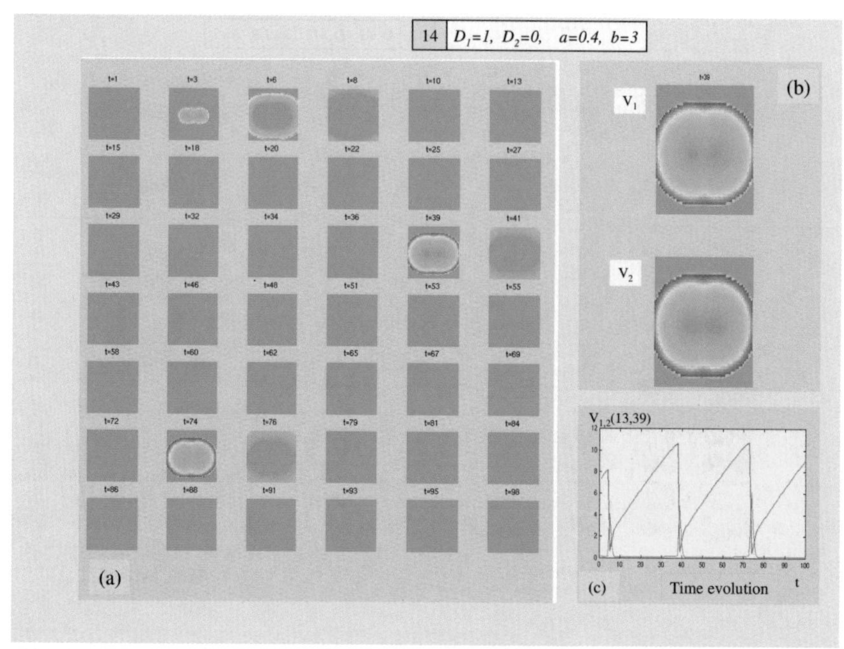

Fig. 29. Dynamic patterns of the Brusselator CNN defined by cell parameter point "14" in Fig. 24; (a) A sequence of 42 time evolution snapshots of the state variable $V_1(j,k)$; (b) A snapshot of the dynamic patterns $V_1(j,k)$ and $V_2(j,k)$, at $t=39$; (c) Time evolution of the state variables $V_1(13,39)$ in red and $V_2(13,39)$ in blue.

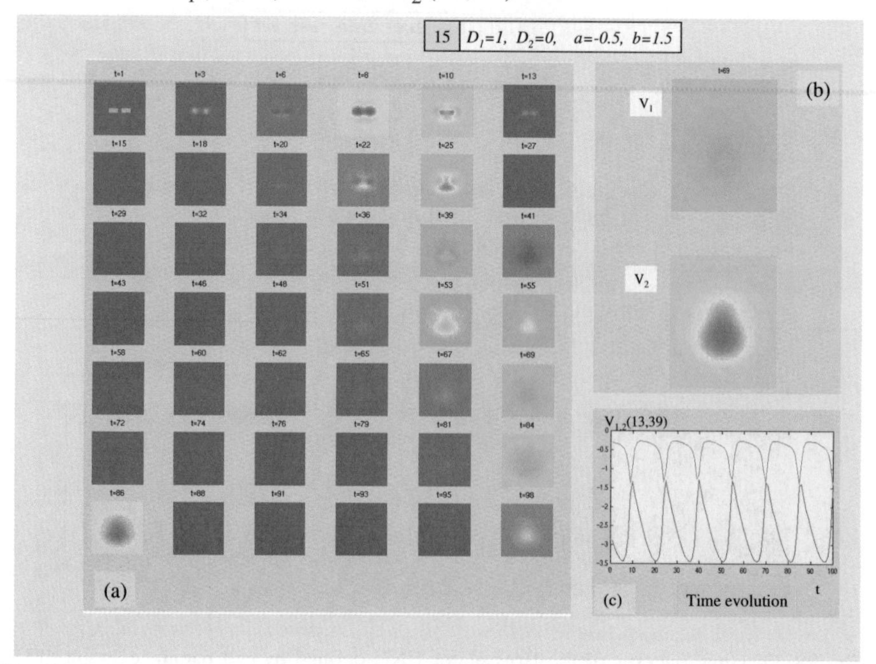

Fig. 30. Dynamic patterns of the Brusselator CNN defined by cell parameter point "15" in Fig. 24; (a) A sequence of 42 time evolution snapshots of the state variable $V_1(j,k)$; (b) A snapshot of the dynamic patterns $V_1(j,k)$ and $V_2(j,k)$, at $t=69$; (c) Time evolution of the state variables $V_1(13,39)$ in red and $V_2(13,39)$ in blue.

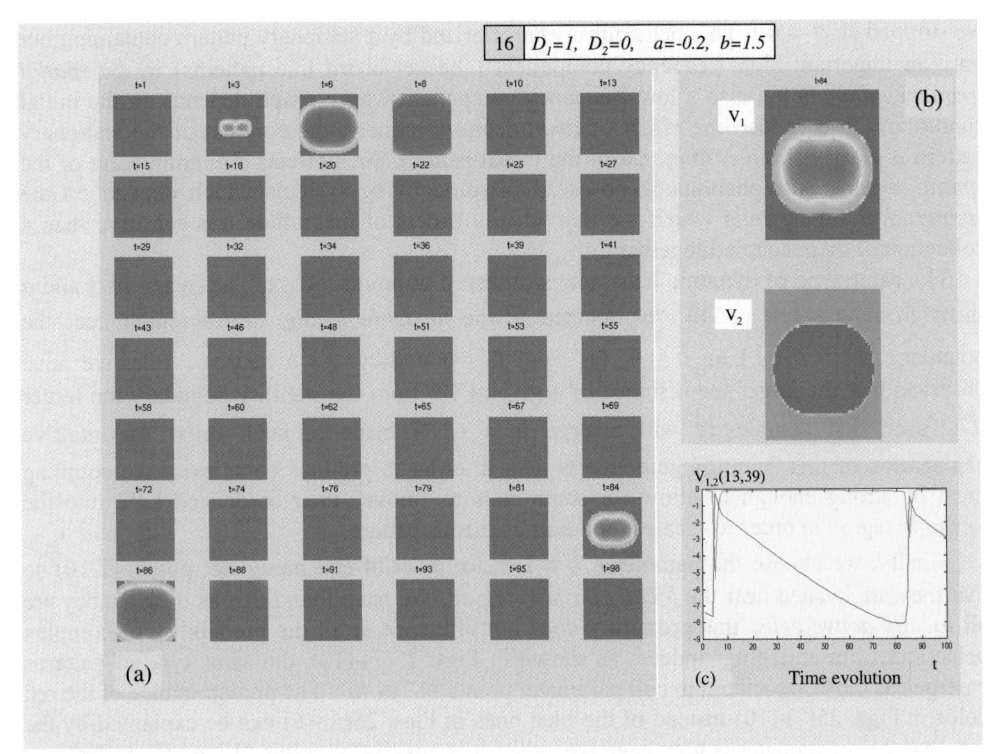

Fig. 31. Dynamic patterns of the Brusselator CNN defined by cell parameter point "16" in Fig. 24; (a) A sequence of 42 time evolution snapshots of the state variable $V_1(j,k)$; (b) A snapshot of the dynamic patterns $V_1(j,k)$ and $V_2(j,k)$, at *t=84*; (c) Time evolution of the state variables $V_1(13,39)$ in red and $V_2(13,39)$ in blue. Due to lack of color resolution in the printer, several arrays appear to be homogeneous in color. The colors of the pixels are nearly equal, *but not identical,* to each other. In fact, the corresponding pattern for $V_2(t)$ has much better contrast (not shown).

Static patterns

Let first choose the cell parameter *b=5,* so that for any *a,* the corresponding CPP (cell parameter point) is far from the boundary between the locally active domain and the locally passive domain. The cell parameter point (CPP) "1" is chosen near the boundary between the stable locally active domain and the unstable locally active domain. After a few trials (with $D_1 = 0$), we found $D_2 = 1$ as a convenient choice for the diffusion coefficient such that non-homogeneous patterns emerge in a CNN made of such locally active cells. We found that for all of the cell parameter points "1" to "10" in Fig.24, complexity emerges when D_2 is greater than some minimum threshold which depends not only on the initial CNN state but also on the location of the CPP. However, when $D_2 \geq 1$ emergent phenomena were found for almost any CPP within the locally active region. Only for a few points (e.g. "4" and "5"), it is necessary to increase this value to $D_2 = 2$. As shown in Figs. 25(1)-(3), the same type of dynamic behavior is displayed when *a* shifts from the *stability boundary* to arbitrarily large values within the *edge of chaos* domain (in this case we stopped at $a = 4$). This behavior is characterized by a stationary pattern containing not only an important high frequency component (checker-board like patterns) in the *spatial frequency domain* but also a low frequency component whose shape depends on the initial condition. Observe also the effect of *amplification*; namely, the contrast of the stationary pattern is clearly stronger than that of the initial pattern. Such effects are reminiscent of the *spatial hyper-acuity* phenomena observed in some living systems which depend on the *cooperation* among cells which are individually orders of magnitude less sensitive than a collection of them coupled together.

The same type of dynamic behavior is observed at points "4"-"6", i.e. when $b=3$ and a shifts from $a = 1.4$ (CPP "6" located in the *unstable locally active* region near the boundary $a^2 = b-1$) to $a = 4$ far from this boundary. As a heuristic rule, we have observed that the larger the distance of a given CPP from the stability boundary the larger D_2 such that complexity will emerge in a CNN made of such cells. An intuitive explanation of this "tuning" guideline is that, in order to produce complexity the coupling must be strong enough to allow for some cells to "move" their associated CPP into the *unstable* region in order to create a non-homogeneous pattern.

Finally, we choose the parameter $b = 1.5$ for a set of cell parameter points (7-10) so that they are located near the *locally passive* region. We have found that as long as they are all *locally active cells,* this proximity does not influence at all the probability of complex behaviors from emerging. Indeed, as shown in Figs. 25(7)-(10), the same type of patterns emerges as those associated to cell parameter points "1" to "6". The predominance of the red color in Figs. 25(7)-(10) instead of the blue ones in Figs. 25(1)-(6) can be explained by the fact that the a parameter is now negative, while it is positive at points "1" to "6". In fact, we have observed that changing the sign of a does not change the shape of the non-homogeneous pattern but only its average value.

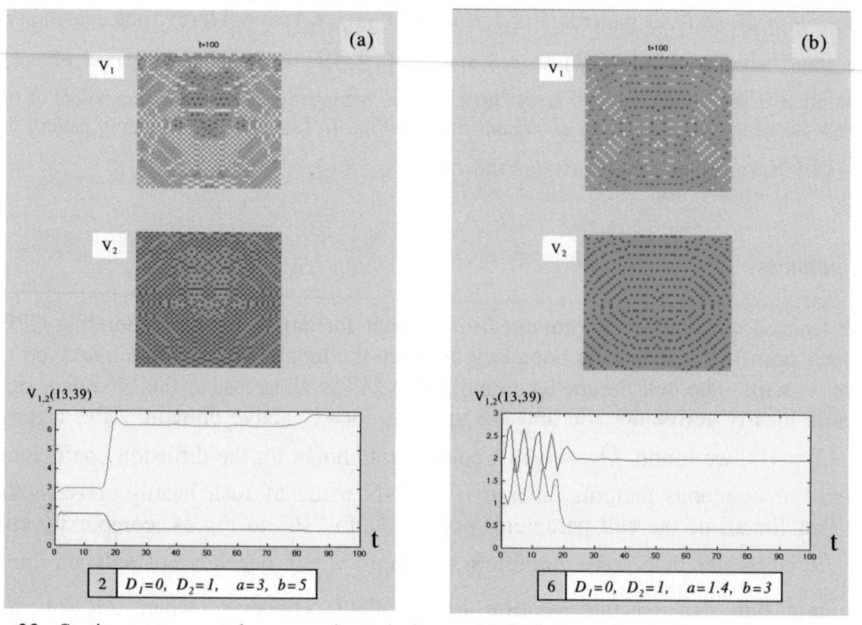

Fig.32. Static patterns and temporal evolution exhibited by a Brusselator CNN at cell parameter points "2", "6", "9" and "10". Each figure displays the stationary pattern in both $V_1(j,k)$ and $V_2(j,k)$ layers at $t=100$ and the temporal evolution for the state variables $V_1(13,39)$ in red and $V_2(13,39)$ in blue; (a) At cell parameter point "2"; (b) At cell parameter point "6"; (c) At cell parameter point "9"; (d) At cell parameter point "10".

For several of the cell parameter points mentioned above, a detailed description of their associated CNN steady states and temporal evolution at an arbitrarily chosen cell are given in Fig.32(a)-(d). It is interesting to observe that for cell parameter points located in the *unstable locally active region* (point "6" in Fig. 32(b), and point "10" in Fig. 32(d)), the temporal evolution exhibits a transient oscillatory behavior, which eventually decays to a static steady state after 4-5 cycles.

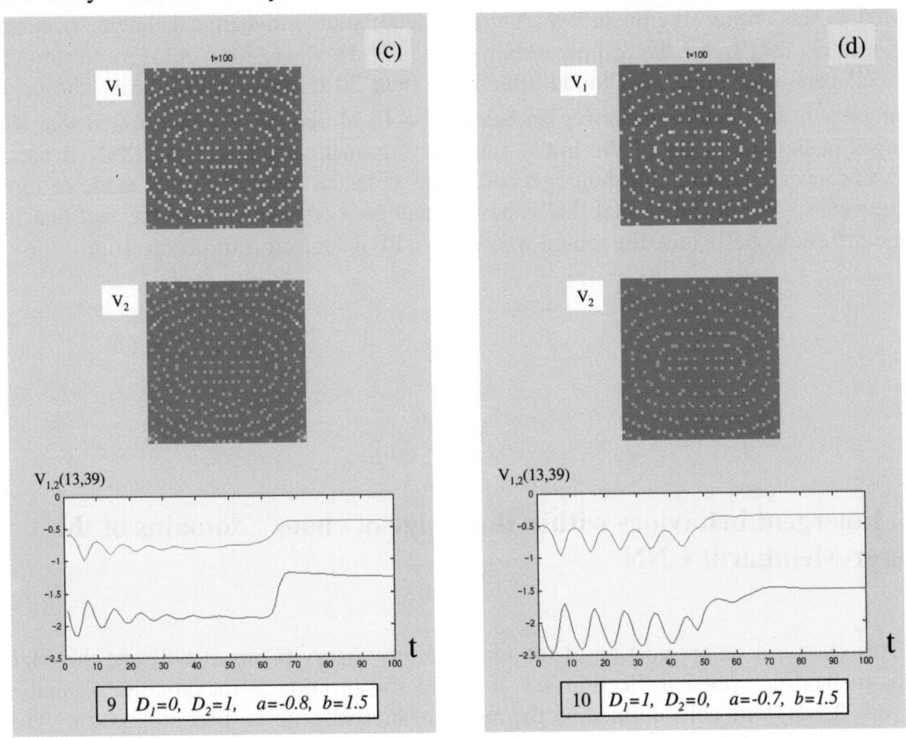

Fig.32. (*Continued*)

The two-diffusion coefficients case

The emergence of static patterns in reaction-diffusion type nonlinear systems is often associated with a mechanism first proposed by Turing in 1952 [Murray, 1989]. It is well known that Turing patterns are stationary patterns produced when both diffusion coefficients are non-zero. However, at cell parameter points "1" to "10", we have seen that *static* patterns can also emerge using only one diffusion coefficient. A similar situation stands for the FitzHugh-Nagumo CNN (Section 4.3.6, Fig.15(5a) and (5b)). It follows that the existence of two non-zero diffusion coefficients is *not* a necessary condition for static patterns to occur. In fact, the situation $D_i = 0$ AND $D_j > 0$, which belongs to the case of a one diffusion coefficient, may be rewritten as $D_i << D_j, D_i \neq 0, D_j \neq 0$ from the perspective of two diffusion coefficients. Indeed, for all cell parameter points located within the portion of the *edge of chaos* domain which is common to both the one diffusion and the two diffusion coefficients cases, we have found that by replacing $D_1 = 0$ with a small positive value (e.g. $D_1 = 0.01$), very similar static patterns would emerge. However, when D_1 is increased above a certain threshold, we have observed that the dynamic pattern usually shifts into a *homogeneous* pattern similar to those observed in CNNs made of *locally passive* cells.

In the case of the FitzHugh-Nagumo cell, Turing-like patterns were found for cell parameter points located in the sub-region of the *edge of chaos* existing *only* in the case of two-diffusion coefficients. As shown in Fig. 24, such a region also exists in the Brusselator CNN. Thus, we expect to observe Turing-like patterns for certain cell parameter points located in the *orange* region below $b=1$. Indeed, such a dynamic behavior occurs, as shown in Fig.25(17), for the cell parameter point "17". However we would like to stress that this cell parameter point was found after more than 20 trials with arbitrarily chosen cell parameters in the *locally active* region below $b=1$. Moreover, we have found that slight changes in the amplitude of the initial state may dramatically affect the CNN dynamics, such as convergence to a dull homogeneous state as in the case of a CNN made of *locally passive* cells. We postulate that this behavior may be a consequence of the fact that for a single diffusion coefficient, the region where this CPP is located is *locally passive.*

* *
*

4.5. Emergent behaviors within the "edge of chaos" domains of the Gierer-Meinhardt CNN

This section is an application of the local activity theory [Chua, 1998] and the edge of chaos method described in Section 4.3. to a specific reaction-diffusion cellular nonlinear network (CNN) with cells defined by the model of morphogenesis first proposed in [Gierer & Meinhardt, 1972]. On the basis of autocatalysis and lateral inhibition, Gierer and Meinhardt proposed a mathematical to explain pattern formation (morphogenesis) in living systems. Using numerical integration, they were able to produce a number of patterns relevant to the formation of biological structures. An analytical treatment of the Geirer-Meinhardt model from the synergetics perspective [Haken, 1994] was presented in [Haken & Olbrich, 1978] using the *order parameter* concept combined with linear stability analysis. While the later approach offers significant contributions in understanding the dynamics of the Gierer-Meinhardt model, it is still too coarse to predict the precise domain in the cell parameter space where emergent behavior may occur.

As shown next, the local activity theory offers a rigorous and effective tool for sharpening existing results in the sense that it can identify more precisely those regions in the cell parameter space which are capable of emergent behaviors.

4.5.1. Local Activity and Edge of Chaos Domains

We will consider throughout this paper the scaled version of the Gierer-Meinhardt model as proposed in [Haken & Olbrich, 1978]. For ease of comparison with our previous works on *local activity* we will replace the notation (ρ,μ) in [Haken & Olbrich, 1978] with (a,b) for the cell parameters. With this substitution, the mathematical model of the scaled Gierer-Meinhardt model is described by the following system of partial differential equations (PDE):

Geirer-Meinhardt PDE with two diffusion coefficients D_1 and D_2:

$$\begin{aligned} \frac{\partial V_1(x,y)}{\partial t} &= a + \frac{(V_1(x,y))^2}{V_2(x,y)} - bV_1(x,y) + D_1\nabla^2 V_1(x,y) \\ \frac{\partial V_2(x,y)}{\partial t} &= (V_1(x,y))^2 - V_2(x,y) + D_2\nabla^2 V_2(x,y) \end{aligned} \tag{43}$$

where V_1 and V_2 are the two state variables characterizing the cell dynamics. The cell parameters are denoted by a, and b, and the spatial coordinates are denoted by x and y. In keeping with the physical meaning of diffusion we will assume the coupling coefficients $D_1 \geq 0$ and $D_2 \geq 0$ throughout this paper.

Step ❶
To apply the concept of *local activity*, the *first step* is to map the above partial differential equations (PDE) into the following associated *discrete-space version*, called a Gierer-Meinhardt CNN:

Gierer-Meinhardt CNN with two diffusion coefficients

$$\begin{aligned} \frac{dV_1(j,k)}{dt} &= a + \frac{(V_1(j,k))^2}{V_2(j,k)} - bV_1(j,k) + I_1(j,k) \\ \frac{dV_2(j,k)}{dt} &= (V_1(j,k))^2 - V_2(j,k) + I_2(j,k) \end{aligned} \tag{43'}$$

where the meaning of the variables was explained in section 4.3.1.

Next we will perform two separate local activity and edge of chaos tests: one for the case *of one diffusion coefficient* ($D_1 \neq 0, D_2 = 0$), and the other for the case of *two diffusion coefficients* ($D_1 \neq 0, D_2 \neq 0$). As explained in Section 4.3.4, we will consider the *restricted local activity – passivity domains.*

Step ❷
The *second step* in applying the local activity theory is to determine the equilibrium points Q_i when $I_1 = I_2 = 0$. In the case of the Gierer-Meinhardt cell, it suffices to investigate only the *equilibrium point* Q_1 defined as follow:

Equilibrium point Q_1 *of the CNN:*

$$V_1(Q_1) = \frac{a+1}{b}, \quad V_2(Q_1) = \left(\frac{a+1}{b}\right)^2 \tag{44}$$

Step ❸
The *third step* consists of determining the four coefficients of the Jacobian matrix $A=\begin{bmatrix} a_{11}(Q_1) & a_{12}(Q_1) \\ a_{21}(Q_1) & a_{22}(Q_1) \end{bmatrix}$ of (43') at the equilibrium point Q_1. For the Gierer-Meinhardt cell they are given by:

$$\boxed{\begin{aligned} a_{11}(Q_1) &= b\frac{1-a}{1+a} \\ a_{12}(Q_1) &= -\left(\frac{b}{a+1}\right)^2 \\ a_{21}(Q_1) &= 2\frac{a+1}{b} \\ a_{22}(Q_1) &= -1 \end{aligned}} \tag{45}$$

As shown in Section 4.3, it is also important to determine the trace $T(Q_1)$ and the determinant $\Delta(Q_1)$ of the Jacobian matrix A :

$$\boxed{\begin{aligned} T(Q_1) &= a_{11}(Q_1)+a_{22}(Q_1)=\frac{2b}{a+1}-b-1 \\ \Delta(Q_1) &= a_{11}(Q_1)a_{22}(Q_1)-a_{12}(Q_1)a_{21}(Q_1)=b \end{aligned}} \tag{46}$$

The information contained in (45) and (46) is then applied as the input of a test algorithm (*step four*) which classifies each cell parameter point (a,b) at the equilibrium point Q_1 into one of the following three disjoint categories:

(a) *Locally Active and Stable* $S(Q_1)A(Q_1)$. In the case of the Gierer-Meinhardt CNN, because there is only one equilibrium point, this region coincides with the *edge of chaos* domain. Following the color code introduced previously this region will be coded in *red* when interpreted from the perspective of "local activity", and in *orange* when applied in the context of "edge of chaos" (see also Fig. 34).

(b) *Locally Active and Unstable* $A(Q_1)U(Q_1)$. This region corresponds to the oscillatory or unstable region of an isolated CNN cell. It is coded in *red* when only *local activity* is considered, i.e. regardless of stability, and in *green* otherwise (see also the legend in Fig.34.)

(c) *Locally Passive* $P(Q_1)$. This region is coded in *blue* and is the region in the cell parameter space where complex phenomena are unlikely to occur. in Reaction-Diffusion CNNs made of such cells.

The test for classifying which one of these 3 categories does a cell parameter point belongs depends on the number of diffusion coefficients.

Step ❹

Local activity and edge of chaos domain, two diffusion coefficients

For the Gierer-Meinhardt CNN cell if the general test in section 4.3.3 is applied, it leads to the following domains:

$$S(Q_1)A(Q_1): \quad b > 0 \text{ AND } \frac{b-1}{b+1} < a < 1 \qquad (47)$$

$$A(Q_1)U(Q_1): \quad b < 0 \text{ OR } \left(-1 < a < \frac{b-1}{b+1} \text{ AND } b > 0\right) \qquad (48)$$

$$P(Q_i): \quad b \geq 0 \text{ AND } |a| > 1 \qquad (49)$$

The subset of the parameters plane (a,b) which satisfies either C3 or C4 coincides with the bifurcation parameter $a_{\max} = \frac{b-1}{b+1}$ derived in [Haken & Olbrich, 1978]. From the perspective of the local activity theory it corresponds to condition $T = 0, \Delta \geq 0$, and is represented in Fig.35 as the vertical boundary between the "red" and the "green" regions. This *bifurcation boundary set* has already been found to play an important role in both the FitzHugh-Nagumo CNN (Section 4.3.) and the Brusselator CNN (Section 4.4.). Indeed, it was found that many interesting complex behaviors are obtained when cell parameter points are located not only within the *edge of chaos domain* but further restricted to a neighborhood of this boundary. According to [Haken & Olbrich, 1978], the inequality $a > a_{\max}$ should be satisfied in order to obtain "soft mode instability", a term used to define emergent complexity in the form of static patterns. A similar condition under a different name is imposed in [Prigogine, 1980] as a prerequisite for emergent stable patterns. Observe that none of these analysis methods can predict an upper bound for the parameter *a* under discussion. For example, in the case of the Geirer-Meinhart model, no information is provided about the existence of an upper bound $a_{MAX} > a$. As we will see in this paper, this upper bound exists $(a_{MAX} = 1)$ and can be determined only via our *local activity theory*. In this case, the *new* information provided by the local activity theory is to identify a part of the neighborhood of this boundary, which need not be checked for complexity because it is locally passive when $I_1 = I_2 = 0$. As long as a cell parameter point is located near the bifurcation boundary $T = 0, \Delta \geq 0$ but in the *locally passive region*, it is highly unlikely for the associated CNN to produce any emergent, non-homogeneous pattern for any initial condition near the equilibrium point. The condition "C2 AND NOT C1" defines not only the edge of chaos domain but it also reveals another important boundary; namely, the one separating the *local activity* domain from the *local passivity* domain. This boundary *cannot* be derived using "stability theory" alone. In fact, this is precisely where the *local activity* theory from [Chua, 1998] becomes indispensable.

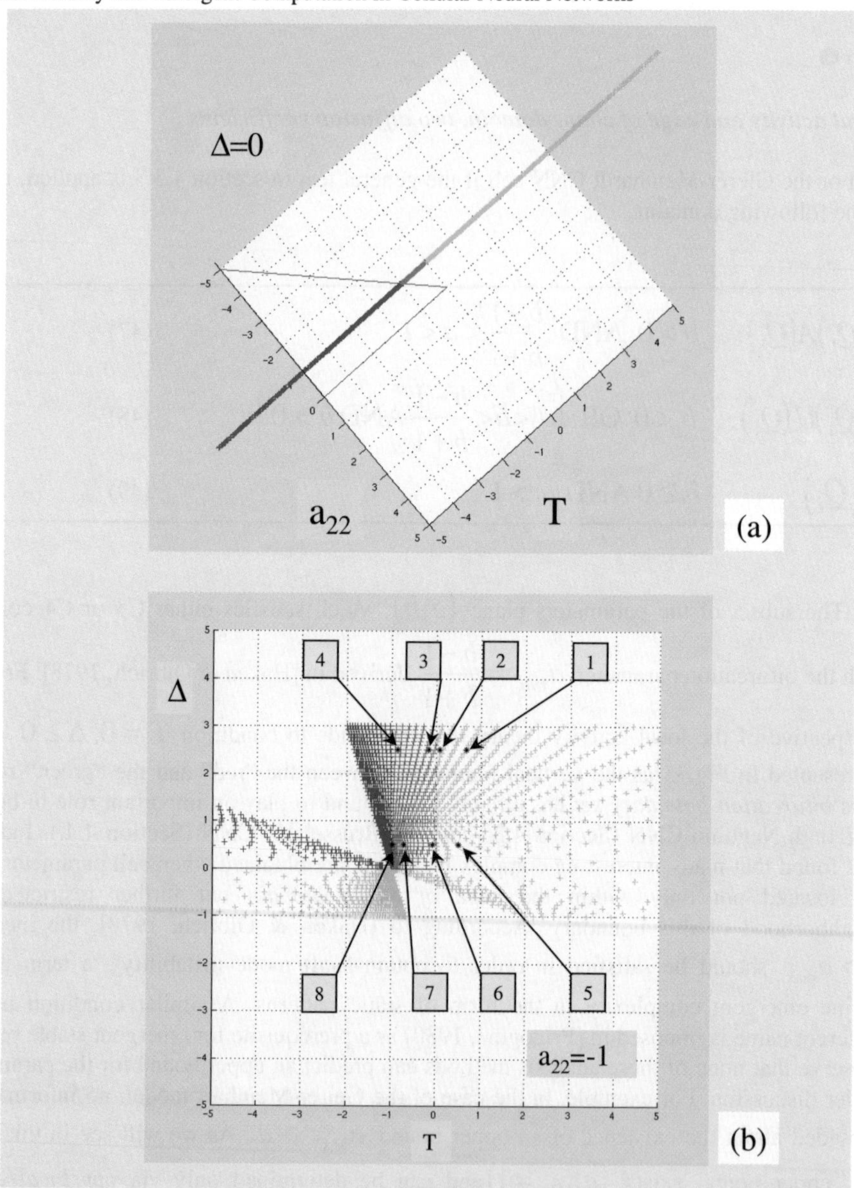

Fig. 33. Cell Parameter Projection Profile $CPPP$ for the Brusselator cell with one diffusion coefficient. The cell parameter domain is represented by a set of 10000 points: $\{(a,b) | \ a=-5+0.1\cdot i; b=-5+0.1\cdot j; i=1,..100; \ j=1,..100\}$. Each cell parameter point (a,b) maps to a $CPPP$ point in the (a_{22},T,Δ)-space. The color at each point indicates its classification into one of the following classes: *locally active and stable (or edge of chaos)* - color *red; locally active and unstable* - color *green*; *locally passive* - color blue; (a) Top view $(\Delta=0)$; (b) Projection onto the $a_{22}=-1$ plane. The location of 8 cell parameter points arbitrarily chosen within the *locally active* domain for one diffusion coefficient is indicated. Each point is assigned a blue box containing its identification label. The same number is used to identify the cell parameter point in the (a,b) parameter plane (Figs. 34, and 35) and in Figs. 36-39.

The cell parameter projection profile (CPPP)

A *cell parameter projection profile* (CPPP) is defined as the projection of any prescribed subset Π of the cell parameter domain of interest into the (a_{22}, T, Δ)-space. In the case of the Gierer-Meinhardt CNN this projection is determined by Eqs. (44) and (45). For $\Pi = \{(a,b) | a \in (-5,5), b \in (-5,5)\}$ the corresponding Gierer-Meinhardt CPPP is displayed in Figs 33(a) and 33(b) as two revealing projections in the (a_{22}, T, Δ)-space, onto the $\Delta = 0$ plane, and the $a_{22} = -1$ plane, respectively. For each of the two parameters, a_{22} and T, the range of variation was quantified into 100 points so that the CPPP visualized in Fig.1 is a discretized version containing 10000 points of the continuous $CPPP(\Pi)$. The colors in Fig. 33 indicate the membership of each cell parameter point (a,b) in one of the three classes defined above: *locally active and stable (edge of chaos)* in *red, locally active and unstable* in *green,* and *locally passive* in *blue.* The projection onto the $\Delta = 0$ plane is shown in Fig. 33(a). Observe that $a_{22} = -1$ independent of the cell parameters, as given by (45). More details concerning the structure of the CPPP are shown in Fig.33(b). As in other cell models (FitzHugh-Nagumo and Brusselator), observe that the "red" region corresponding to the *edge of chaos* is a relatively small subset of the entire CPPP domain. This observation indicates the relevance of the local activity theory, particularly for discarding the uninteresting cell parameters located in the "blue" subdomain. The location of 8 interesting cell parameter points in the (a_{22}, T, Δ)-space is specified in Fig. 33(b). All of these points share a common property: They are all located within the *locally active* domain and thus it follows from the local activity theory that CNNs made of such cells can exhibit emergent behaviors. The dynamic simulations to be presented next will confirm this prediction for the locally active Gierer-Meinhardt cells.

Bifurcation diagrams

The local activity domain and the edge of chaos domains are shown in Fig. 34 (the upper row) for the case $a \in [-5,5]$, and $b \in [-5,5]$. It is interesting to observe how the cell parameters were chosen in the original paper on the Gierer-Meihardt model. As in the case of the reaction-diffusion model reported in [Prigogine, 1980], the cell parameter points are located in the neighborhood of the bifurcation boundary between the stable and the unstable domains. Indeed, from most papers on reaction-diffusion systems it appears that this bifurcation boundary is the only region of interests in searching for interesting parameters. However, using the local activity theory, we can obtain a more exact profile of the cell parameter space and eventually may choose cell parameter points far from the bifurcation boundary while keeping the resulting CNN system in an emergent computation regime. For example, in [Gierer & Meinhardt, 1972] several calculations were reported to demonstrate the "pattern formation" capability of their model. In terms of our scaled model, the four cell parameter points investigated in Fig.1 of [Gierer & Meinhardt, 1972] have the following corresponding coordinates: Cell parameter point $A(a = 0.96,\ b = 5)$ corresponds to the simulation results reported in Fig.1(b) of [Gierer & Meinhardt, 1972]; cell parameter point $B(a = 0.2133,\ b = 0.778)$ corresponds to the results shown in Fig.1(c)-(h); cell parameter point $C(a = 0.096,\ b = 1)$ corresponds to Fig. 1(i), and cell parameter $D(a = 0.128,\ b = 0.67)$ corresponds to Fig. 1(k)-(m) of [Gierer & Meinhardt, 1972].

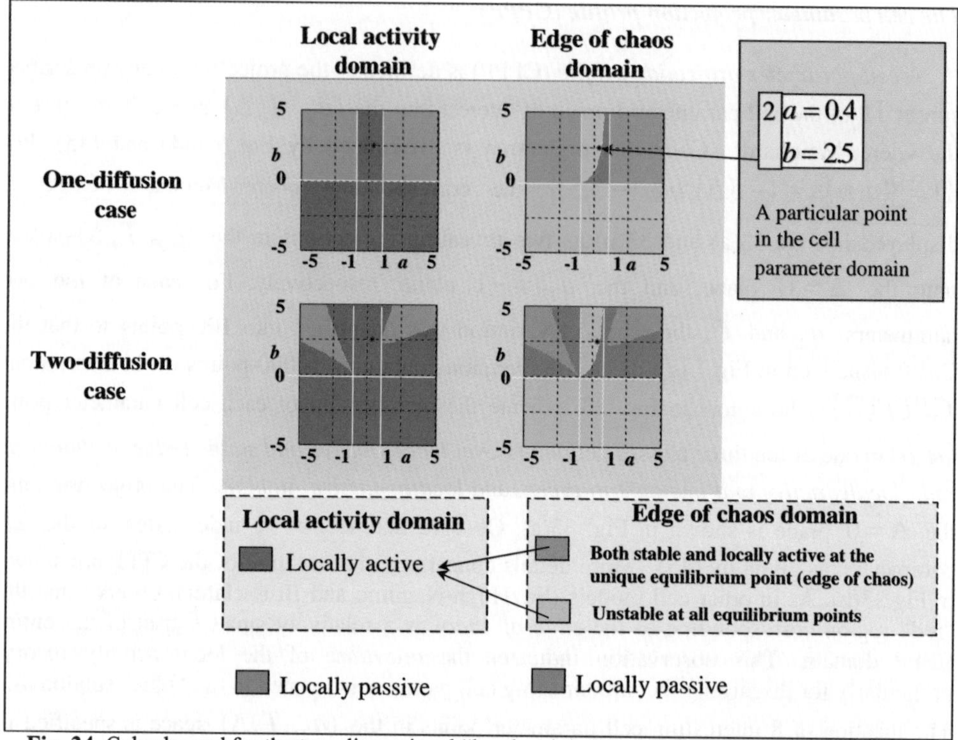

Fig. 34. Color legend for the two-dimensional "local activity" and "edge of chaos" bifurcation diagrams in Fig.35: Each selected point of interest in the $(a-b)$ parameter plane is assigned a numerical label (highlighted with a thin light-blue box next to a light-yellow "caption box" specifying the parameters (a,b)) so that it can be easily identified from the corresponding CNN dynamic simulation shown in Figs.36-39. For example, the parameter point labeled "2" in this figure has its associated CNN dynamics presented in Fig. 36(2).

Note that, no explanation is provided in [Gierer & Meinhardt, 1972] about the reasons why these cell parameter points and not others were selected. It can be easily verified that all of these points lie within the *edge of chaos* region as shown in Fig.35 where they are marked with yellows dots and labels. Many other additional cell parameter points leading to similar behaviors can now be selected within the *edge of chaos* region, determined on the basis of the *local activity theory,* to be illustrated in Section 4.5.2.

Thus our local activity based analysis provided not only a systematic and efficient way to determine the cell parameters via a *unified* approach which is applicable to any type of cells, but additional information on what regions in the parameter space can be ignored by simply identifying the *local passivity* domain; namely, $b>0 \text{ AND} |a|>1$ in the case of the Gierer-Meinhardt model. The boundary between the *local passivity* domain and the *local activity* domain *cannot* be determined via a local stability analysis, since cells defined by parameter points located in both $S(Q_1)A(Q_1)$ and $P(Q_i)$ regions share the same stability property, i.e. they are always *stable*. To illustrate the lack of complexity in such a region, we have randomly picked several cell parameter points in the locally passive domain $(|a|>1, b>0)$ and have found no emergent behavior in the associated CNN when $D_2=0$ (condition for which the result of this test applies).

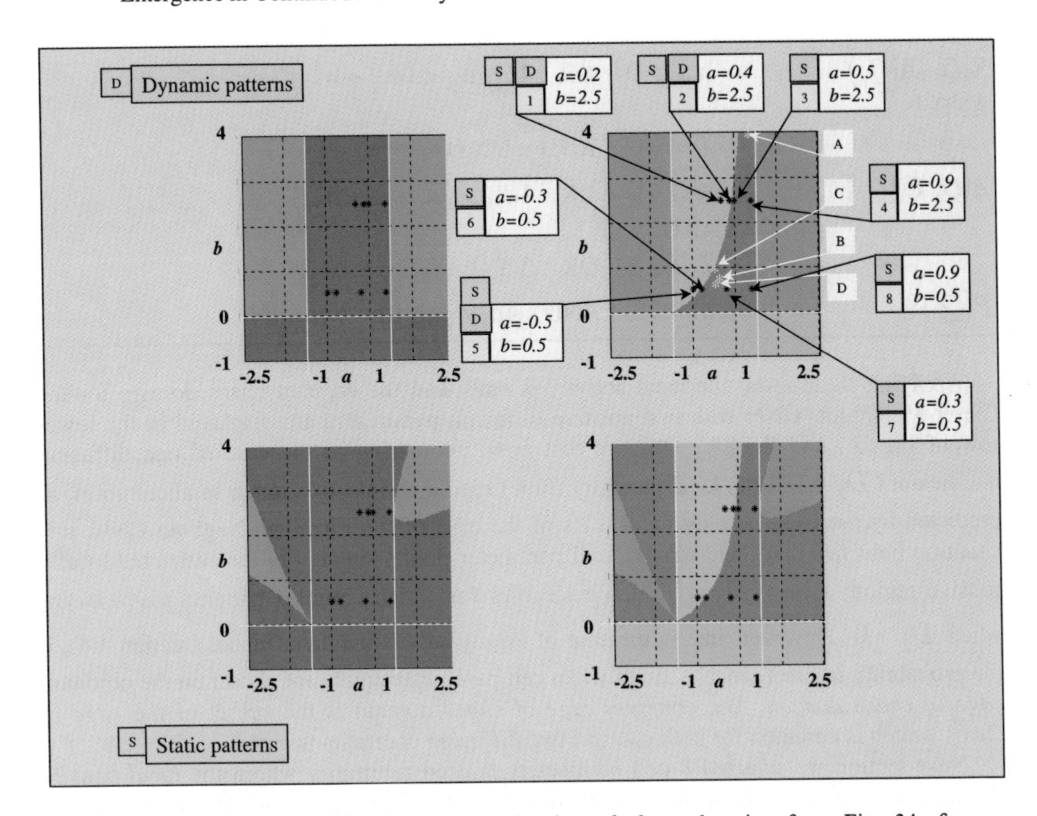

Fig. 35. Details of the *local activity* and *edge of chaos* domains from Fig. 34, for $a \in [-2.5, 2.5]$, and $b \in [-1,4]$. Eight cell parameter points (1 to 8) are indicated, each one is assigned an additional literal label ("S" or "D") identifying the type of dynamical behavior exhibited by the corresponding CNN. The corresponding results of the CNN dynamic simulations are shown in Fig. 36 for the case of static patterns, and in Figs. 37-39 for the case of dynamic patterns.

The case of one diffusion coefficient presented above assumes the coupling is made via the first cell state variable. However, we may also use the second state variable for the coupling. In [Dogaru & Chua, 1998b], it is proved that for any cell having two state variables, the *edge of chaos domain,* the *local passivity domain,* and the *unstable local activity* domain do not depend on whether $D_1 \neq 0$ or $D_2 \neq 0$. Thus, for the *Gierer-Meinhardt* case, we can state explicitly that the tests (47), (48), and (49) yield the same classification for each cell parameter point, independent of the choice of the diffusion coefficient.

Local activity and edge of chaos domain, two diffusion coefficients

For the Gierer-Meinherdt CNN if the general test in section 4.3.3. is applied, it leads to the following domains:

$S(Q_1)A(Q_1)$:	$a_{22} > 0$ OR $4a_{11}a_{22} < (a_{12} + a_{21})^2$
AND	
	$T < 0$ AND $\Delta > 0$
$A(Q_1)U(Q_1)$:	$a_{22} > 0$ OR $4a_{11}a_{22} < (a_{12} + a_{21})^2$
	AND
	$T \geq 0$ OR $\Delta \leq 0$
$P(Q_i)$:	$a_{22} \leq 0$ AND $4a_{11}a_{22} \geq (a_{12} + a_{21})^2$

Applying these tests, the local activity domain and the edge of chaos domain for the Gierer-Meinhardt CNN with two positive diffusion parameters are displayed in the lower row of Fig. 34 and Fig.35. Observe that now, comparing to the case of one diffusion coefficient ($D_2 = 0$) the local passivity (blue) region shrinks to a much smaller subset, as predicted by the local activity theory. As in the case of the FitzHugh-Nagumo CNN, this situation indicates that there may be cell parameter points situated in the *restricted* locally passive region (when $D_2 = 0$) derived earlier for which complex patterns may emerge when $D_2 > 0$. However, like in the case of Prigogine's Brusselator model (section 4.4), it the probability of emergence is small when cell parameter points are chosen in the common *edge of chaos domain.* The common *edge of chaos* domain is the subset of the *edge of* chaos, which is common for both one and two diffusion coefficients cases.

Next section we selected 8 cell parameters located arbitrarily within the *local activity domains* common to both the one-diffusion coefficient case and the two-diffusion coefficients case, and demonstrate by computer simulations that emergence is characteristic for these points. These examples clearly illustrate the "fine-tuning" ability of the local activity theory.

4.5.2. Emergence at the edge of chaos

As shown previously, using the analytical techniques described in extenso in Section 4.3, the cell parameters associated with each CNN cell are mapped into a subset called the *cell parameter projection profile* (CPPP) in the (a_{22}, T, Δ)-space, as shown in Fig. 33. In the Gierer-Meinhardt case, both *red* and *green* regions are found in the CPPP. The location of these cell parameter points (CPP) and their membership in either the *locally active (stable and unstable) domain,* or the *locally passive* domain, are shown in Fig. 35 following the legend in Fig.34. Let us now examine the various patterns generated by the Gierer-Meinhardt CNN. The boundary condition is assumed to be toroidal (i.e. periodic) for all examples.

The one-diffusion coefficient case

In this case, according to the theory of local activity, cell parameter points situated in the region $|a| \geq 1, b > 0$ are unlikely to exhibit complexity. Many points were chosen arbitrarily in the *locally passive* region, and the corresponding Gierer-Meinhardt CNNs were simulated starting from various non-homogeneous initial states near the equilibrium point. In all of these cases, the dynamic evolution converges to a stable state characterized by the same equilibrium point (defined by Eq. (44) in Section 4.5.1) for each cell.

This result is of course predicted by the theory of local activity. Observe that no other theory based only on stability criterion could predict this result. In contrast, the stable cells located in the *edge of chaos* region, as well as cells located in the *locally active and unstable* domain nearby the *edge of chaos*, are likely to exhibit complex behaviors. In order to verify this prediction, we have explored the *local activity* and *edge of chaos* profile of the Gierer-Meinhardt CNN in Fig. 35 and have adopted the following strategy: Two arbitrary values of the parameter *b* were selected ($b \in \{2.5, 0.5\}$) and for each value of the *b* parameter, several arbitrary values were assigned for the parameter "*a*" so that we could explore both *stable* and the *unstable* regions of the *local activity domain* . At least one parameter "*a*" was chosen near the separating boundary between the *stable* and *unstable* regions. As a result, a set of 8 cell parameter points (labeled from "1" to "8" in Fig.35) were generated. The points "1","2" and "5" are located in the "green" region; i.e., they correspond to "unstable" uncoupled cells. As shown in the dynamical simulations in Figs. 37-39, they are responsible for producing both dynamic and static patterns, depending on the couplings.

The remaining cell parameter points ("3","4", and "6"-"8") are located strictly inside the *edge of chaos* region (shown in "orange" in Figs. 34 and 35). They are responsible for producing emergent behaviors (non-homogeneous static patterns in these cases) similar to those described in [Gierer and Meinhardt, 1972], and in [Haken and Olbrich, 1978]. The dynamics of their associated CNN is displayed in Fig. 36(x) where "x" is the integer number (enclosed within a "pink" background) which identifies the corresponding CPP in Fig. 35. For example, the CNN dynamics corresponding to CPP "2" is displayed in Fig.36(2). For the same cell parameter point, different sets of coupling coefficients and/or initial states may lead to completely different behaviors. In order to differentiate such cases, a letter is appended near the identification number of the cell parameter point. For example, two different dynamical behaviors can be observed in Fig. 36(3a) and Fig. 36(3b), respectively, where the different behaviors are only the consequence of choosing different sets of coupling coefficients. The initial states used in all simulations presented in Fig. 36 (with a few exceptions which are described in the text) are chosen near the cell equilibrium point and consist of a pattern of "small" magnitudes whose "color codes" are defined in the legend at the bottom of Fig. 36. Only three cells have slightly perturbed values, and as shown in the simulations presented below, such a small perturbation is enough to trigger various emergent phenomena when the cell parameter points are located within or nearby the *edge of chaos*.

It is interesting to observe that for all points located in the unstable region ("1", "2", and "5"), the corresponding emergent phenomenon consist of *stationary patterns* when $D_2 \neq 0$, or in general when $D_2 >> D_1$ while *dynamic patterns* are obtained only when choosing $D_1 \neq 0$ or, in general, $D_1 >> D_2$. The dynamic patterns associated with points "1" "2", and "5" are shown in Figs.5, 6 and 7, respectively. This behavior has been reported also for different models of cells described by a second order model (Sections 4.3 and 4.4), leading to the conjecture that it may represent a generic property (independent of the cell model). When $D_1 >> D_2$, no emergent patterns were observed for cell parameter points located within the *edge of chaos region.* This seems also to be a generic phenomenon for any of the cell models previously investigated using the local activity theory. Figure 36 presents the emergence of non-homogeneous static patterns at each of the 8 cell parameter points identified in Fig.35.

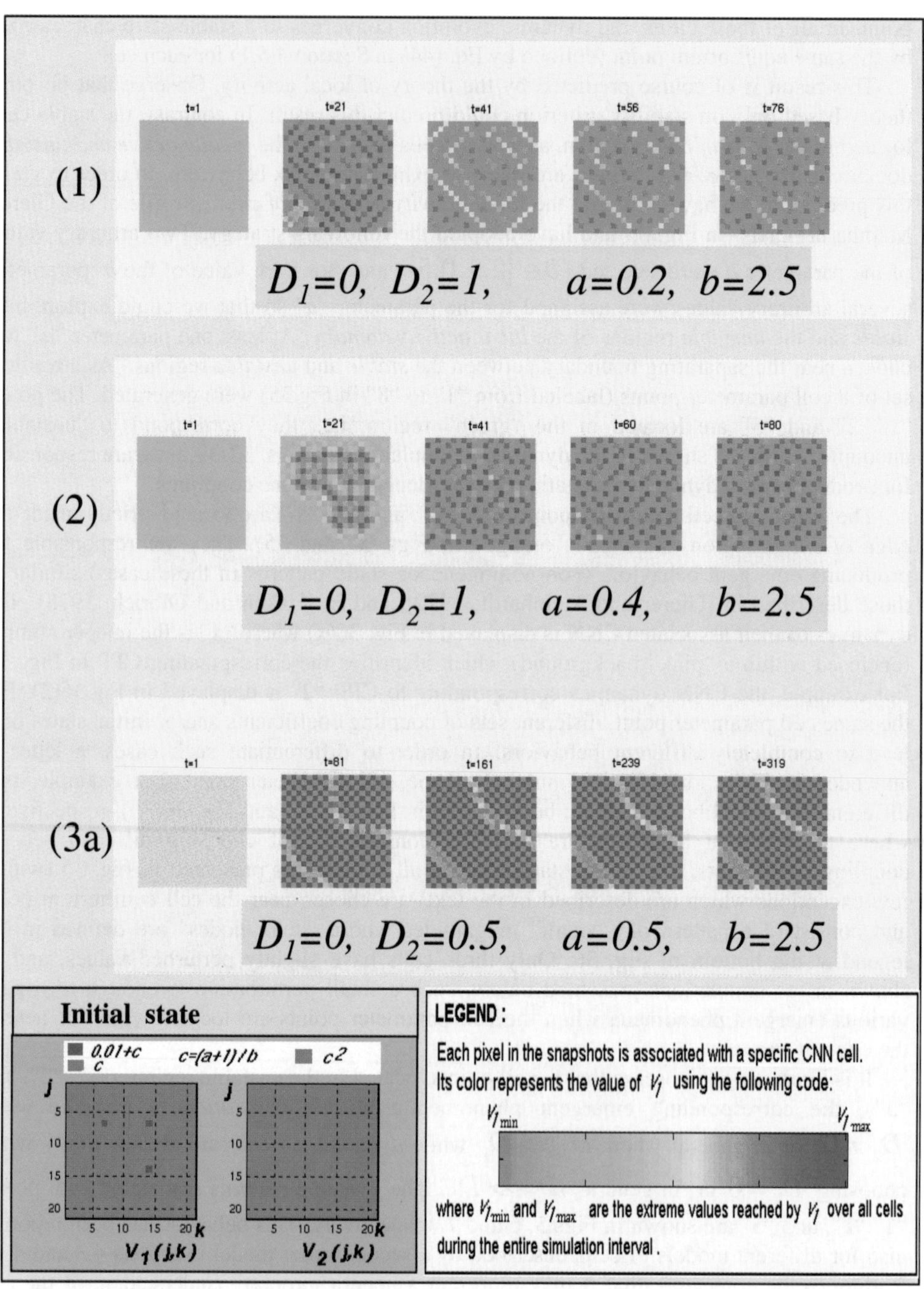

Fig. 36. Examples of dynamic simulations at different cell parameter points responsible for producing static patterns. The color code and the initial state are defined in the legend. The label assigned in the upper left (light-blue margin) of each sequence of 5 snapshots consists of a number indicating the cell parameter point in one of the previous figures. An additional letter is used in some cases to differentiate among different couplings and/or initial states for the same cell parameter point. For example, two different couplings were considered for the cell parameter point "3" shown in Fig. 36(3a), and in Fig.36(3b), respectively.

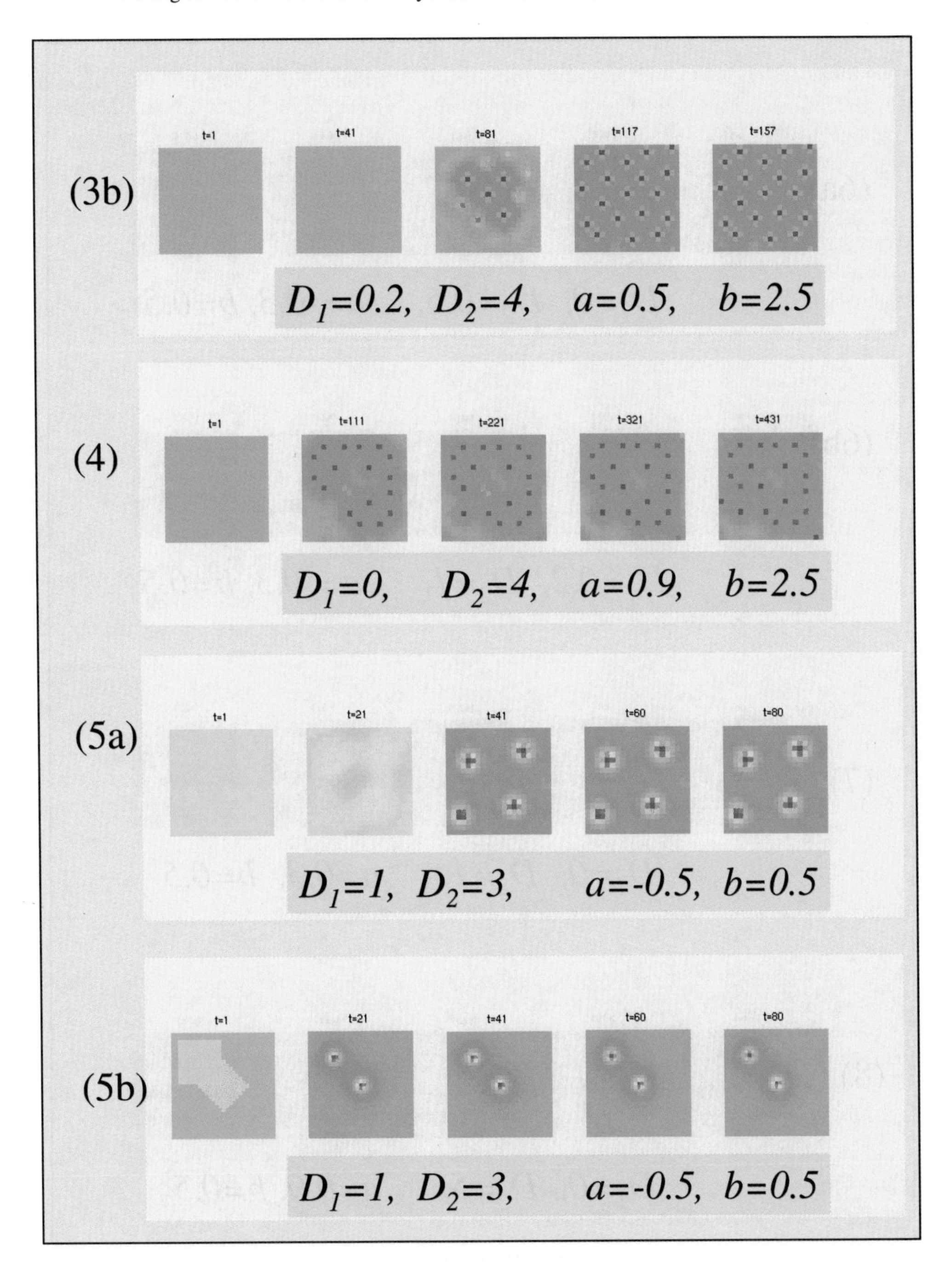

Fig. 36. (*Continued*)

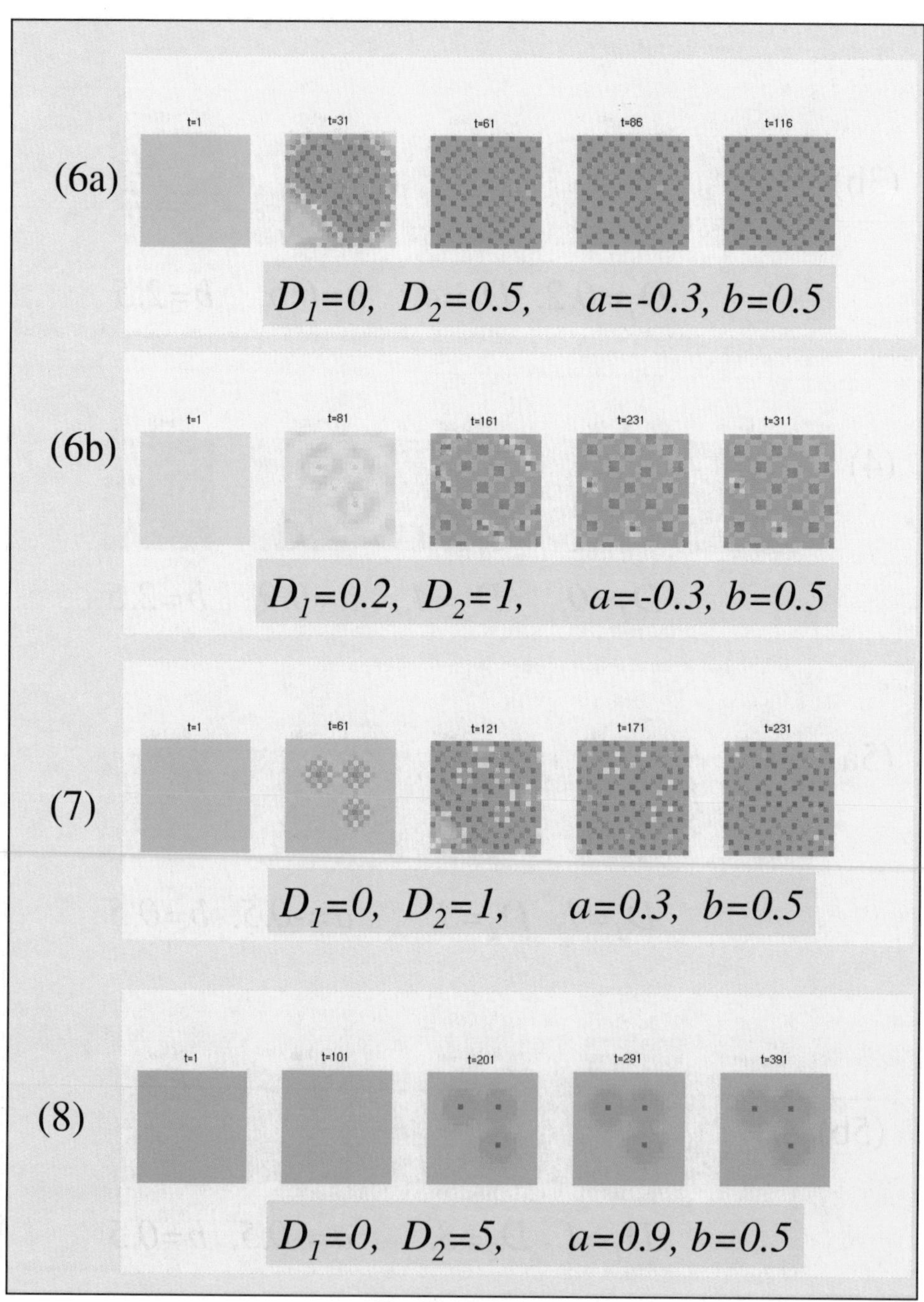

Fig. 36. (*Continued*)

Static patterns

Let us first choose the cell parameter *b=2.50.* The cell parameter point (CPP) "1" is chosen near the boundary between the stable locally active domain and the unstable locally active domain. After a few trials (with $D_1 = 0$), we found $D_2 = 1$ to be a convenient choice for the diffusion coefficient, which allows robust non-homogeneous patterns to emerge in CNNs made of such locally active cells. As in the case of other cell models, we have found that for all cell parameter points "1" to "8" in Fig.35, complexity emerges when D_2 is greater than some minimum threshold. As shown in Figs. 36(1),36(2), 36(3a), and 36(4), qualitatively the same type of dynamic behavior is displayed when a varies from the "green" domain (where the uncoupled cell is unstable) to the right boundary between the "orange" and the "blue" regions in Fig. 35. It is interesting to observe that even though the first two cell parameter points "1", and "2" correspond to *unstable* uncoupled cells, the coupling using $D_2 \neq 0$, had stabilized the dynamics of the CNN system to form stable patterns. Nothing changes in this scenario when the value of the parameter b changes. Indeed, as shown in Fig.36(6a), Fig. 36(7), and 36(8), qualitatively the same type of static patterns emerges, as in the case $b = 2.50$. As observed in the case of the Brusselator CNN, the closer a cell parameter point is to the bifurcation boundary between the "green" and the "orange" sub-domains, the smaller is the amount of coupling (D_2) required for the emergence of a static, non-homogeneous pattern. However, the cell parameter points *do not* necessarily need to be located on, or nearby this boundary. In fact, cell parameter points "4" and "8" are good examples of situations where complexity emerges from the "edge of chaos" region, relatively far from the "stable-unstable" boundary but relatively close to the "active-passive" boundary. It can be observed that in both cases, relatively large magnitudes of the coupling coefficient D_2 were required in order to trigger the development of non-homogeneous patterns. The transient time is also larger than for cell parameter points located near the "unstable-stable" bifurcation boundary. However, if the cell parameter point is located in the "blue" (restricted locally passive) region, no emergent phenomena can be triggered, as predicted by the local activity theory [Chua, 1998], and confirmed by many simulations with $a = 1.10$. Such a prediction would not be possible in the framework of any of the previous theories used to explain emergence in reaction-diffusion systems.

Dynamic patterns

Let us now choose $D_2 = 0$, and $D_1 > 0$. In this case, dynamic non-homogeneous patterns are generated when the cell parameter points are located within the *green* region in Fig.35, which corresponds to the *locally active and unstable* domain. The same situation had been observed for the Brusselator CNN cell (Section 4.4). A "non-emergent" behavior in this case will correspond to cells oscillating *independently* of each other with not-correlated phases. In a visual representation of the type shown in Figs. 37-39 such a *non-synergetic* behavior would be manifested as a change in the colors of the cells while the shape of the initial pattern would remain unchanged during the temporal evolution. At a glance, one can see that this is clearly not the case in Figs. 37-38 and therefore in all of these cases we have obtained an emergent, complex behavior. A notable exception, discussed in detail in the next paragraph, is shown in Fig. 39 for cell parameter point "5". Except for the example shown in Fig. 38, the same initial state specified in the legend of Fig.36 was considered in all of the remaining examples of dynamic patterns.

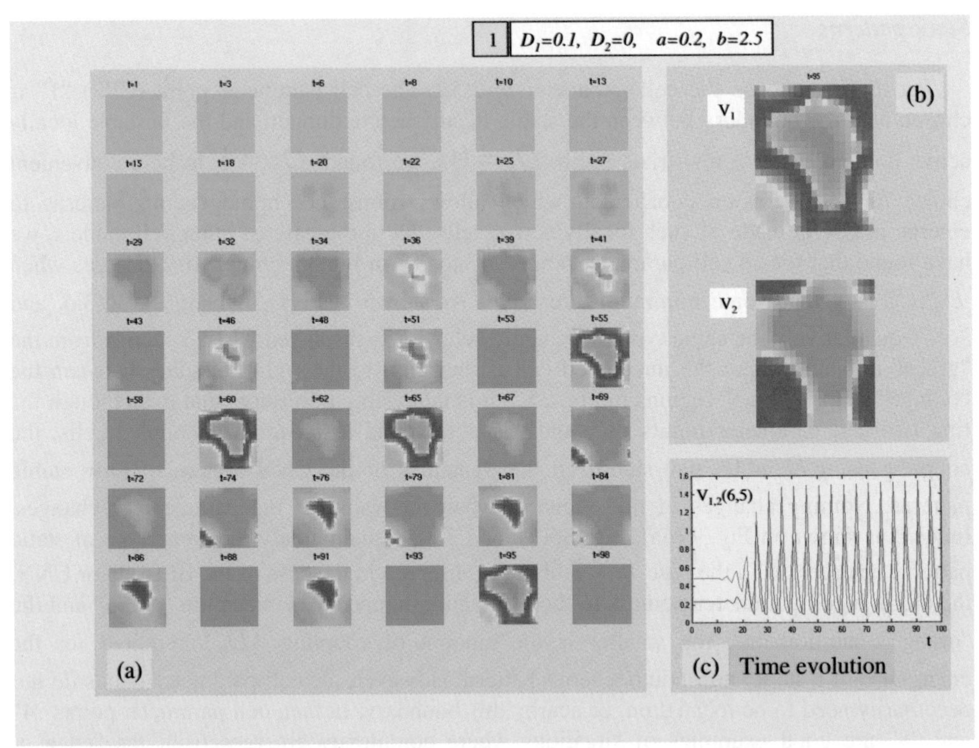

Fig.37. Dynamic patterns of the Gierer-Meinhardt CNN defined by cell parameter point "1" in Fig. 35; (a) A sequence of 42 time evolution snapshots of the state variable $V_1(j,k)$. Due to lack of color resolution in the printer, several arrays appear to be homogeneous in color. They are in fact non-homogeneous where the colors of the pixels are nearly equal, but *not identical*, to each other; (b) A snapshot of the dynamic patterns $V_1(j,k)$ and $V_2(j,k)$, at *t=95*; (c) Time evolution of the state variables $V_1(6,5)$ in red and $V_2(6,5)$ in blue.

Let consider first the cell parameter point (CPP) "1" $(a = 0.2,\ b = 2.5)$, which is located in the "unstable" domain, near but not very close to the *stability boundary* (the boundary between the "green" and the "orange" sub-regions in Fig. 35. The temporal evolution shown in Fig. 37(c), as well as the 42 dynamic snap-shots shown in Fig. 37(a) illustrates the features of a specific spatio-temporal dynamic pattern. Here groups of cells are acting in synchrony by being simultaneously excited (red color in Fig. 37(a)) or inhibited (blue color in Fig. 37(a)). Such a synchronization phenomena was observed in various biological systems (e.g. in the superficial layers of the brains' cortex [Calvin, 1996]) and it appears to be generic for systems made of coupled oscillatory cells. Previous investigations on other types of Reaction-Diffusion CNNs [Dogaru & Chua, 1998a,b] using the local activity theory gave similar results when the cell parameter points were located in the unstable region, near but not very close to the stability boundary. By increasing $a = 0.40$ while keeping the parameter *b* unchanged $(b = 2.5)$ we still keep our cell parameter point in the "green" (unstable) sub-domain, but in a closer neighborhood to the *stability boundary*. A dynamical behavior similar to that presented in Fig. 37 is also observed, as shown in Fig. 38 for the cell parameter point "2". However, now the oscillations are taking place around a steady state, which is different from the uncoupled cell equilibrium point. This is why the snapshots presented in Fig.37(a) have a lower contrast. It was also found that a much larger perturbation in the initial state is required in order to "trigger" dynamic patterns made of clusters of cells oscillating in synchrony.

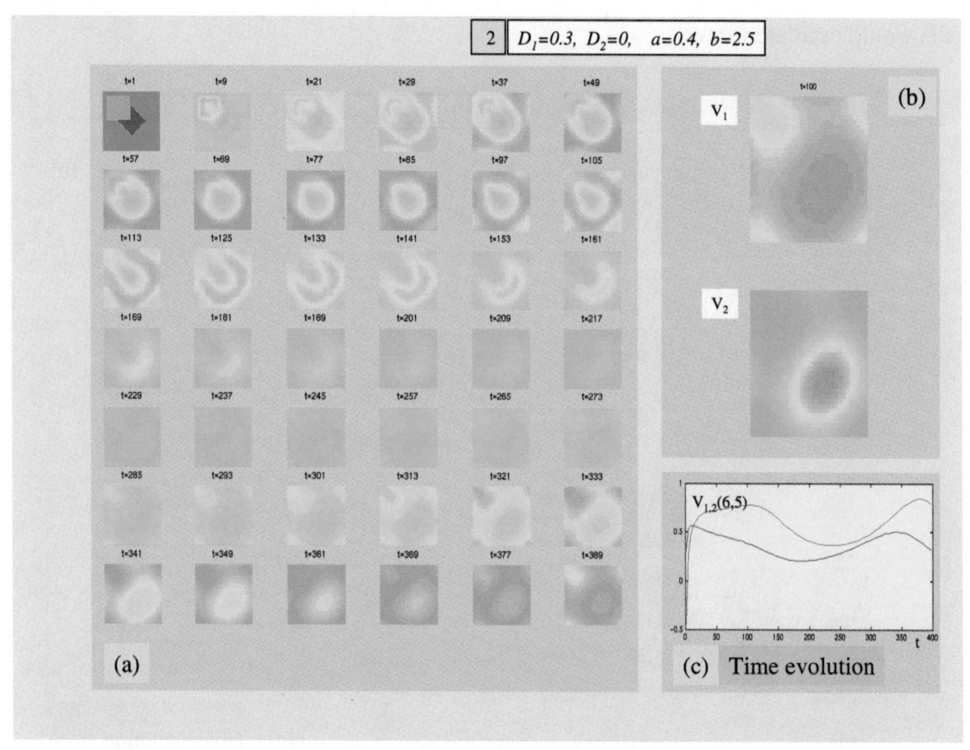

Fig.38. Dynamic patterns of the Gierer-Meinhardt CNN defined by cell parameter point "2" in Fig. 35; (a) A sequence of 42 time evolution snapshots of the state variable $V_1(j,k)$. Due to lack of color resolution in the printer, several arrays appear to be homogeneous in color. They are in fact non-homogeneous where the colors of the pixels are nearly equal, but *not identical*, to each other; (b) A snapshot of the dynamic patterns $V_1(j,k)$ and $V_2(j,k)$, at *t=100*; (c) Time evolution of the state variables $V_1(6,5)$ in red and $V_2(6,5)$ in blue.

A very interesting phenomenon is observed at the cell parameter point "5", as exhibited in Fig. 39. In this case, a small fluctuation in the initial state (3 cells have slightly different magnitudes from the equilibrium state of the uncoupled cell - as shown in the legend of Fig. 36) is able to trigger a clustering phenomenon (groups of cells having the same type of activation) similar to those described for the previous cell parameter points. However, in this case, the dynamic patterns persist only for a short transient period $(26 < t < 36)$ (Fig.39a). Thereafter, in contrast to the previous examples, the CNN decays towards a "lower magnitude" steady state. This global equilibrium state characterized by $V_1(j,k) = V_2(j,k) = 0$ corresponds to a homogeneous pattern but which *cannot be reached* in a system made of uncoupled cells. Indeed, as discussed in Section 4.5.1, for an uncoupled cell, $V_1 = V_2 = 0$ is an unstable, degenerate, equilibrium point. Therefore, we can speak of an emergent phenomenon in this case, even if it occurs only in the form of a short lasting (ephemeral) non-homogeneous (transient) pattern. Any living system is characterized by a finite lifetime. This example exhibits a qualitatively similar phenomenon in the sense that a non-homogeneous pattern had emerged and evolved during its life time $(26 < t < 36)$ but which eventually "dies" when the corresponding CNN had reached a state characterized by its lowest possible energy. Without coupling, the same collection of cells would oscillate forever.

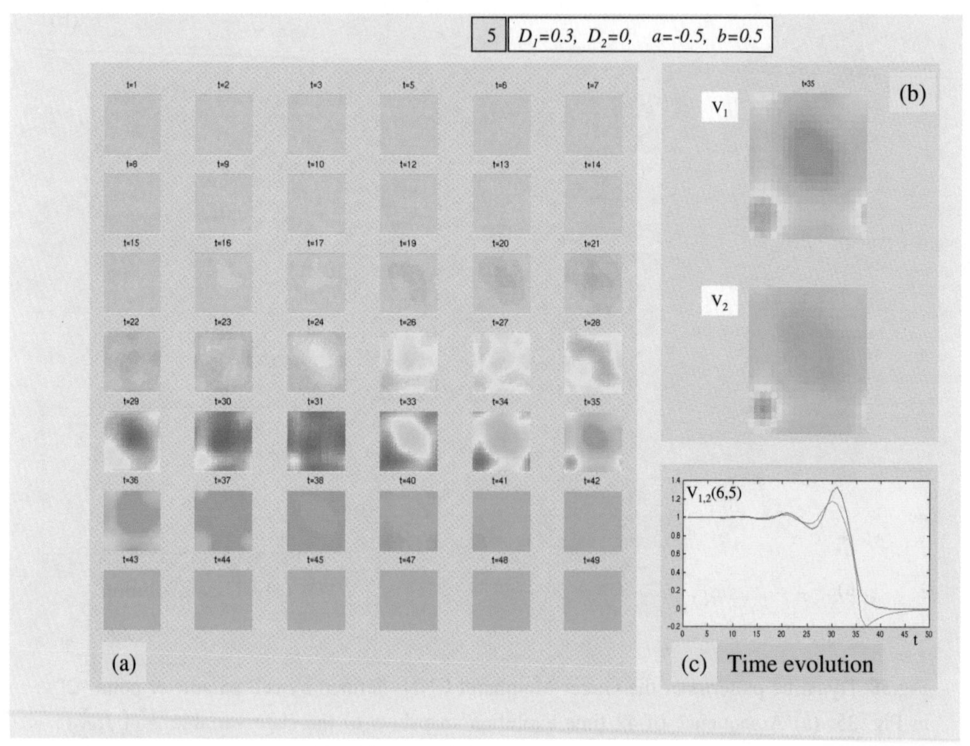

Fig.39. Dynamic patterns of the Gierer-Meinhardt CNN defined by cell parameter point "5" in Fig. 35; (a) A sequence of 42 time evolution snapshots of the state variable $V_1(j,k)$; (b) A snapshot of the dynamic patterns $V_1(j,k)$ and $V_2(j,k)$, at $t=35$; (c) Time evolution of the state variables $V_1(6,5)$ in red and $V_2(6,5)$ in blue.

The two-diffusion coefficients case

The emergence of static patterns in reaction-diffusion type nonlinear systems is often associated with a mechanism first proposed by Turing [Turing, 1952]. It is well known that Turing patterns are stationary patterns produced when both diffusion coefficients are non-zero. However, at cell parameter points "1" to "8", we have seen that *static* patterns can also emerge with a single diffusion coefficient. A similar situation has been reported in previous sections for the FitzHugh-Nagumo CNN, and for the Brusselator CNN. It follows that the existence of two non-zero diffusion coefficients is *not* a necessary condition for static patterns to emerge. In fact, the situation $D_i = 0$ AND $D_j > 0$, which belongs to the case of one diffusion-coefficient, may be rewritten as $D_i << D_j, D_i \neq 0, D_j \neq 0$ from the perspective of two diffusion coefficients. Indeed, for all cell parameter points located within the region of the *edge of chaos* domain which is common to both the one diffusion and the two diffusion coefficients cases, we have found that by replacing $D_1 = 0$ with a small positive value (e.g. $D_1 = 0.01$), very similar static patterns would emerge.

In the case of the Gierer-Meinhardt CNN it is found that static patterns will emerge for two non-zero coefficients with $D_2 > D_1$, when the cell parameter points are located in a relatively narrow neighborhood around the bifurcation between the "green" and the "orange" domains (the *stability boundary*). This observation is consistent with the experimental results reported in the Gierer-Meinhardt literature. However, we must stress that by replacing $D_1 = 0$ with a very small value (e.g. $D_1 = 0.0001$) the results reported previously for one-diffusion coefficient are still valid (but now there are two non-zero diffusion coefficients). Thus, static patterns would emerge always when cell parameter points are located in the "orange" region, and in a small neighborhood of it which extends into the "green" region, as long as $D_2 >> D_1$. However, only within a relative narrow neighborhood of the *stability boundary* can the condition $D_2 >> D_1$ be relaxed to $D_2 > D_1$ (e.g., $D_2 = 3D_1$ instead of $D_2 = 1000D_1$), so that stationary patterns will emerge for comparable values of the two diffusion coefficients. This is the case for cell parameter points "3", "5", and "6", illustrated by CNN simulation examples presented in Figs. 36(3b), 36(5a), 36(5b) and 36(6b). Note that the cell parameter point "5" is located in the "green" (unstable) region. Yet, when these cells are coupled, the CNN behavior is similar to the one obtained when locally active and stable cells are coupled. Similarly to the patterns reported in Fig.7 of [Meinhardt & Gierer, 1974], the patterns shown in Figs. 36(3b), 36(5a) and 36(6b) are characterized by the emergence of an almost regular grid of peaks of activation (called "bristle" patterns in of [Meinhardt & Gierer, 1974]). Depending on the initial condition, such patterns may have a computational meaning similar to Fig. 36(5b) where the output can be interpreted as an extraction of the gravity centers of two regular objects in the input pattern.

4.6. Concluding remarks and engineering perspectives

This chapter focused on methods of applying the local activity theory [Chua, 1998] to Reaction-diffusion CNN systems determining those domains of the cell parameters such that emergence is likely to occur. We have extensively described in Section 4.3 an "edge of chaos" procedure, as a corollary to the local activity theory (see also [Dogaru & Chua, 1998a,b]). As case models we considered first the FitzHugh-Nagumo CNN (section 4.3), then the Brusselator CNN (section 4.4), and finally the Gierer-Meinhardt model (section 4.5).

It is shown that a *precisely defined* subset of the local activity domain, called the *edge of chaos*, can be identified in the cell parameter space, using simple and fast algorithms independent of the diffusion coefficients. Since the choice of cell parameters within or nearby this domain leads to complex dynamics (including spiral waves, Turing-like patterns, information computation behaviors and chaotic dynamic patterns) in the corresponding Reaction-Diffusion CNN, our definition of "edge of chaos" shares the same properties first advanced by [Packard, 1988], and [Langton, 1990]. However, when comparing our approach to previous works on *edge of chaos*, there are several major differences:

❶ First, our theory is *continuous in both time* and *state* while previous works on the "edge of chaos" deal exclusively with discrete time, discrete state systems. In our theory, cellular automata is just a special case of a much more general "analog" dynamical system called *Generalized Cellular Automata* [Chua, 1998]. In particular, as shown in Chapter 3, any local rule defining a cellular automaton can be realized by simple piece-wise linear functions. For example, the local rules of the well-known Game of Life[10], can be simply implemented using the nonlinear function $y = sign(u_0 + a - |b + cu_\sigma|)$, where a, b and c are *real* parameters, u_0 denotes the output of the central cell, and u_σ represents the sum of the remaining eight neighboring cell outputs at the previous discrete-time moment. When these parameters are *continuously* varied, the resulting Boolean local rule (here, the two state levels are coded with -1 and +1, respectively) changes, so that, the rule associated with the Game of Life, is realized within a *compact domain* of parameters (see more details in Chapter 5). One particular point in the cell parameter domain of the "Game of Life" is, for example, $(a = 3,\ b = 5,\ c = 2)$. Such a parameter domain can be explicitly defined by its *failure boundary* [Chua, 1998] and thus it can be identified using nonlinear techniques. In contrast, the transition point identified within the domain of variation of the λ parameter in [Langton, 1990] can be determined only by brute force computer experiments and thus there is no guarantee that it is generic. This chapter describes in detail how *edge of chaos* domains of three different Reaction-Diffusion CNN models can be identified within the failure boundaries of the cell parameter domain. The same algorithm is applicable to any other *reaction-diffusion* equations and can be generalized to any other complex systems modeled by a CNN.

❷ Second, as pointed out in the work of [Mitchell *et.al.*, 1993], where the experiments described in [Langton, 1990] and [Packard, 1988] were reconsidered, there is no evidence of a generic relationship between λ, (the parameter defined via the discrete structure of the transition function), and the computational ability in the Cellular Automata. In contrast, the *edge of chaos* is defined precisely in [Chua, 1998] as the cell parameter domain for which the uncoupled cells are *not only locally active, but also stable*. It is the *local activity* property, tempered by stability constraint, which makes it possible to perform non-trivial computations and display self-organization.

Although the "edge of chaos" domain as defined within this chapter is giving a much narrow domain than the local activity domain, its definition cannot still *guarantee* the computational emergence (static or dynamic). The *edge of chaos* membership of a cell parameter point is a *necessary but not sufficient* condition for emergence. But this is reasonable since we perform all our analysis using the *uncoupled* cell, therefore ignoring the coupling. On the other hand a mathematical analysis of the whole CNN systems (coupled cells) is prohibitive and difficult to carry.

In addition there are some heuristics based on observations coming from the "edge of chaos" analysis that we have performed on several second order Reaction-Diffusion systems. Such heuristics can be used to choose better inside the edge of chaos domains and they also give indications on how to choose the diffusion coefficients such that emergence will occur.

Heuristics in determining cell parameter points leading to emergence

First, it was clearly determined that *richer emergent behaviors occur in cells having more equilibrium points at the "edge of chaos"*. Let us recall the FitzHugh-Nagumo model where the most interesting emergent phenomena occur in the edge of chaos region in a sub-region characterized by three stable equilibrium points and in the nearby of the cell

[10] The "Game of Life" is an example of local rules belonging to the "complexity" class IV [Wolfram, 1984], or within the *edge of chaos* [Langton, 1990].

parameter points satisfying both conditions C3 and C4 in the set of local activity conditions. Such behaviors as *spiral patterns* were not observed in systems characterized by a unique equilibrium point.

There are several more *common* aspects, which appear to be *independent* of the cell types:

① In the case of two-diffusion coefficients, "Turing-like" static patterns can emerge for $D_1 << D_2$. In the case of one-diffusion coefficient, "static patterns" can still emerge for cell parameter points located in the *edge of chaos* region when $D_1 = 0$ in the above inequality. The further a cell parameter point is from the boundary between the (red) *edge of chaos* and the (green) *locally active and unstable* domain, the larger should be chosen the coupling coefficient D_2 in order to spawn emergent behaviors.

② Turing-like patterns can also emerge in CNNs made of cells with their parameters located within the *edge of chaos* domain existing *only for the two diffusion case* (the same domain for one-diffusion coefficient case is typically *locally passive*). However, in such cases, the probability of emergence is smaller than that when the cell parameter points are located within *edge of chaos* domains valid for both one and two diffusion coefficients. In fact, for the Gierer-Meinhardt cell, within the one-diffusion coefficient *edge of chaos* no example of emergent behavior was found, but this does not necessarily means that it does not exists.

③ Dynamic patterns are typically associated with cell parameter points located near the *edge of chaos domain* but within the *locally active and unstable* region, and they usually emerge for $D_1 \neq 0$ in the case of one-diffusion coefficient, or $D_2 >> D_1$ in the case of two-diffusion coefficients.

④ The boundary $T = 0, \Delta \geq 0$ which separates the *edge of chaos region* by the *locally active and unstable* region (also called a *stability boundary*) plays an important role in that for the cell parameter points located within the *edge of chaos* domain, or in the *locally active and unstable* domain but very close to this boundary, rich complex behaviors consisting of either static and dynamic patterns usually emerges. *Chaotic dynamic patterns* for example were usually found for cell parameter points located within the *locally active and unstable region* but specifically very close to the *edge of chaos* sub-region. Such examples were provided by the cell parameter point "7" for the FitzHugh-Nagumo cell in Section 4.3.6, or by cell parameter point "11" for the Brusselator cell in Section 4.4, or by the cell parameter points "1" and "2" in Section 4.5 (the Gierer-Meinhardt cell).

⑤ Cell parameter points in the restricted *locally passive* region typically produces no emergent behaviors in CNNs made of such cells, a result which is in fact anticipated (for initial conditions chosen near the equilibrium point) by the local activity theory. It is important to stress that a stability analysis *cannot* determine the "edge of chaos" region in the cell parameter domain where the cells are *stable* but *locally active*.

Uncertainties in locating emergent behaviors

The "edge of chaos" domain as defined herein is a *precisely defined* set only under the assumption $\bar{I}_1 = \bar{I}_2 = 0$. Therefore it should be viewed as a minimal set where there is a potential for complexity, in fact Chua's local activity theory clearly states that any locally active cell has a potential for emergence and complexity when coupled with similar cells. Assuming a wider range for the coupling currents, one should then consider a larger domain called an *extended edge of chaos,* which will *always* include the precisely defined *edge of chaos*. Such a domain is determined by the union of all *perturbed edge of chaos* domains determined according to the algorithm presented in Section 4.3 where $\bar{I}_1, \bar{I}_2$ are allowed to vary within a given range. Observe that a precise definition of the *extended edge of chaos* is impossible, since the $\bar{I}_1, \bar{I}_2$ variation range depends on various factors difficult to be taken in consideration, such as the initial state of the CNN, the degree of coupling between CNN cells and the entire CNN dynamics. However, the assumption $\bar{I}_1 \approx \bar{I}_2 \approx 0$ (or equivalently, an infinitesimal variation range for $\bar{I}_1, \bar{I}_2$) stands for most of the cases of interest. In this case, the *extended edge of chaos* set is almost identical with the *edge of chaos* domain. It adds only a thin but *ambiguously* specified domain, expanding within the neighborhood of the *edge of chaos* frontiers. This is the reason why, we discussed as having potential for emergence cell parameter points which are not only inside but also *near* the *edge of chaos domain*, henceforth specifying the *extended edge of chaos* by the word *near*. While a cell parameter point situated within the *edge of chaos* domain always belongs to the *extended edge of chaos domain* the reverse may not be true. As shown in Section 4.3, for several parameter points (e.g. "6","7","8","10","13") complex behaviors characterized by non-homogeneous patterns were observed in the corresponding one diffusion coefficient CNN. It is an example confirming that such points belong to the *extended edge of chaos* domain while they are *not* within the *edge of chaos.* As expected, such points are situated *near* the boundary of the *edge of chaos* domain for the one-diffusion coefficient case.

A question of interest regarding the *edge of chaos* domain is the following: Compared to the entire cell parameter space, is this a small or a large sub-domain? Depending from the perspective, we may admit both alternatives as valid answers. Indeed, compared to a purely random search in the cell parameter space, the *edge of chaos* is very small. For example, taking the case of the FitzHugh-Nagumo cell if one spans the parameter range $a \in [-20,20]$, $b \in [-20,20]$, $\varepsilon \in [-4,4]$ with a resolution of $N = 40$ for each parameter range, one can easily check that there are 106880 corresponding points in the (a_{22}, T, Δ) space*. The *edge of chaos* is a well-defined sub-set of the (a_{22}, T, Δ) space. It was found numerically that only 46 out of the above 106880 cell parameter points are included in this region, representing only 0.043% of the entire cell parameter space. Thus, the simple algorithm for determining the *edge of chaos* discarded 99.967% of the cell parameter points as useless since they surely do not produce emergent behaviors in the corresponding Reaction-Diffusion CNN. Still, as discussed earlier, within this small subset not all parameter points within the *edge of chaos* were found interesting in terms emergent behaviors. Within a pool of 9 randomly selected cell parameter points from the previously mentioned 46 points, and for the fixed values of diffusion coefficients $D_1 = 0.1$, and $D_2 = 0.5$, only 2 out of 9 cell parameter points were found to lead to emergent behaviors (non-homogeneous patterns) in a one-diffusion coefficient Reaction-Diffusion CNN. In a Reaction-Diffusion CNN with two coefficients, 4 out of the above 9 cell parameter points located within the *edge of chaos* were found responsible for complex (emergent) behaviors. From this second perspective, the *edge of chaos* is still large and we are interested to find methods for an even finer location of the emergent cell parameter points. For reasons explained above (the *edge of chaos* method does not consider cell coupling in its definition) we believe it would be hard to narrow this domain even further using the local activity theory alone. But, looking at the above results one should be optimistic. Indeed, although roughly only about 30% or less of the cell parameter points within the *edge of chaos* effectively lead to emergent behavior, this still indicates a remarkable capability of the edge

---

* For some points in the cell parameter space, there are three corresponding points in the (a_{22}, T, Δ) space, associated with three equilibrium points.

of chaos method to locate cell parameter points with potential for emergence. Without the "edge of chaos" method, using exhaustive search, one should search for those cell parameter points in a space which is 2325 (i.e. $\frac{1}{0.00043}$) times larger! This is indeed like searching a needle in a haystack.

Engineering perspectives and potential applications of the edge of chaos analysis

Within this chapter a methodology to locate narrow domains with potential for emergence within the cell parameter space of a Reaction-Diffusion CNN was presented. The method has the advantage that only the uncoupled cell is studied leading to fast numerical algorithms to locate the interesting cell parameter domains, but it cannot predict what is the exact value of the diffusion coefficients for which emergent phenomena will occur. However, as various experiments presented above show, there is little experimental effort to determine such values of the diffusion coefficients.

The whole methodology exposed in this chapter is applicable only for second order cells (i.e. defined by 2-state variables). In a practical implementation these corresponds to maximum two layers of resistive grids connecting the cells (i.e. the implementation of the diffusion local connectivity). The "reaction" part is in fact the locally active cell, which can be easily implemented as a nonlinear electronic (or possibly molecular) circuit.

Although a similar methodology based on the local activity can be developed for higher order cells (see for example [Min, L., *et.al.*, 2000a,b]), from an engineer's perspective it is more interesting to focus on 2-layers Reaction Diffusion CNNs since more than 2 resistive grid layers being difficult to implement. It is quite easy to design and model mathematically simple nonlinear circuit cells where a few parameters are tunable. Certain dynamic phenomena can be identified afterwards using the analytic techniques of local activity and edge of chaos exposed herein. Then one can use such circuits for various tasks, for example as an intelligent and programmable image sensor where several basic processing functions can be easily integrated on the same chip including the sensing device.

Integrated circuit designs to implement a second order Reaction-Diffusion CNN were already reported in [Arena *et al.,* 1998], [Serrano and Vázquez, 1999], where some basic emergent behaviors were also reproduced. The techniques proposed in this chapter may be used to improve the designs providing an easiest way to identify the useful regions of parameters for which emergent computation occurs.

Another area of interest for applying the edge of chaos based methods is the emerging area of *nanotechnology*. Recently, very high densities of active devices with nano-meter dimensions were demonstrated in laboratories and in a near future such systems will enter in the industrial realm. One such successful device is the Resonant Tunneling Diode (RTD) [Mizuta and Tanoue, 1995], which can be operated at room temperature. Although several schemes of combining RTDs to produce basic logic gates were already reported, an interesting direction for the use of such nanotechnology devices is the emergent computation in arrays based on nano-sized active cells. This direction is the only one to allow very high densities of cells since chemical processes such those leading to the formation of Turing patterns can be effectively used to "program" very tiny islands of active devices. Instead, classic litographic methods may fail at such very small device sizes. Then the next step would be to control the cell parameters in such a way to produce emergent computation in the array of cells and here is where our method could help.

Chapter 5
Emergence in Discrete-Time Systems: Generalized Cellular Automata

In the previous chapter we investigated the emergent computation in a Reaction-Diffusion cellular architecture. The most important result of this chapter is the *"edge of chaos"* method, an extension of the *local activity theory* [Chua, 1998] applied to the mathematical model of the cell to locate potentially emergent sub-domains in the parameter space. Such sub-domains can be directly related to various types of dynamic behaviors (e.g. passive, active and unstable or active and stable). Particularly important are the "edge of chaos" sub-regions, being proved that choosing cell parameter points within these sub-regions it is likely that emergent phenomena occur in the corresponding cellular nonlinear network. Some of these emergent phenomena may have computational relevance, e.g. in various image-processing tasks.

The Reaction-Diffusion CNN systems are continuous-state, continuous-time systems. Still, in many situations one deals with discrete-time cellular systems such as cellular automata with discrete or continuous state cells. In [Chua, 1998] the concept of *a generalized cellular automata* was introduce to describe unitary more types of discrete-time cellular systems including the binary cellular automata, the continuous state cellular automata as well as some *generalized cellular automata* without an equivalent in the previous literature.

This chapter provides some insights and practical methods for locating interesting cell parameter points such that emergent computation will occur in discrete-time cellular systems.

To identify the emergent behaviors, we have introduced a non-homogeneity measure, called *cellular disorder measure,* inspired from the *local activity theory*. Based on its temporal evolution, we are able to partition the cell parameter space into a class U (unstable-like) region, a class E (edge of chaos-like) region, and a class P (passive-like) region. The similarity with the "unstable", "edge of chaos" and "passive" domains defined precisely and applied to various reaction-diffusion CNN systems [Dogaru & Chua, 1998b,c] opens interesting perspectives for extending the theory of local activity to discrete-time cellular systems with non-linear couplings.

Another characteristic of our method is the choice of a parametric cell described trough a nonlinear equation instead of a transition table often used in cellular automata literature. Such cells were described in detail in Chapter 3. The theory of the universal piece-wise linear cells is here exploited to provide a set of parameters which can vary within certain domains while different types of dynamic behaviors can be identified and classified using the *cellular disorder measure* much like we did for the Reaction-Diffusion systems in Chapter 4.

The novelty of our approach consists in a method for *precisely* partitioning the cell *parameter space* into sub-domains via the *failure boundaries* of the piecewise linear CNN (cellular neural network) cells [Dogaru & Chua, 1999a] of a *generalized cellular automata.* Instead of exploring the rule space via statistically defined parameters (such as λ in [Langton, 1990]), or by conducting an exhaustive search over the entire set of all possible local Boolean functions, we explore a *deterministically structured* parameter space built around certain cell parameter points. These cell parameter points are chosen such that their associated cell implement "interesting" local Boolean logic functions (or transition tables). The meaning of "interesting" here is that such cells were already proved to generate some form of emergent behavior.

In section 5.1 the well-known "Game of Life" [Berlekamp *et al.*,1982] cellular automata is reconsidered to exemplify our approach and its advantages. Starting from a piecewise-linear representation of the "Game of Life" local Boolean function, and by introducing two new cell parameters that are allowed to vary continuously over a specified domain, we are able to draw a "map-like" picture consisting of planar regions which cover the cell parameter space. A total of 148 sub-domains and their failure boundaries are precisely identified and represented by colored *paving stones* in this mosaic picture (see Fig. 1), where each stone corresponds to a specific local Boolean function in cellular automata parlance. Except for the central "paving stone" representing the "Game of Life" Boolean function, all others are *mutations* uncovered by exploring the entire set of 148 sub-domains and determining their dynamic behaviors. Some of these mutations lead to interesting emergent behaviors.

In section 5.2 we report a novel class of dynamical behaviors observed in *a continuous state uncoupled* generalized cellular automata (see Chapter 2 for its simulation) with piecewise-linear (PWL) cells. Unlike the binary uncoupled cellular automata investigated in Section 5.1, this cellular system is the equivalent of *continuous-state* cellular automaton. However, both the binary and the continuous-state cells are described by an almost identical piecewise-linear equation. In fact, the equation of the cell is a *mutated version* of the one used to implement several mutations of the binary "Game of Life" cells where the binary output function is removed. Starting from an almost homogeneous initial condition self-making (*autopoietic* in the sense of [Varela *et al.*, 1974]) patterns reminiscent of simple living systems, emerge as a result of *nonlinear* coupling among cells. Similar to patterns of organization characterizing living systems, our patterns display features such as growth, maturity and death. The discovery of such patterns was possible via *mutations* in several piecewise CNN cell realizations for the Game of Life [Berlekamp *et al.*, 1982].

In the last section of this chapter we will apply again the idea of a slight *mutation* in the cell equation, obtaining emergent computation in a *coupled* generalized cellular automata. Emergent computation is here successfully applied to restore meaningful shapes from highly corrupted images (in our example the amplitude of noise is 10 times larger than of the useful signal).

Although the methods used in this chapter to locate the "interesting" regions in the parameter space are not entirely analytic they provide an effective way to detect sets of parameters responsible for emergence. The examples in this chapter have all a common root; namely, the universal CNN cell equation which implements the well-known local rule (seed) of the "Game of Life" cellular automata. However the same techniques can be applied to any other "seed". Our experience suggests that a wide variety of emergent behaviors are obtained when mutating *seeds* that already proved to posses some emergent computation properties when used for cellular automata cells. The theory in chapter 3 provides an efficient way to build nonlinear equation descriptions of any seed previously described in the form of local rules or transition tables.

5.1 Emergent phenomena in binary cellular automata

5.1.1. Introduction

It is well known that under certain circumstances, "computation" emerges in systems composed of coupled cells. Many examples, such as Conway's Game of Life [Berlekamp *et al.* ,1982], Langton's [Langton, 1986], Reggia's [Reggia *et al.*,1993] and Sipper's [Sipper, 1995] self-replicating loops, Varela's autopoietic networks [Varela *et al.*, 1974], [McMullin & Varela, 1997] and others, are based on cellular structures represented via a cellular automata [Toffoli & Margolis, 1987] formalism, where the cell model is defined by a transition table, or a set of local rules. In recent years, novel dynamic regimes called "edge of chaos" [Packard, 1988], [Langton, 1990], have been conjectured to have special significance from the perspective of complexity and emergence. Unfortunately, due to the lack of a mathematical definition, the term "edge of chaos" remains a fuzzy jargon in the cellular automata literature. It is interesting to observe that most systems reported so far to be operating in the "edge of chaos" regime have their cells described by models where some "common sense" *human knowledge had been abstracted into a set of local rules* describing the transition table. One such example is the well-known Conway's "Game of Life" [Berlekamp *et al.* ,1982]. On the other hand, if the transition tables defining the local rules are randomly generated over the entire space of local Boolean functions[1], emergent behaviors having computational potentials[2] are much less likely to occur. Therefore, the basic problem of cellular computing is to identify and invent cells such that emergent behaviors with computational potentials can occur in the corresponding cellular system.

Recently, an evolutionary approach based on Genetic Algorithms was used to invent cells (often defined in the form of transition tables or local Boolean functions) so that the resulting cellular automata would perform some prescribed task. Such examples are presented in [Sipper, 1995], [Mitchell *et al.* 1996] where the goal was to evolve CA's capable of exhibiting simple computation tasks (e.g., to decide whether the number of cells in state "one" is larger than the number of cells in state "zero"). A similar approach was reported by [Lohn, 1996] for the task of discovering structures capable of self-replication in more general cellular space models. Often, mutations and evolutionary techniques applied directly to the huge space of Boolean functions lead to excessive computation times when performed on conventional computers. This situation raises the question of whether seeking computational properties via an exhaustive search through the entire space of local transition functions can be replaced by a faster search algorithm over a smaller subset of representative local Boolean functions. In this section we show that such a confined family of functions can be precisely determined by exploring the well-defined parameter domain of a piecewise-linear CNN universal cell (chapter 3). Starting from a prescribed local Boolean function (e.g., the one defining Conway's "Game of Life"), henceforth called a *seed function*, an entire *family* of new local Boolean functions is generated. This is done by introducing a 2-parameter unfolding of the Game of Life and performing mutations around the cell parameter point representing the local Boolean "seed" function. Analytical methods can be used to locate each sub-domain in the cell's parameter space associated with the corresponding member of the function *family*.

In [de la Torre & Martin, 1997], the problem of investigating "game of life"-like cellular automata is studied by choosing a finite set (corresponding to 1296 games) of parameters (*fertility* and *survival*). These parameters differ slightly in value from those normally used to define the local rule of the "Game of Life". Investigating the relationship between the morphology of the games and two *ad-hoc* statistical measures (*density* and *activity*) the authors concluded that: "..no systematic correlation between the morphology and the parameters of the game has been found", and "no game was found with features as fascinating as those of the game of life..".

Comparing to previously mentioned works our contribution has a character of universality and it is applicable not only to the "Game of Life", but also to any cell performing a local Boolean rule. Our search parameters are not related to a specific definition of some local rule (like *survival* and *fertility*) but rather they came from the formalism of *universal CNN cells* presented in detail in Chapter 3. Moreover, by expanding

---

[1] In the case of binary cellular systems with a von Neumann neighbourhood (9 inputs per cell) there are $2^{2^9} \approx 1.3 \cdot 10^{154}$ distinct cell transition tables (i.e., Boolean functions of 9 variables).

[2] In the sense of human perception, we may consider computation as a kind of selective resonance to dynamical behaviours, which are relevant to the survival of the species.

the concepts of *local activity, local passivity* and *edge of chaos* to discrete-time systems, such as *generalized cellular automata,* we can define a quantitative measure of complexity which reveals a strong correlation between its parameter values and the morphology of the observed patterns. Using this measure it is possible to locate a well-defined region called an "edge of chaos" in the parameter space where complexity abounds. When the "game of life" is chosen to be the seed game, within its edge of chaos region some other interesting "games" were found. Moreover, in Section 5.2 it is shown that interesting complex behaviors with *continuous* (non-binary) outputs can be obtained by mutating the universal CNN cell whose parameters are centered at the "game of life" local Boolean function.

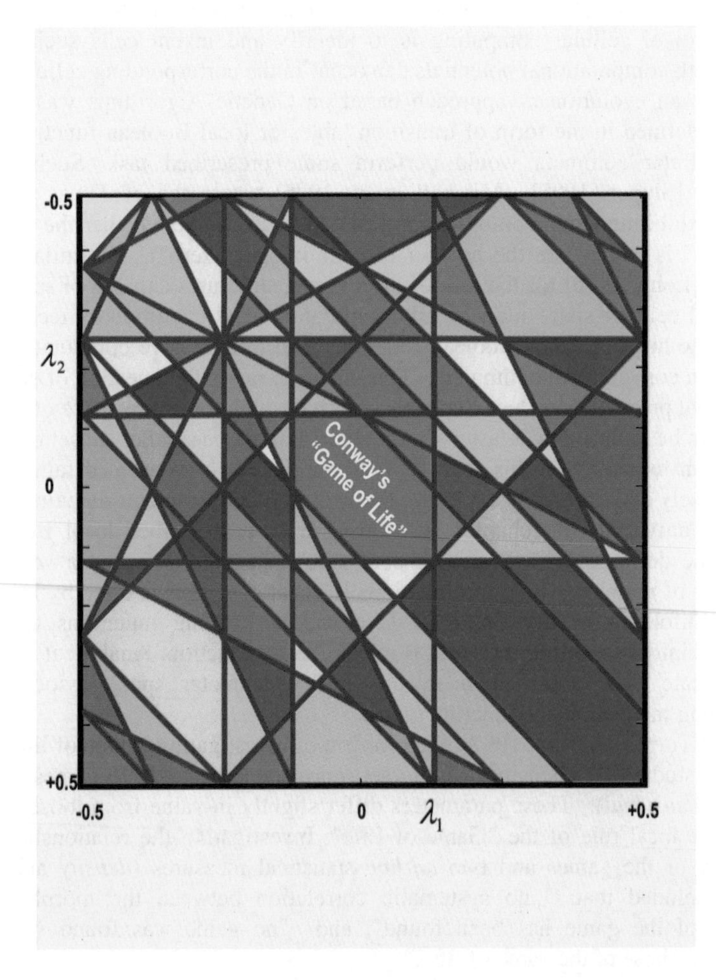

Fig.1. The failure boundaries of the unfolded 2-parameter family of piecewise-linear CNN cell realizations $w = 1.5 + 0.5u_5 - |2.5 + (1+\lambda_1)(u_7 + u_6 + u_4 + u_1) + (1+\lambda_2)(u_9 + u_8 + u_3 + u_2)|$ are shown as dark blue line segments inside the "square" parameter space $(\lambda_1, \lambda_2) = [-0.5, 0.5] \times [-0.5, 0.5]$. Each of the 148 sub-domains represented here as a colored "paving stone" corresponds to a distinct local Boolean function. Each cell parameter point (λ_1, λ_2) inside the central hexagonal "paving stone", including the point $(\lambda_1, \lambda_2) = (0, 0)$ realizes one and the same local Boolean function called the "Game of Life". It is only when one crosses the boundary of this hexagon that Conway's Life mutates into another Boolean function.

In our approach, a binary cellular automata is replaced by its functional equivalent called a *generalized cellular automata* (GCA) [Chua, 1998], which is built around a *cellular neural network* kernel. The key component of the cellular neural network is a *deterministic* dynamical system called a *cell.* In Chapter 3 it was shown that any local Boolean function admits a realization in the form of a piecewise-linear cell and the specific function is determined by a set of parameters called a *gene.* Section 5.1.2 introduces the *generalized cellular automata* model and a particular cell realization for the local Boolean function "Life". Since the cell equations are *deterministic* and defined by a piecewise-linear nonlinearity, the defining equations of the *failure boundaries* [Chua, 1998] which enclosed a sub-domain of the cell parameters space for the Boolean function "Life" can be determined analytically, as shown in Section 5.1.3. The result is a partition of the cell parameter space into sub-domains where each sub-domain corresponds to the realization of a uniquely defined local Boolean function. Such a partition is represented in this paper as a mosaic made of colored *"paving stones",* as shown in Fig. 1. The name "paving stones" is used here as a suggestive alias for cell parameter sub-domains. Since our CNN cell is designed as a realization of a particular *seed* function, there is a *paving stone* associated with this *seed* function. The remaining *"paving stones"* represent a confined set or a *family of* Boolean functions (henceforth also called mutations) built around the *seed* by mutating the cell parameters. In order to distinguish among different *paving stones,* or their associated *cell realizations*, an integer label (From 1 to 148 in our example) is attached to each stone.

It is interesting to determine how *mutations* represented by different "paving stones" within the cell parameter space determine the dynamic behavior of the corresponding generalized cellular automata. Therefore, in Section 5.1.4 we introduce a set of criteria to characterize the global behavior of the resulting generalized cellular automata. In particular, a set of 12 "*paving stones*" has been selected as representative illustrations of various qualitatively different dynamical behaviors. Snapshots of these dynamical behaviors are presented and described in detail in the same section. By applying the above cited criteria each "paving stone" can be assigned a color code (see Figs. 2-3) which reveals an organized structure in the cell parameter space.

Each paving stone, or sub-domain, is associated with one of three types of dynamical behaviors, namely a *class U* (unstable-like) behavior, a *class P* (passive-like) behavior, or a *class E* ("edge of chaos"-like) behavior. The most interesting behaviors (including the "Game of Life" and several newly discovered behaviors) correspond to "paving stones" situated in the "E" zone which is topologically a domain separating the "U" domain from the "P" domain. A similar situation had been theoretically predicted for continuous time cellular systems via the mathematically rigorous local activity theory [Chua, 1998]. A set of methods for determining various behavioral zones for different Reaction-Diffusion CNN systems, such as the "edge of chaos" region, the "unstable" region, and the "passive" region, were presented in Chapter 4.

5.1.2. Generalized Cellular Automata

This is a remainder of the notions introduced in Chapter 2. A generalized cellular automaton (GCA) is composed of a CNN lattice with an additional discrete time loop. Therefore the output y_{ij} of the CNN cell C_{ij} is sampled at the discrete time moments $t_n = n\Delta T$, $n = 1,2,...$, and then fed into the cell input at the beginning of a new cycle at t_{n+1} of the dynamical CNN evolution. The pair of integers (i, j) locates the CNN cell within a rectangular grid. The duration of each cycle of the CNN evolution is ΔT, where ΔT is chosen so that all transients had settled down and the CNN has reached a steady state before a new cycle begins. When implemented on a state-of-the art silicon chip, ΔT is approximately 10^{-9} seconds. The state space dynamics over each cycle $t_n \le t \le t_{n+1}$ in the case of a piecewise-linear (PWL) cell is characterized by the following ordinary differential equation:

$$x_{ij}(t_n) \overset{\Delta}{=} x_0,$$

$$\dot{x}_{ij}(t) = -x_{ij}(t) + \sum_{k,l \in \mathrm{N}} a_{kl}\, y_{kl}(t) + w(u_{kl}(t)),\ \ t \in (t_n, t_n + \Delta T) \qquad (1)$$

$$y_{kl}(t) = f(x_{kl}(t)) = 0.5(|x_{kl}(t)+1| - |x_{kl}(t)-1|), \qquad (2)$$

where $x_{ij}(t)$ is the *state* of cell C_{ij} at location site (i, j) at time *t*, and $w(u_{kl})$, called the *discriminant function,* is a piecewise-linear function of 9 arguments $u_{kl}(t)$ representing the cell inputs from all cells located within the sphere of influence N of the cell C_{ij}, where $u_{kl}(t) \overset{\Delta}{=} u_{kl}(t_n^+) = y_{kl}(t_n^-)$ for $t_n \le t < t_n + \Delta T$. Here ΔT is chosen greater than the time where all transients have vanished. The variables $t_n^- = \lim_{\substack{\varepsilon \to 0 \\ \varepsilon > 0}} (t_n - \varepsilon)$ and $t_n^+ = \lim_{\substack{\varepsilon \to 0 \\ \varepsilon > 0}} (t_n + \varepsilon)$ denote the *left-hand* limit and *right-hand* limit, respectively, at $t = t_n$. Observe that $u_{kl}(t)$ is in general a *square wave* whose amplitude in general jumps discontinuously from one value to another, depending on the output $y_{kl}(t)$ at $t = t_n^-$.

The additional GCA *loop* is described by:

$$u_{ij}(t_n^+) = y_{ij}(t_n^-) \qquad (3)$$

for the general case where the output is fed back to the input, which is the case throughout this chapter. When implemented via a CNN *universal chip* [Roska and Chua, 1993] endowed with the desired *discriminant function* $w(\bullet)$, equation (3) is programmed as a *single instruction*. We assume next that $x_0 = 0$, and choose a periodic CNN boundary condition [Chua, 1998] so that corresponding boundaries of the CNN grid are identified. The size of the grid is $N \times M$ and the cells are indexed by $i, j \in \{1,2, \ldots, N\} \times \{1,2, \ldots, M\}$. The distribution of cells around the central cell C_{ij} is denoted by

$$u_{kl}\big|_{kl \in \mathrm{N}} = \begin{array}{|ccc|} u_{i-1,j-1} & u_{i-1,j} & u_{i-1,j+1} \\ u_{i,j-1} & u_{i,j} & u_{i,j+1} \\ u_{i+1,j-1} & u_{i+1,j} & u_{i,+1j+1} \end{array} \equiv \begin{array}{|ccc|} u_9 & u_8 & u_7 \\ u_6 & u_5 & u_4 \\ u_3 & u_2 & u_1 \end{array} \qquad (4)$$

where the notation on the right is a simplified abbreviation. The same notation applies also for the other cell variables (state, output, and feedback coefficients a_{kl}).

Observe that there are two levels of coupling among the cells. The first level of coupling is *continuous* in time and is specified by the feedback coefficients a_{kl}. The second level of coupling is *discrete* in time since $u_{kl}(t)$ is *held constant* throughout each cycle, and is specified by adding the GCA feedback loop. The nonlinear, feed-forward discriminant function $w(u_{kl})$ is responsible for this coupling.

Uncoupled GCA

Throughout Sections 5.1 and 5.2 we consider the case of *uncoupled CNN cells*, defined in [Chua, 1998] as cells where there is no recurrent connection between the state x_{ij} of a cell and its neighbors x_{kl} (i.e. $a_{kl} \neq 0$ only if $kl \equiv ij$). Moreover, for all generalized cellular automata with *uncoupled CNN cells* we will assume $a_{ij} = 2$ throughout this paper so that the CNN cell output always converges towards a stable *binary* state (i.e., either -1 or +1) at the end of each cycle of duration ΔT [Chua, 1998]. With these constraints, our generalized cellular automaton is functionally equivalent to a binary cellular automaton except that the cell *is not defined via a transition table, or a set of rules, but via a deterministic non-linear equation with some prescribed parameters.* In this special *uncoupled* case, the dynamics of our binary GCA can be written as a *discrete time* equation in the *discrete time variable t* marking the beginning of a new cycle of duration ΔT :

Generalized Cellular Automata (GCA): *non-recurrent (uncoupled) case*

For $t = 1, 2, \ldots, \infty$

$$u_{ij}(t) = y_{ij}(t-1) \tag{5}$$

$$y_{ij}(t) = \text{sgn}(w(u_{kl}(t))), \tag{6}$$

End

When the standard CNN cell is used in an implementation, the sign function in Eq. (6) is *not* a built-in cell output function, as in a *perceptron,* but rather it is the asymptotic (steady state) response of the *continuous* time nonlinear CNN dynamics defined by Eqs. (1), and (2). However one may use different other circuit techniques to implement directly equation (6) and (5) while removing equations (1) and (2).

Equation (5) and (6) are a different way to define a *cellular automaton.* The major difference is that instead of a transition table, the nonlinear function in (6) is now used to define the cells. This approach has several important implications:

(i) First, it provides a convenient practical method for implementing local logic. As shown in Chapter 3, for many local Boolean functions of interest compact piece-wise linear representations exist which can be easily implemented as electronic circuits. Moreover it is guaranteed that such a nonlinear representation exists for any arbitrary transition function.

(ii) Second, it is much easier to make a change in the transition function (6). Indeed, instead of modifying an entire table, one could simply tune one or more (gene) parameters.

Although in equations (5) and (6) above the case of *binary* states is considered, by removing the function sign one can easily get the representation of a cellular automaton with a continuum of states as used in Sections 5.3. and 5.4. Finite states cellular automata (with a number of states greater than 2) can be also easily obtained by replacing the sign function with a proper quantifier function in (6).

The specific cell for realizing a particular Boolean function depends on the choice of the discriminant function $w(\bullet)$ whose parameters can be determined by one of the methods described in Chapter 3. A piecewise-linear CNN cell has an important advantage over the smooth CNN cell because it allows an efficient determination of the *failure boundaries* between sub-domains (henceforth called *paving stones*) representing different local Boolean functions.

In addition to the special case of a generalized cellular automata with *uncoupled* (i.e. non-recurrent state) *CNN cells,* which have always an *equivalent* cellular automata, in Section 5.3 we will consider the case of a *generalized cellular automata* with *a coupled* (i.e. state recurrent) *CNN cell.* This is a much more general construct (having a *continuum* of states) than any binary or multi-state cellular automata [Chua, 1998]. In this case, Eq. (6) no longer holds, and must be replaced by the original output equation (2).

Generalized Cellular Automata for the "Game of Life"

Using the notation in (4) the following 3 discriminant functions are all valid cell realizations of the local Boolean function associated with the Game of Life:

Realization 1:

$$w = 1.5 + 0.5u_5 - \left| 2.5 + (u_1 + u_2 + u_3 + u_4 + u_6 + u_7 + u_8 + u_9) \right| \tag{7}$$

Realization 2:

$$w = 1.5 - \left| 2.5 + (u_1 + u_2 + u_3 + u_4 + u_6 + u_7 + u_8 + u_9) + 0.5u_5 \right| \tag{8}$$

Realization 3:

$$w = 1 + 0.5u_5 - \left| 3 - (u_1 + u_2 + u_3 + u_4 + u_6 + u_7 + u_8 + u_9) + 0.5u_5 \right| \tag{9}$$

The first "Life" realization was obtained by training a piecewise-linear structure called a rectification neural network [Dogaru & Chua, 1998a], and the second is the result of applying a systematic design procedure presented in Chapter 3 which is valid for arbitrary local Boolean functions. The third "Life" realization is adapted from results independently proposed in [Mar & St. Denis, 1996] where the set of linguistic rules describing the "Life" local Boolean function were reinterpreted from the perspective of fuzzy logic (in this case the two binary states are {0,1}). Among these realizations, (8) is the simplest since it requires a minimum number of arithmetical operations. However, one can see that (7) has the same arithmetic complexity; what differs is the location of u_5 within the formula. In this paper we will choose the first realization and introduce an unfolding of the "Game of Life" in the $\lambda_1 - \lambda_2$ parameter plane:

2-Parameter Unfolding of the local Boolean function "Game of Life" :

$$w = 1.5 + 0.5u_5 - \left| 2.5 + (1+\lambda_1)(u_7 + u_6 + u_4 + u_1) + (1+\lambda_2)(u_9 + u_8 + u_3 + u_2) \right| \tag{10}$$

Comparing (10) with (7), we see that the local Boolean *seed* function "Life" is recovered at the cell parameter point $(\lambda_1 = 0, \lambda_2 = 0)$. Because of the strong nonlinear nature of the sign function in (6), the domain of the cell parameters for realizing the "Life" Boolean function is much larger than the single parameter point (0,0). In fact it covers an area represented as a hexagonal *paving stone* located at the center of the mosaic in Fig. 1. It can be shown analytically via the failure boundary method in [Chua, 1998] that *all* points (λ_1, λ_2) located within this "Life" paving stone give valid realizations for "Life". Hence the above "unfolding" uncovers a *continuum* of *distinct* piecewise-linear CNN cells, each one being a valid realization of Conway's Game of Life.

In the next section a method is given to find the boundaries between various cell parameter sub-domains of Eq. (10) and therefore identify the *paving stones* of the Boolean function *family* formed around the *seed* by mutating the cell parameters λ_1 and λ_2 over any specified domain of interest. For example, in the next we will choose the following domain $(\lambda_1, \lambda_2) = [-0.5, 0.5] \times [-0.5, 0.5]$ as our domain of interest. For clarity, we have restricted ourselves to a two-dimensional unfolding. However, one can perform a similar unfolding in higher dimensions, for example, by replacing the constant 2.5 in (10) by $2.5 + \lambda_3$, thereby adding a new dimension to the parameter space.

5.1.3. Failure boundaries and *paving stones* in the cell's parameter space

In order to determine the *failure boundaries* let us consider the following reasoning. Equation (6) gives a realization of some *Boolean* function, if and only if, for any of the 2^9 possible combinations of binary inputs, the discriminant function gives either $w > 0$, or $w < 0$. Consequently, if there is at least one combination of binary inputs u_{kl} such that $w = 0$, the cell output in that case will be 0 instead of -1 or +1.

Therefore, the *failure boundaries* (separating two local Boolean functions) in the cell parameter space can be easily determined by solving $w(\lambda_1, \lambda_2) = 0$ for each of the 512 possible combinations of binary inputs. For each combination of inputs (labeled J, $J = 1, 2, \ldots, 512$), the result is a compact manifold representing a set of points $\Lambda_J = \{(\lambda_1, \lambda_2) \mid w(\lambda_1, \lambda_2, u_1^J, u_2^J, \ldots, u_9^J) = 0\}$, which belongs to one or more failure boundaries. The entire set of failure boundaries is thus defined by $\Lambda = \bigcup_{J=1}^{512} \Lambda_J$.

As long as the prescribed cell has a piecewise-linear realization, the equations $w(\lambda_1, \lambda_2) = 0$ can be easily solved, the results being linear segments when only two parameters are considered, and hyper-planes of dimension $m-1$ in the general case of m parameters. The above analytical approach makes it clear why the choice of a piecewise-linear cell realization is superior to other nonlinear descriptions of the cell.

Let use the above procedure to determine the failure boundaries of Eq. (10), by determining Λ_J where J is chosen in this illustration to correspond to the case when all inputs are equal to -1. Substituting $u_1 = u_2 = \ldots = u_9 = -1$ in (10), the equation $w(\lambda_1, \lambda_2) = 0$ can be rewritten as $1.5 + (-0.5) - |\, 2.5 + (1 + \lambda_1)(-4) + (1 + \lambda_2)(-4) \,| = 0$, which reduces to:

$$1 = |\, 5.5 + 4(\lambda_1 + \lambda_2) \,| \tag{11}$$

To solve this equation, the following two cases must be considered:

Case 1: $(\lambda_1 + \lambda_2) > \dfrac{-11}{8}$ (positive argument of the absolute value function);

From (11) it follows that $(\lambda_1 + \lambda_2) = \frac{-9}{8}$ which is a valid solution since $\frac{-9}{8} > \frac{-11}{8}$.

Case 2: $(\lambda_1 + \lambda_2) \leq \frac{-11}{8}$ (negative argument of the absolute value function);

From (11) it follows that $(\lambda_1 + \lambda_2) = \frac{-13}{8}$ which is again a valid solution since $\frac{-13}{8} < \frac{-11}{8}$. Therefore, the subset Λ_1 of the failure boundary is a set formed by the following parallel lines:

$$\lambda_2 = \frac{-9}{8} - \lambda_1 \text{, and } \lambda_2 = \frac{-13}{8} - \lambda_1$$

Similarly, the remaining 511 subsets of the set of failure boundaries can be determined, the results being presented in Fig. 1, where the failure boundaries are shown in dark blue color for enhanced visibility. Within the domain of investigation $(\lambda_1, \lambda_2) = [-0.5, 0.5] \times [-0.5, 0.5]$ these failure boundaries determine a partition composed of 148 *"paving stones"*, each of which is associated with a distinct local Boolean function. Different colors were chosen in Fig.1 to identify the "paving stones", or sub-domains, of the two-dimensional $\lambda_1 - \lambda_2$ cell parameter space. For convenience, each paving stone in Fig.1 is assigned an integer label (from 1 to 148), which is different from the corresponding truth table ID code [Chua, 1998], the latter requiring 155 digits.

Observe that each of the 148 Boolean functions is uniquely specified via (10) and the *2 real parameters* (λ_1, λ_2) corresponding to each cell parameter point within the "paving stone" associated with that function. For some representative local Boolean functions their integer labels are indicated in the corresponding "paving stones" in Fig. 2. A complete table with the representative cell parameter points for each of the 148 functions had been determined so that one can easily implement the associated cell described by (10). An excerpt from this table containing only the 11 local Boolean functions chosen in Section 5.1.4 for illustrations is shown in Table 1.

Table 1. Cell parameter points for a set of 11 local Boolean functions from a *familiy* of 148 functions generated by mutations of the piecewise-linear cell (10) which represents an unfolding of the Conway's game of life.

Numerical label	5	12	17	23	32	37	45	55	61	114	122
$128\lambda_1$	-1	-31	3	-25	-15	17	35	37	-45	-33	-47
$128\lambda_2$	-29	-31	-15	-5	19	17	-63	-61	-57	3	17

The value of λ_1 and λ_2 corresponding to each of the 11 "mutated" cells in Table 1 is obtained simply by dividing the two members in each column by 128.

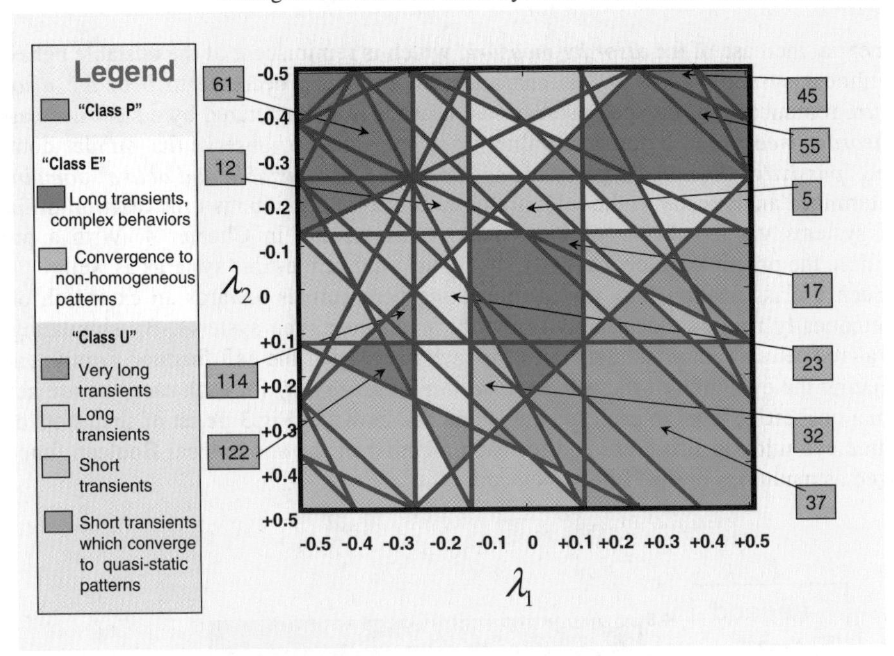

Fig.2. A detailed view of behavioral domains with 11 representative qualitatively distinct dynamic behaviors described in the text below, and identified here by numerical labels attached to their corresponding "paving stones" within the cell parameter space. Failure boundaries are enhanced for visibility and shown in dark blue. The color code of the behavioral sub-domains is defined in the legend.

5.1.4. Locating emergent behaviors and their dynamics

The failure boundaries in Fig. 1 indicate that a great variety of distinct local Boolean functions can be realized by mutating two real parameters around the central point $(\lambda_1, \lambda_2) = (0,0)$, which corresponds to a piecewise-linear realization of the local-Boolean *seed* function "Life". This diversity is evident from the many colors assigned to the "paving stones" in Fig. 1. However, mosaics such as Fig. 1 can capture only the *static* aspects of the CNN cell. It is of great interest to label each "paving stone" with additional information concerning the qualitative aspects of the emergent *dynamics* exhibited in the generalized cellular automata defined by Eqs. (5) and (6), where the cell parameters are situated within that "paving stone". For example, it is well known that the local-Boolean *seed* function "Life" is capable of universal computation and self-reproduction [Berlekamp *et al.* ,1982]. Are there any similar behaviors to be found in the other members of the 148-member family of functions generated by parameter mutations? What other types of dynamic behaviors can exhibit the generalized cellular automata corresponding to members of this family? Can the parameter space be partitioned into regions, each one exhibiting the same dynamical behavior?

To answer these questions, we have adopted the following strategy: First, we define a non-homogeneity measure *m(t)* of the instantaneous state $x_{ij}(t)$ at time *t* of a generalized cellular automata, henceforth called its *cellular disorder measure.* Based on the temporal evolution of the *disorder measure m(t),* we assign the corresponding dynamic behaviors to one of three distinct qualitative classes; namely, an "Unstable-like" or *"U" class*, a "Passive-like" or *"P" class,* and an "Edge of chaos"-like or *"E" class*. Class U is characterized by an exponential increase of the *disorder measure*, which is reminiscent of the unstable behaviors in nonlinear systems. Class "P" is characterized by a fast decrease to 0 of the *disorder measure* reminiscent of damped oscillations. Class E is characterized by a slow decrease of the *disorder measure* to a non-zero value. It is interesting to observe that similar domains (called *"passivity" domain*, *"edge of chaos" domain*, and *"unstable and active" domain* can be determined analytically in the cell parameter space for continuous-time reaction-diffusion CNN systems via the theory of local activity, as we did in Chapter 4. With a proper definition, the principle of local activity should act on discrete-time systems as well.

Such a classification may reveal interesting perspectives towards an extension of the mathematically rigorous local activity theory to discrete time systems. By simulating the generalized cellular automata dynamics for each member of the 148-function family, and by calculating the evolution of the *cellular disorder measure m(t)* for each case, we are able to assign a qualitative label to each "paving stone" as shown in Fig.3. A set of snapshots of the dynamic evolution is also presented for each member of the above local Boolean functions realized as mutations of the "Life" CNN cell.

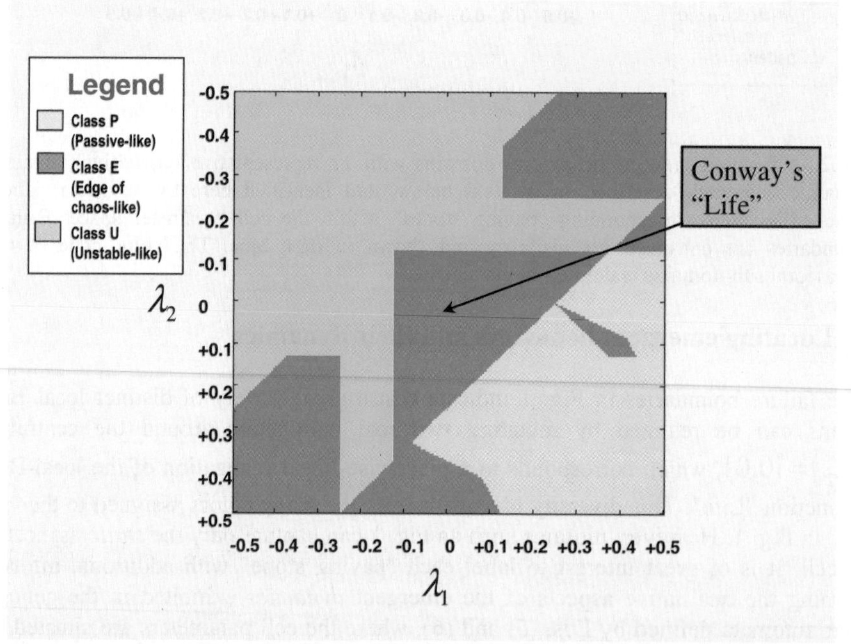

Fig. 3. Behavioral domains in the cell parameter domain; Color *blue* corresponds to cell parameter domains leading to a passive-like Class P "passive-like" behavior of the corresponding generalized cellular automata. Domains exhibiting an *active-like* behavior are split into two sub-categories: an "edge of chaos"-like domain E depicted in *red*, and an unstable-like domain U depicted *green*. Observe that the "edge of chaos"-like domain includes all cell parameters leading to the realization of the local Boolean function called "Game of Life".

The dynamic simulations of the CNN evolution are shown in Figs. 5-7 for a CNN grid having 50 × 80 cells (with two exceptions, shown in Fig.6(d) and Fig.6(e), where the CNN grid has 80 x 120 cells). In each simulation an *initial state of the generalized cellular automata* is chosen to be a matrix of outputs $\mathbf{Y} = \{y_{ij}(0)\}_{i,j\in\{1,2,\ldots,50\}\times\{1,2,\ldots,80\}}$, and each new iteration corresponds to incrementing the discrete time variable t by 1. The CNN outputs at iteration t are displayed as snapshots depicted here in a "film-like" frame sequence to dramatize its dynamic evolution in time (Figs. 5-7). The cell output is coded so that a negative output value of -1 is represented by the color *light yellow*, while the color *red* is used to code a positive output value of +1. The time variable *t* associated with each snapshot is printed in its upper left corner.

Cellular Disorder Measure

To measure the cellular disorder, or non-homogeneity, quantitatively of an evolving pattern within a cellular grid at any discrete-time moment *t* we have developed the following algorithm:

For each cell in the grid we assign an integer code J equal in value to the decimal representation of the 9 bits formed by the outputs of all cells within its neighborhood and ordered as in equation (4). It follows that each cell is assigned to exactly one of the $2^9 = 512$ possible codes: 0,1,2,...,511. Then count the number of cells n_J assigned to each code $J = 0,1,...,511$. It follows that the total number of cells *N* in the grid is equal to $N = \sum_{J=0}^{511} n_J$. Observe that there may be codes assigned to no cell, i.e., in general, there exists J such that $n_J = 0$. Let us define the *cellular disorder measure "m"* to be the normalized sum of 512 local probabilities $p_J = \frac{n_J}{N}$; namely,

$$m = \frac{-1}{9} \sum_{J=0}^{511} p_J \log_2 p_J \tag{12}$$

It can be easily checked that $0 \le m \le 1$. The case $m = 0$ corresponds uniquely to a perfectly homogeneous pattern within the CNN grid. Indeed, if all cells have the same output value (-1 or +1), there is only one binary code (0 or 511) uniformly distributed to all cells. Therefore there is only one term in (12) with $p_J = 1$, so it follows that $m = 0$. If at least onc cell within the grid violates the homogeneity property there will be more than 1 codes represented, each with $p_J > 0$, and hence $m > 0$. Since a homogeneous pattern was demonstrated in [Chua, 1998] to be the result of *local passivity*, a *cellular disorder measure* equal to 0 can be interpreted as a measure of passivity-like behaviors in our discrete-time case. The extreme value $m = 1$ corresponds to the highest degree of disorder or non-homogeneity. In this case, all codes are equally represented by $p_J = \frac{1}{512}$ among cell neighborhoods so that $\sum_{J=0}^{511} p_J \log_2 p_J = -\log_2 512 = -9$, and hence *m=1* in this case.

We should stress on a subtle but important difference between our definition of *cellular disorder measure* and the conventional definition of cell *entropy* used in other works. In the latter case the neighborhood is ignored and the entropy is computed as $h = p_1 \log_2 p_1 + p_{-1} \log_2 p_{-1}$ where p_1 is the fraction of cells having the outputs +1 and $p_{-1} = 1 - p_1$ is the remaining fraction of cells having -1 outputs. Therefore, both a highly irregular chaotic pattern and a very regular "chess-board"-like pattern may assume the same maximum cell entropy $h = 1$, even though the chess board pattern is a highly regular one with only 2 distinct cell neighborhoods. Using our non-homogeneity measure the chaotic pattern will give $m \approx 1$, while the chess board pattern will give $m = 1/9$ indicating a much more ordered structure. Thus, the inclusion of neighboring cells is essential for measuring the degree of order, or disorder, over the entire CNN grid.

The definition (12) can be easily extended from binary to multi-state case. Let us consider the number of distinct states 2^q. The only change is in the coding of each potential cell neighborhood where not only 1 but q bits are now allocated to each of the 9 positions. The result is a $9q$ binary word coding up to 2^{9q} possible cell neighborhoods. For example, one may want to evaluate the magnitude of cellular disorder for the array output of continuous state cellular automata. The first operation will be to quantify the outputs with a reasonable number of bits, say $q=8$. Then the magnitude of cellular disorder is evaluated with the generalized formula:

$$\boxed{m = \frac{-1}{9q} \sum_{J=0}^{2^{9q}-1} p_J \log_2 p_J} \qquad (12')$$

Three Qualitative Classes of Dynamic Behaviors

By plotting the temporal evolution of the *cellular disorder measure m(t)* one can identify the following three major classes of dynamic behaviors by examining the increase, or decrease, in the disorder measure *m(t)* with respect to time over a sufficiently long interval.

Class P Behavior

"Passive-like" or "P" behavior is associated with an initially *rapid decrease* in *m(t)* as a function of time, followed by a convergence towards $m = 0$. One such example is the time evolution of the local Boolean function "37" depicted in Fig. 4.

Class E Behavior

"Edge of chaos"-like or "E" behavior is associated with a general tendency of *a decrease* in the local maxima of *m(t)* as a function of time. In this case, the evolution of *m(t)* is, in general, not a monotonic function of time. However, asymptotically the *cellular disorder measure m(t)* will converge in this case to either a constant (dc) steady state value, or to relatively small oscillatory fluctuations (as shown in Fig. 4 for the local Boolean functions "32","17","45", and "55".)

Class U Behavior

An "Unstable-like" or "U" behavior is associated with a *general tendency of an increase* in the local maxima of *m(t)* as a function of time assuming that the initial disorder measure *m(0)* is relatively small so that the increasing process will not be limited from the beginning. As in the case of the "E" behavior, the increasing process is non-monotonic in general. However, asymptotically the disorder measure *m(t)* will tend to either a constant value, or to some small oscillatory fluctuations such as that shown in Fig. 4 for the local Boolean functions "5","12","61","114", and "23". For convenience, we have used a logarithmic time scale in Fig. 4.

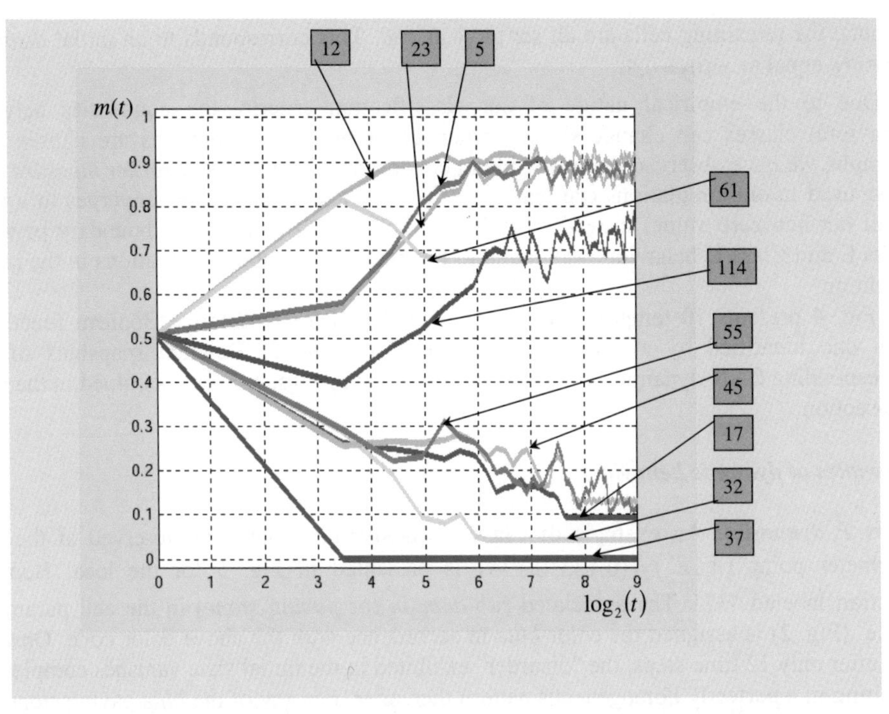

Fig. 4. Time evolution of the *cellular disorder measure m(t)* plotted from a set of 10 representative cell realizations. A numerical label is attached to each function *m(t)* over $2^9 = 512$ time steps. Observe that *m(t)* can decrease rapidly to 0, indicating a "passive-like" Class P behavior as in the case of cell realization "37", or it can decrease slowly (cell realizations "32", "17", "45", "55") indicating an "edge of chaos"-like Class E behavior, or it can eventually increase above *m(0)* (cell realizations "114", "5", "23", "12", "61") indicating an "unstable like" Class U behavior.

An interesting feature of the Class U behavior which makes it easy to identify is that starting from a very small cellular disorder measure *m(0)* (such as the one shown for all simulations Fig.7), and after a small period of time, the entire grid is filled with a highly non-homogeneous pattern characterized by a large *m*. Since Class P behaviors are even easier to detect, it follows that if a behavior is found to be neither Class U or Class P, then it should be classified as Class E. This simple criterion was used to fill in the paving stones in Fig. 3 with colored labels associated with the above three classes of behaviors of *m(t)*. Following the color code previously used in Chapter 4, we have used the color *blue* to label a Class P behavior, the color *green* to label a Class U behavior, and the color *red* for Class E behavior.

Taking into consideration the finer features and additional results from dynamic CNN simulations, many more subclasses have been identified within each class, and depicted in Fig. 2. A detailed description of each sub-class is given in the following sub-section where several representative dynamic behaviors are discussed.

It follows from the above classification criteria that one should choose the initial state so that the value of $m(0)$ reflects a compromise. This is a compromise between a relatively small *disorder measure* (required for identifying Class U behaviors), and a sufficiently large *disorder measure* which is required for detecting those decreasing tendencies which are characteristics of Class E and Class P behaviors. In our example such a goal was achieved by choosing a 51×51 CNN grid, where except for a small 31×31 square (located in the middle of the grid) with cells having a random equally-probable distribution of -1 and +1 outputs, the remaining cells are all set to -1 at $t=0$. This corresponds to an initial *disorder measure* equal to $m(0) \approx 0.5$.

Due to the empirical nature of our classification criteria, the boundaries between behavioral classes can change slightly when different initial conditions are chosen. For example, we have observed that using initial conditions with a higher *disorder measure* than those used in our simulations can lead to a "weak" Class E (where m converges to a very small but non zero value) instead of a Class P behavior. However, the boundary between Class E and Class U behaviors was found to be much more stable to variations in the initial condition.

Fig. 4 presents 10 temporal evolutions of m for 10 representative Boolean functions, each one identified by a corresponding label from Fig. 3. Detailed snapshots of the corresponding CNN dynamic evolutions are presented in Figs. 5-7 and examined in the next sub-section.

Examples of dynamic behaviors

Class P dynamics: An example of Class P (passive-like) behavior observed at the cell parameter point $(\lambda_1, \lambda_2) = (0.133, 0.133)$ is presented in Fig. 5 for the local Boolean function labeled "37". The associated *sub-domain* (or *paving stone*) in the cell parameter space (Fig. 2) is assigned the color *blue* in accordance with the above color code. Observe that after only 12 time-steps, the "disorder" exhibited in the initial state vanishes completely, resulting in a perfectly homogeneous pattern thereafter. For any of the *blue* paving stones in Fig. 2 a similar behavior is observed. As shown in Fig. 4, the *cellular disorder measure* in this case decreases rapidly to 0.

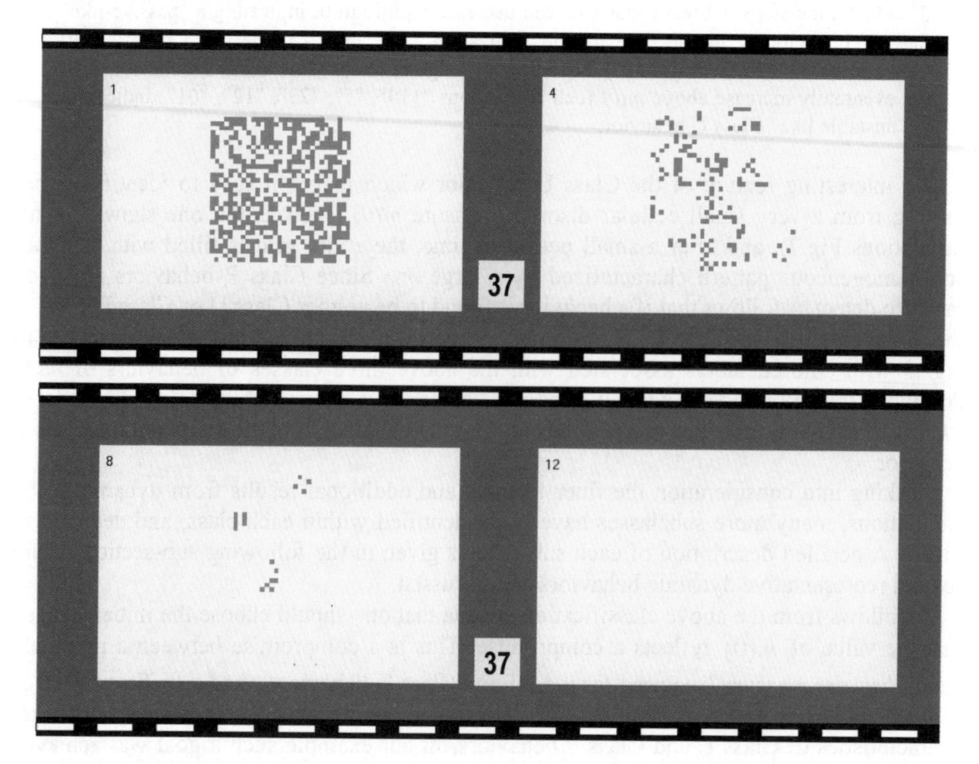

Fig. 5. Snapshots of dynamic evolution from cell realization "37" illustrating a typical "passive-like" Class P behavior.

Class E dynamics: Observe from Fig. 2 that depending on the qualitative aspects of the dynamics, two sub-classes of class E ("edge of chaos"-like) behavior can be distinguished:

- Sub-class 1 corresponds to those *paving stones* colored with *magenta*. Its characteristic is a relatively slow convergence towards a non-homogeneous static (fixed-point) or dynamic (low-period oscillations) pattern having a small *disorder measure.* An example belonging to this subclass is shown in Fig. 6(a) for the local Boolean function labeled "32". Observe that compared to the Class P "passive-like" behavior described above, 44 time steps are now required, instead of 13 time steps, to reach steady state. As shown in Fig. 4, this steady state corresponds to a non-homogeneous static pattern with a small *disorder measure,* $m \approx 0.05$. From a *computation* perspective, such behaviors can be exploited to extract certain features from an initial pattern. It represents a potentially meaningful transformation of an initial state pattern, and is functionally equivalent to the class of "Information Computation" dynamic behaviors observed from CNNs operating in the "edge of chaos domain" [Dogaru & Chua, 1998b,c].

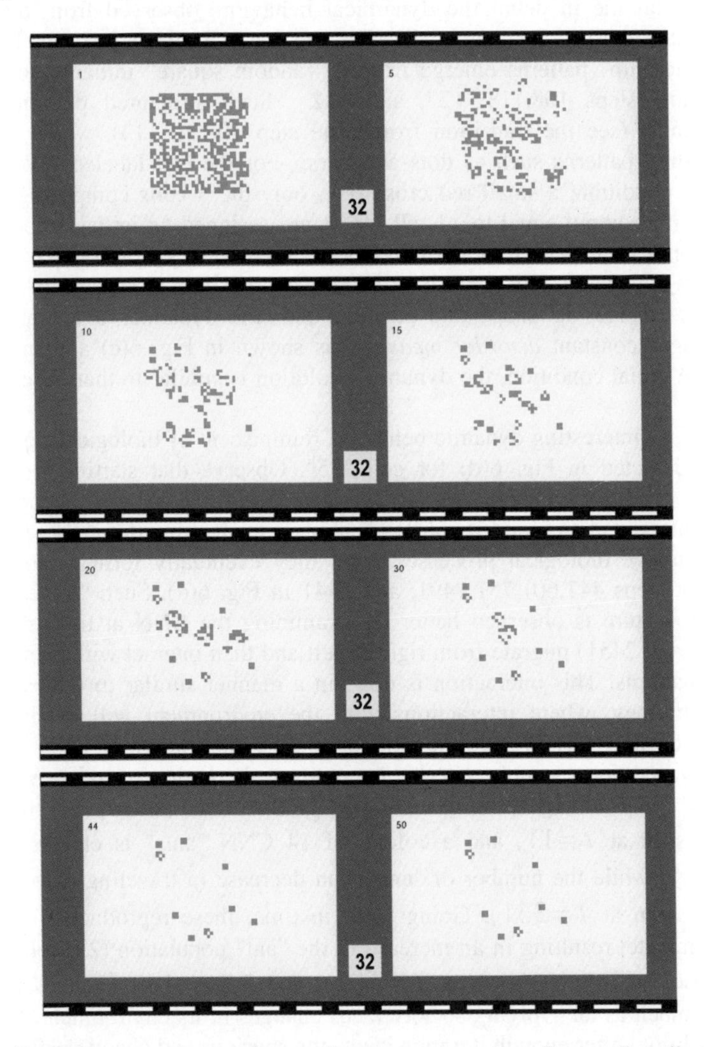

Fig. 6(a). Evolution towards a static, non-homogeneous pattern from cell realization "32". Information computation functions such as "corner detection", or "edge detection", are behaviorally similar to this pattern of evolution.

- Sub-class 2, which is reminiscent of the "Class IV" cellular automata proposed in [Wolfram, 1984] or of the "edge of chaos" proposed in [Langton, 1990], corresponds to those *paving stones* colored in *red.* Its characteristic is long transients (with a decreasing tendency in *m*) and the emergence of regular interacting patterns. Observe that the central paving stone associated with the "Life" function belongs to this subclass. Since the dynamic behaviors of "Life" are well known, in this paper we will illustrate only those dynamic behaviors resulting from *mutations of* the *seed* function "Life"; namely, the Boolean functions labeled "122", "17", "55", and "45". The positions of the associated paving stones are identified in Fig. 2 and representative snapshots of the dynamics of their corresponding generalized cellular automata are displayed in Fig. 6(b)-(e). Observe from Fig. 4 that for three of these cells, the rate of decrease in *m* tends to be slower and the average disorder measure at steady state (after a sufficiently large number of iterations) tends to be larger ($m \approx 0.1 \div 0.15$) than those from sub-class 1.

Let now examine in detail the dynamical behaviors observed from the generalized cellular automata defined by these cells: For the local Boolean function labeled "122", "travelling spaceship" patterns emerge from a "random square" initial state, as shown in Fig.6(b) at time steps 108,113, 122, and 312. Such structured patterns evolve and sometimes vanish (see the transition from time step 108 to 113), while surrounded by simpler travelling patterns such as dots and bars. For the cell labeled "17", instead of a random initial condition, a small red cross (i.e., only the 5 cells composing the cross are assigned an initial output equal to +1, all others are assigned an initial output equal to -1) initial condition was chosen. Instead of an indefinite growth (typical for Class U "unstable-like" behaviors), such an initial condition typically evolves into a "spaceship"-like organized pattern, which changes its shape and position until the dynamics eventually settles into patterns having a constant *disorder measure,* as shown in Fig. 6(c) and in Fig.4. For a random square initial condition, the dynamic evolution is similar to that observed from cell "122".

Another very interesting dynamic behavior, reminiscent of biological reproduction and evolution, is depicted in Fig. 6(d) for cell "55". Observe that starting from a "random square" initial condition, a small regular 2-legged insect-like pattern (henceforth called a CNN "*ant*") emerged at *t=11* in Fig. 6(d). Then, more such structures emerge, similar to the typical reproductive biological processes, until they eventually form "colonies", such as those shown at steps 441,601,751,1491, and 2441 in Fig. 6(d). Such "travelling colonies" (the travelling feature is observed better by examining the CNN at time steps 1491 and 1501, or 2441 and 2451) migrate from right to left and then interact with a small number of unstructured patterns. This interaction is done in a manner similar to typical evolutionary processes in biology, where interactions with the environment will eventually produce mutations and changes. As a result, the number and the distribution of CNN "ants" within a colony may vary. For example, at the beginning where the reproductive tendency is dominant, new "ants" emerge from of a completely disordered pattern (1 well formed CNN "ant" can be seen at $t=11$, and a colony of 14 CNN "ants" is clearly observable at $t=51$). After a while the number of "ants" can decrease (a traveling colony of 11 CNN "ants" can be seen at $t=331$). Going forth in time, these reproductive processes will eventually dominate, resulting in an increase in the "ant" population (23 well formed CNN "ants" compose the colony shown at $t=601$), and followed by changes in the colony configuration much as an evolving society faces changes in its environment. Depending on the initial condition, after enough iteration steps, the spurious and chaotic patterns seen from the earlier time frames in Fig. 6(d) will eventually vanish and evolve into a stable configuration of west-travelling CNN "ants" (shown in Fig. 6(d), at time steps 2441, and

2451). The larger the CNN grid is, the longer is the duration of the transient from a completely disordered pattern to a stable travelling colony of CNN "ants". Note that the very long transient in this case is reflected in Fig. 4 as well.

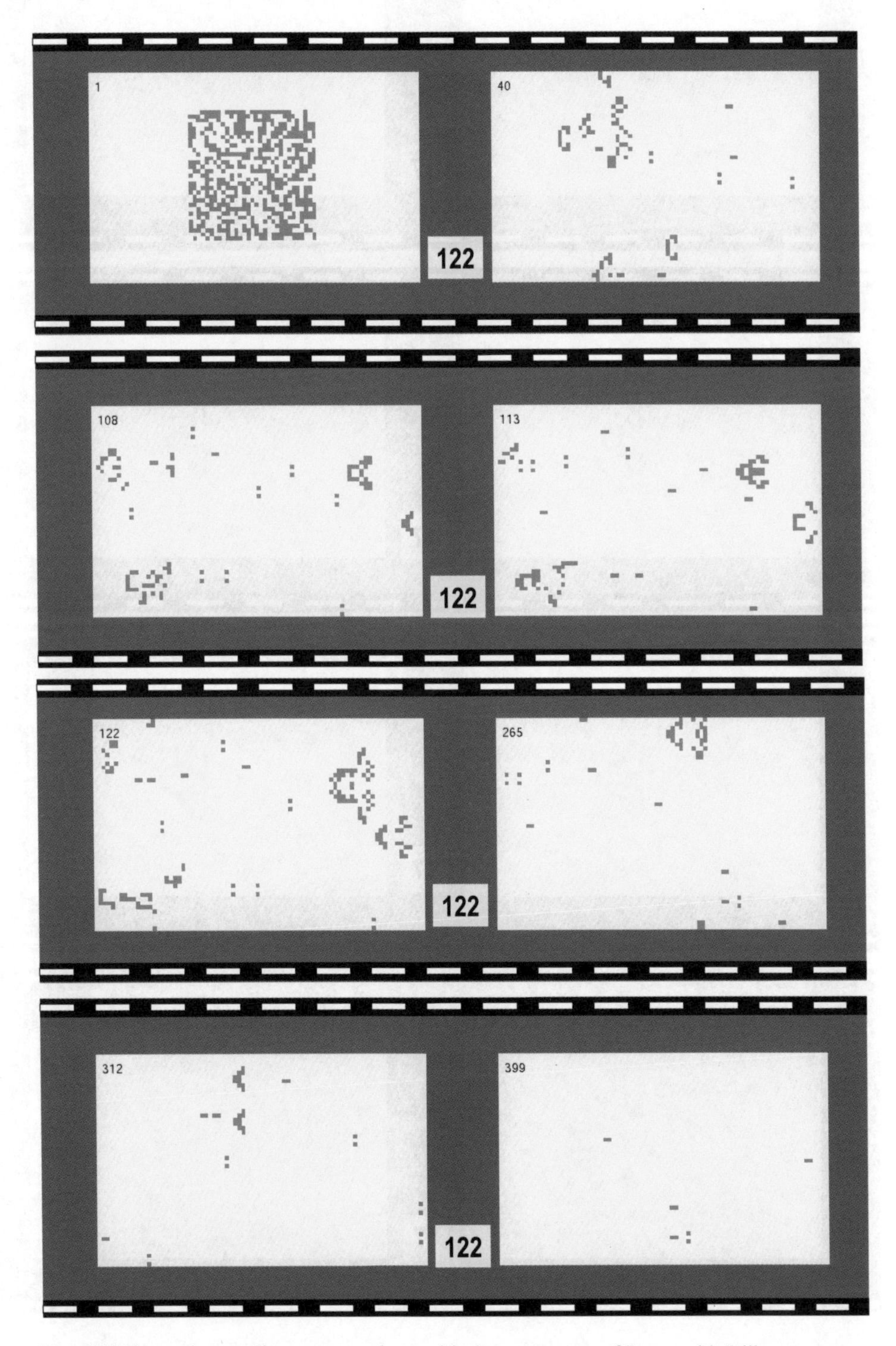

Fig. 6(b) Dynamic traveling patterns along with the emergence of "space ship"-like structures observed from cell realization "122".

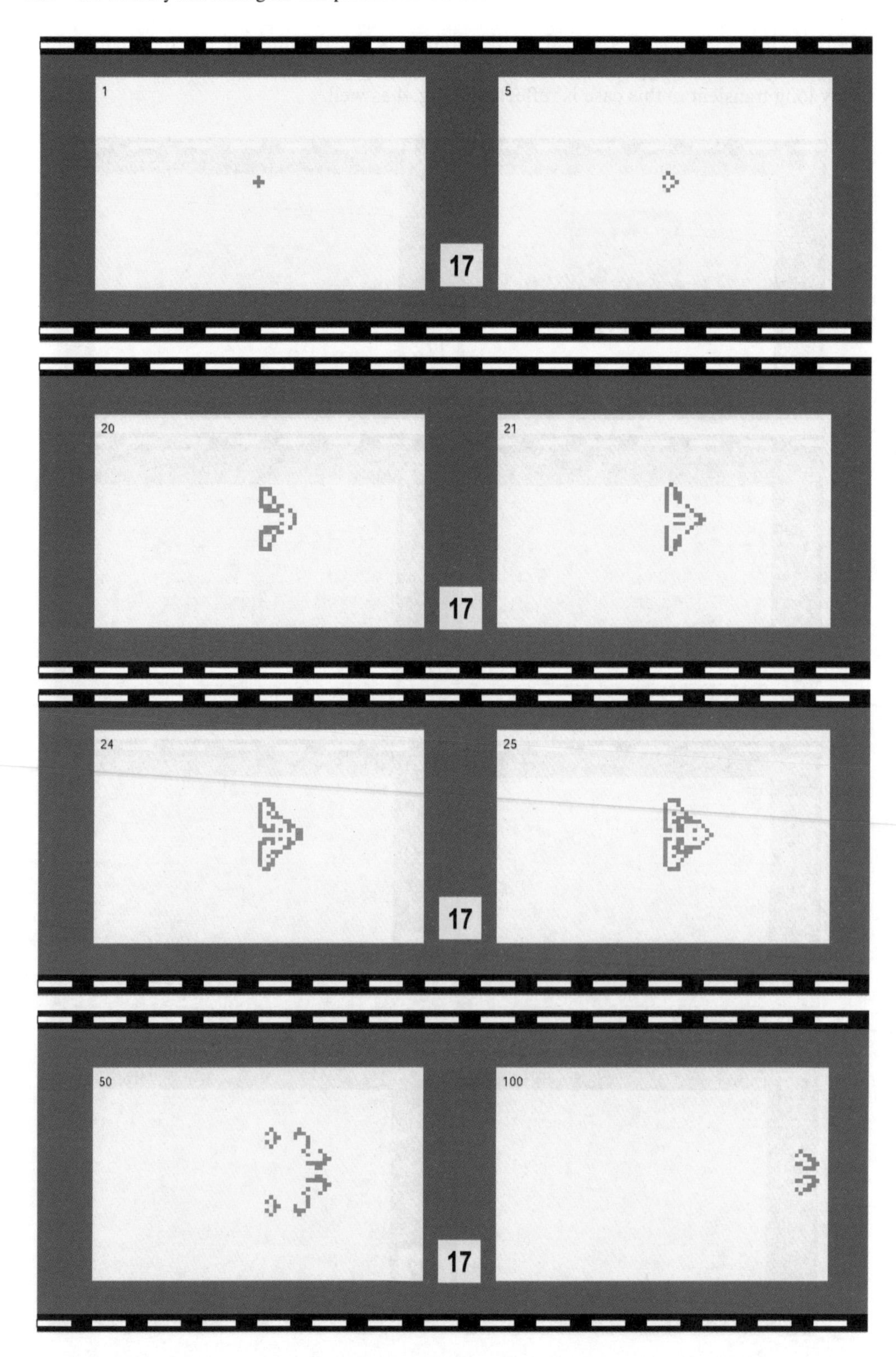

Fig. 6(c) Emergence and metamorphosis of "space ship" patterns from a small "red cross" initial state observed from cell realization "17".

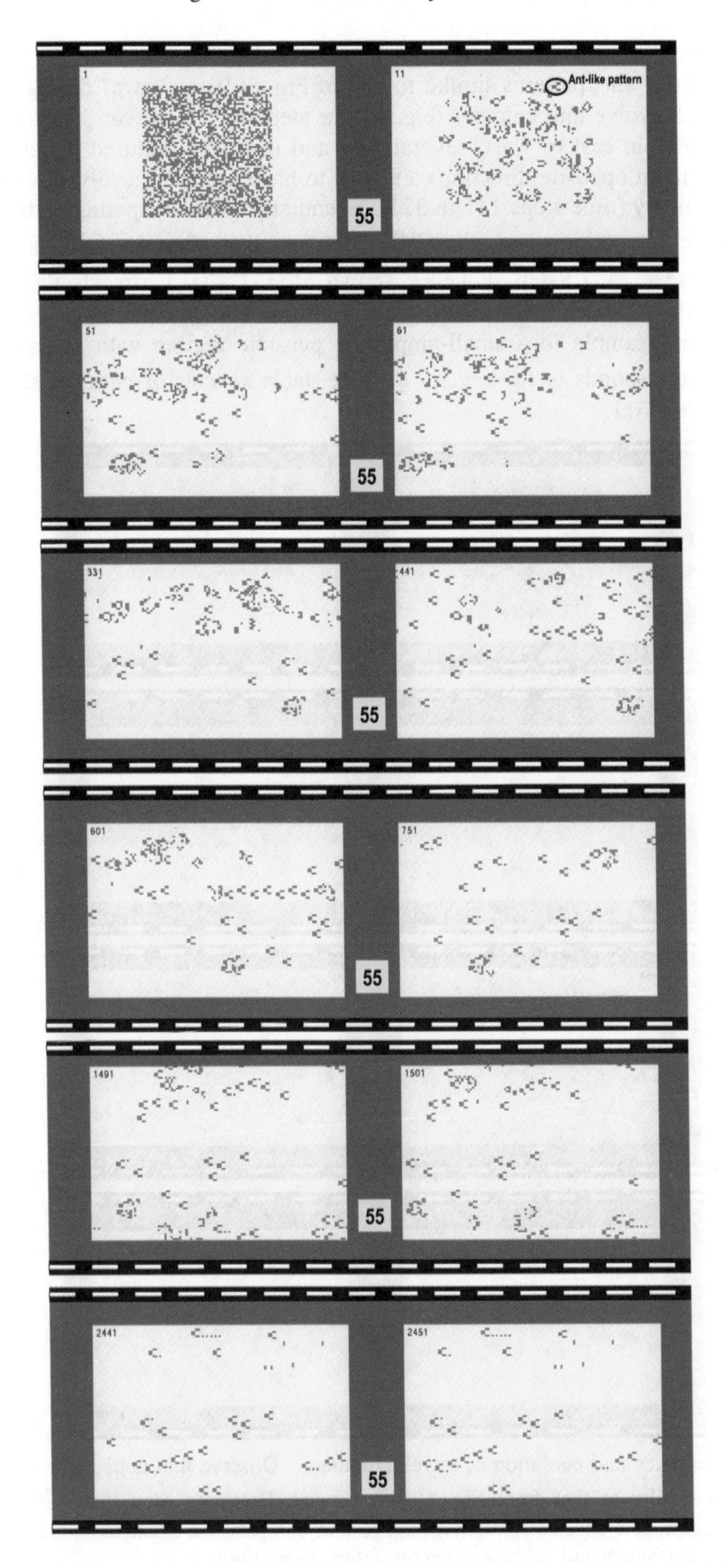

Fig. 6(d) Emergence and evolution of an "ant colony" from a "random square" initial state observed from cell realization "55".

An even more interesting behavior can be observed in Fig. 6(e) for the Boolean cell "45". In this case, "ant" patterns similar to that of Fig. 6(d) are "born" during the first 150 cycles, and which evolve into colonies (e.g. at time step 141). However, besides these CNN "ants" travelling from east to west, several new and distinct structured patterns (many of them travelling in an opposite direction) are seen to have emerged, evolved, and interacted with the "ant" colony (time steps 141 to 321). Eventually these competing patterns are seen to have *won* the competition with the CNN "ants" so that after $t = 611$ no CNN "ant" pattern can be seen. A 10-step sequence shown after $t = 611$ reveals the diversity of "patterns" and their dynamic evolution. Observe from Fig. 4 that the *disorder measure* converges in this example to a small-amplitude periodic regime with the average value $m \approx 0.13$. It corresponds to the several distinct stable *species of patterns* shown in Fig. 6(e), after time step 611.

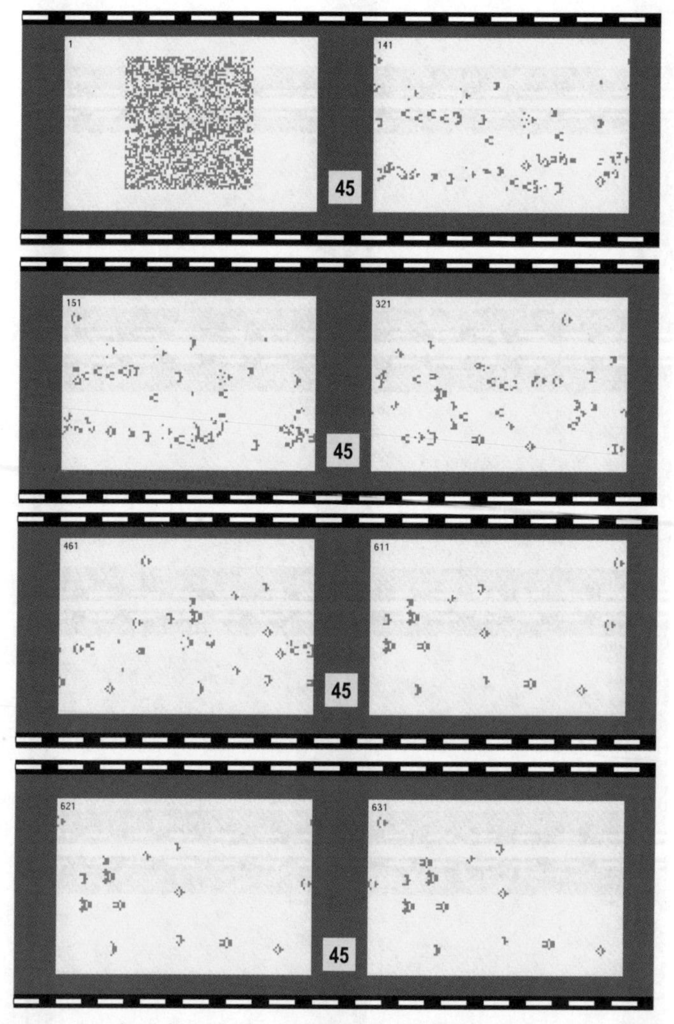

Fig. 6(e) Emergence and evolution of traveling patterns. Observe that until iteration 461 an "ant" pattern similar to that from Fig. 6(d) is present. However, after $t = 321$ several different types of new species (patterns) emerge and compete for existence until the "ant" species (patterns) vanish and become extinct in future generations.

Class U dynamics: This case corresponds to the *green* paving stones in Fig. 3. Within this domain an even finer classification based on the qualitative behaviors observed during our simulations can be made:

- A first group of three sub-classes (colored in *orange* and two shades of *light green* in Fig. 2) can be defined from the transient duration observed during the process of increasing *disorder measures:*

The *orange* color corresponds to *very long transients* as observed from cell "114". In fact, as seen in Fig. 2, such behaviors are rather rare, where only 7 out of 148 *paving stones* are assigned to this sub-class. Observe also from Fig. 4 that in this case the *disorder measure m(t)* even decreases initially as in Class E ("edge of chaos"-like) behaviors, but after a while the dynamics eventually switches to a steady increase in the local maxima of *m(t)* towards $m \approx 0.80$. This interesting type of Class U ("unstable-like") behavior is presented in Fig. 7(a), where large patterns grow slowly from a small initial red cross pattern in a fashion reminiscent of the stretching of a piece of rubber, or of a plant growing from its seed.

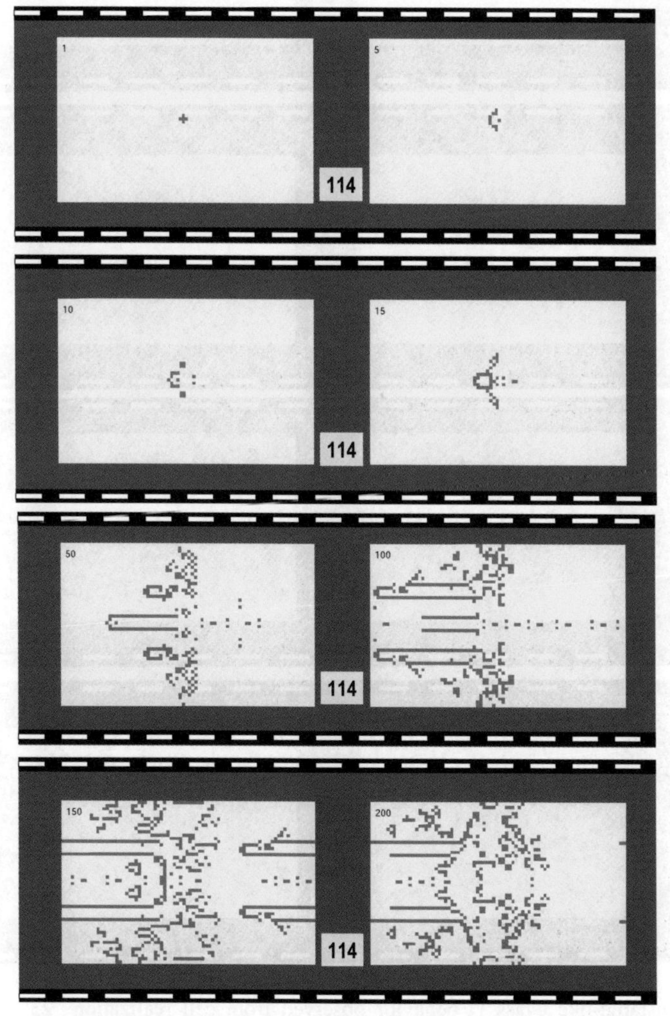

Fig. 7(a) Emergence of a growing complex pattern via a very slow transition process (see also Fig. 4, cell realization "114"). The degree of non-homogeneity increases slowly as more complicated pattern emerges.

A representative cell which exhibits a *long-transient* unstable behavior (but shorter than the preceding case) can be observed from cell "23", where such behaviors are labeled in Fig. 2 with a *yellowish green* color. As shown in Fig. 4, for this cell the increase in the *disorder measure* is faster than in the previous case. Snapshots of the dynamical evolution are presented in Fig. 7(b). Observe how the small cross initial state evolves towards a disordered pattern, and eventually spans the entire CNN grid after a sufficiently large number of time-steps. Note that small regularities still can be seen in the final pattern as a consequence of the stronger order and symmetry of the initial state pattern. Simulations with small random square initial conditions gave a completely disordered pattern after 200 or more time steps.

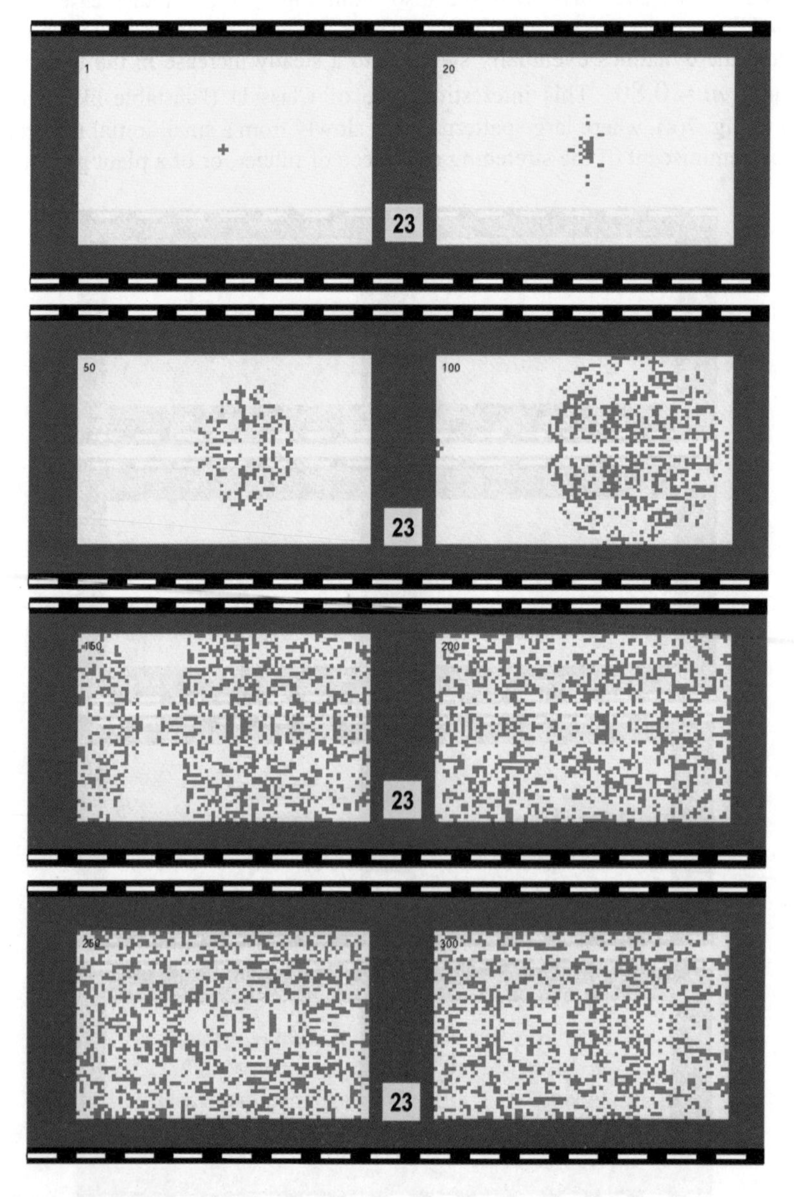

Fig. 7(b) Unstable-like Class U behavior observed from cell realization "23". A growing pattern emerges which eventually covers the entire frame. Observe the evolution towards a chaotic dynamic pattern.

Finally, the sub-class of *short* transitions labeled in *light-green* color in Fig. 2 is represented by dynamic evolutions of the Boolean functions "5" and "12". Within this sub-class it is obvious from Fig. 4 that the Boolean function "12" leads to faster transitions to a high *disorder measure* when compared to that of Boolean function "5". In fact, the above sub-classification into, respectively, *very long, long,* and *short* unstable-like behaviors is somewhat subjective, and this is the reason why we have classified all of them into the same group. The "arabesque" pattern formation shown in Fig. 7(c) is qualitatively similar to that shown in Fig. 7(b) for Boolean function "23".

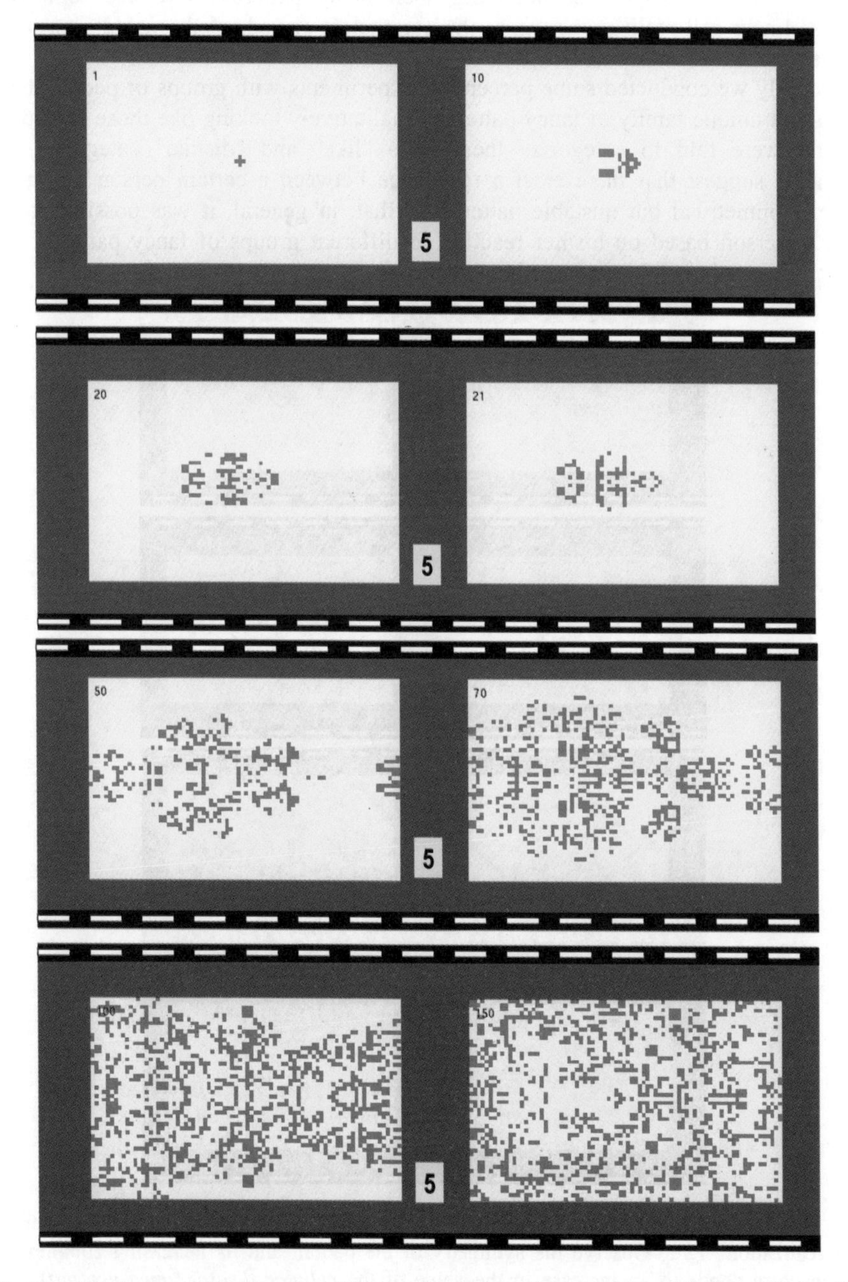

Fig 7(c) Similar unstable-like class U behavior observed from cell realization "5". In this example we focused on a narrower region within the CNN grid. Observe that the initial state "explodes" into a complex pattern and continues to increase in both size and complexity.

The evolution depicted in Fig. 7(d) for cell "12" exhibits symmetrical patterns with increasing complexity and *disorder.* This symmetry can be simply explained upon observing that the paving stone associated with the Boolean function "12" lies on the diagonal line $\lambda_1 = \lambda_2$ which corresponds to *semi-totalistic* Boolean functions, as can be easily inferred from equation (10). Semi-totalistic local Boolean functions belongs to a class of functions, which includes the "Life" *seed* function, that are insensitive to the positions of the cell inputs in the surrounding neighborhood, except that associated with the center of the neighborhood. In fact, a simple "recipe" for obtaining such fancy patterns is to use semi-totalistic piecewise-linear cell realizations where the parameters are adjusted to obtain an *unstable-like* behavior.

Recently we conducted some perception experiments with groups of people that were exposed to a unique family of fancy patterns (qualitatively looking like those shown in Fig. 7(d)) and were told to categorize them (into "like" and "dislike" categories). These experiments suggest that there exist a resonance between a certain person and a certain group of symmetrical but unstable patterns so that, in general, it was possible to clearly identify a person based on his/her reaction to different groups of fancy patterns. More details on this topic will be discussed in Chapter 6.

Fig. 7(d) Emergence of "fancy" patterns in the unstable-like class U regime observed from cell realization "12". Observe the symmetry of the pattern and its increasing complexity, which corresponds to an increase in the value of the *cellular disorder measure m(t)* as *t* increases.

- A second group of Class U ("unstable-like") behaviors is depicted in Fig.2 corresponding to those *paving stones* colored in *dark green*. They are characterized by an initial monotonic increase in the *disorder measure m(t)* and followed by a slow monotonic decrease in *m(t)* which tends asymptotically to a constant value. This steady state value of *m(t)* corresponds to a non-homogeneous "labyrinth-like" pattern with a stationary or a low-period dynamics, as shown in Fig. 7(e), at $t = 80$ and $t = 100$ for cell "61", where except for a few cells in the center all others converge to a stationary regime. The cause for this saturation is the finiteness of the CNN grid, and the only difference between this type of dynamical behavior and the ones mentioned in the other group of unstable-like behaviors is the richness in the dynamics observed in the permanent regime. Although in this case (*dark-green paving stones*) the large value of *m(t)* corresponds to a quasi-static pattern, in all other cases, a large disorder measure is always accompanied by qualitatively distinct changes at *each* cell as a function of time.

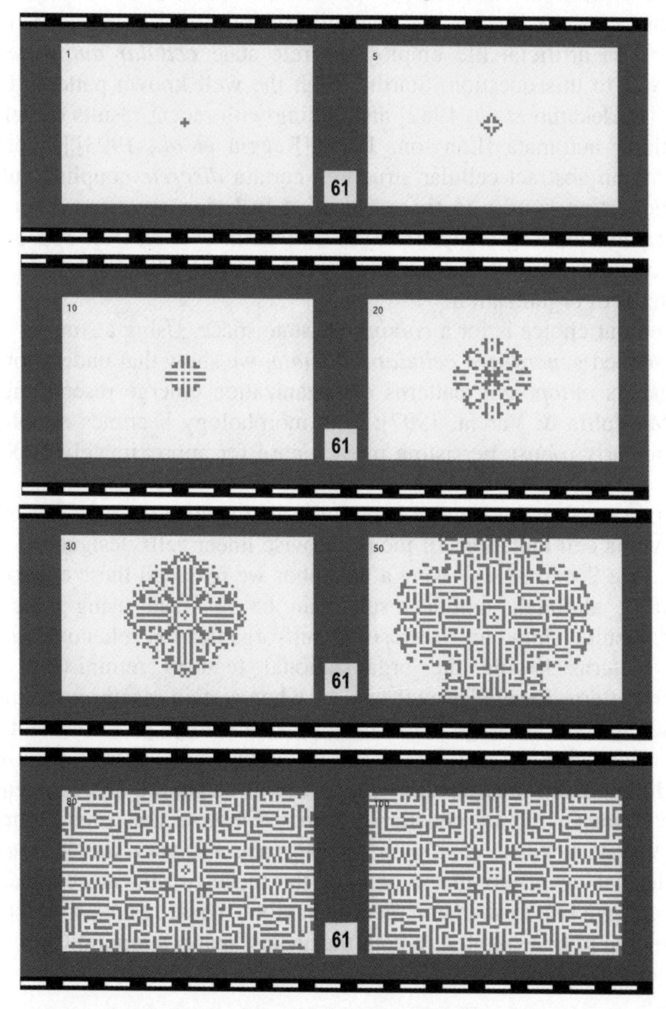

Fig. 7(e) Emergence of complex but quasi-stable labyrinth-like patterns from cell realization "61". Observe the strong symmetry, which has the same explanation as that of Fig.7(d). However, unlike the time evolution observed from the previous examples which continue to vary with time, the evolution here converges to a quasi-static pattern with a large disorder measure (see the nearly horizontal asymptote in *m(t)* in Fig.4). This labyrinth-like pattern is reminiscent of the well-known Turing patterns observed from reaction-diffusion CNN equations.

5.2. Emergent phenomena in continuous state cellular automata

5.1.1. Introduction

According to modern theories of life [Kauffman, 1993,1995], [Capra, 1996], living organisms can be viewed as highly structured *patterns* emerging from the *coupling* of simple chemical elements in auto-catalytic networks. Varela [Varela & Maturana, 1974] defines *autopoiesis* as the self-making *pattern of organization* characterizing living systems, emphasizing that *independent* of their physical embodiment, such patterns are a signature for life and cognition. While biological systems can be viewed as *patterns of organization* emerging within a *chemical* structure (the physical embodiment of the system's pattern of organization), one may ask if there is a general principle for the emergence of life. To date most researchers on artificial life employ discrete state *cellular automata* aiming at an affirmative answer to this question. Starting with the well known patterns from Conway's "Game of Life"[Berlekamp *et al.,* 1982] and ending with recent results on self-reproducible patterns in cellular automata [Langton, 1996],[Reggia *et al.*, 1993],[Sipper, 1995] it is proved that given an abstract cellular structure, certain *discrete* coupling rules can induce patterns of organization similar to those observed in living systems. Formal models of *autopoiesis* were developed [Zeleny, 1977] in a *discrete-time discrete-space* cellular model and recent simulation results based on such models [McMullin & Varela, 1997] exhibit autopoietic patterns of organization.

In this section our choice is for a *continuous* state space. Using a simpler cellular model; namely, an uncoupled *generalized cellular automata,* we show that under a proper choice of the gene parameters *autopoietic* patterns of organization emerge resembling the behavior described as [McMullin & Varela, 1997]: "..a morphology becomes established which is apparently particularly robust, persisting in each case for approximately 1000 time steps of the model".

The results presented in this section exploit the analog state space of *generalized cellular automata* (GCA) via cell mutations of the piece-wise linear cells designed to implement the "Game of Life" (see Section 5.1). Using a metaphor we may call these experiments "analog mutations in life", although similar results can be obtained using different *seed* cell functions. The results show an interesting self-organization phenomena; namely, the emergence of patterns having the organizational features reminiscent of unicellular organisms. Such patterns have a "growth" stage, when certain *membranes* emerge, isolating different chaotic modes. This stage is followed by a "maturity" phase, when the membranes evolve without major changes and, finally, by an "aging" and "dying" stage when both the membranes and the pattern organization vanish. In this stage, the GCA system enters into a stationary state characterized by chaotic dynamics. Observe that such complex dynamics occur in a GCA system with a *continuous*[3] state space and very simple *analog* cells. It is truly remarkable that *mutations* in the gene of the piecewise-linear cell realization of the Game of Life can give rise to such emergent behaviors. These results suggest that there is a tremendous potential to be exploited in both the analog spatio-temporal GCA dynamics, and in the mutation of CNN genes.

[3] In fact, in digital computer, the state space contains a large number of discrete states.

5.2.1. The cellular model

Since the CNN used in our simulations is an uncoupled one, the model can be simplified as shown in Section 5.1. The only difference is that in our *continuous state* model the sign function in (6) is now removed. It follows:

Continuous State Generalized Cellular Automata (GCA): *non-recurrent (uncoupled) case*

For $t = 1, 2, \ldots, \infty$

$$u_{ij}(t) = y_{ij}(t-1) \quad (13)$$

$$y_{ij}(t) = w(u_{kl}(t)), \quad (14)$$

End

Where all notations follow the description in Section 5.1. Let us now consider several simulation results in a CNN grid with 201 × 201 cells. In each simulation an *initial state of the generalized cellular automata* is considered as a matrix of outputs $\mathbf{Y} = \{y_{ij}(0)\}_{i,j\in\{1,2,\ldots,N\}\times\{1,2,\ldots,N\}}$, and each new iteration corresponds to incrementing n by 1. The CNN outputs at iteration n are displayed as snapshots depicted here in a "film-like" frame sequence to dramatize its dynamic evolution in time (Figs. 9-13). The magnitude of the cell outputs is coded with pseudo colors following the legend in Fig. 8. Note that the most negative value output (-1) is represented by the color *blue,* while the color *red* is used to code the most positive output value (+1).

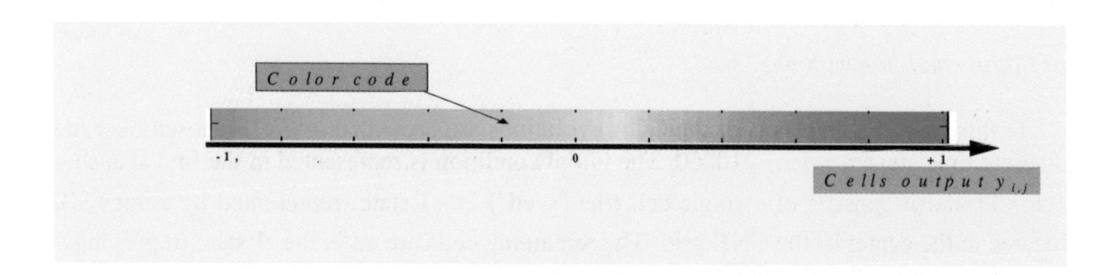

Fig.8. The color code for interpreting the CNN cell output. Blue color represents the most negative (-1) output while the most positive output is represented by the red color.

5.2.2. Emergence of autopoietic patterns by mutations in the "game of life"

Let us consider first the seed cell, which is another piecewise-linear realization for the Game of Life local logic. The function w in (14) of the cell is defined by:

$$u_\sigma = \underset{**}{1} \cdot (u_1 + u_2 + u_3 + u_4 + u_6 + u_7 + u_8 + u_9) \quad (15)$$

$$w(\mathbf{u}) = \underset{*****}{sign}\left(\underset{***}{0.8} + 0.3u_5 + 0.3u_\sigma - \left|2 + 0.3u_5 + \underset{*****}{0.82}u_\sigma\right| + 0.5\left|u_\sigma\right|\right) \quad (16)$$

where $\mathbf{u} = [u_1, u_2, ..., u_9]$ is defined in Section 5.1.2, equation (4).

The above cell corresponds to *binary* generalized cellular automata, and the equations (15) and (16) are just another alternative to equations (7), (8), and (9) in Section 5.1.

Let us now consider the following *mutation* of the previous Generalized Cellular Automata:

Analog Mutation (1) of the
Generalized Cellular Automata cell (version 1) for "Life" (step n)

$$u_\sigma = \underset{++++}{0.5} \cdot (u_1 + u_2 + u_3 + u_4 + u_6 + u_7 + u_8 + u_9) \quad (17)$$

$$w(\mathbf{u}) = \underset{++++}{0.9} + 0.3u_5 + 0.3u_\sigma - \left|2 + 0.3u_5 + \underset{++++}{0.6}\, u_\sigma\right| + 0.5|u_\sigma| \quad (18)$$

Observe that the CNN cell model of the GCA (17)-(18) is even simpler than the previous model because the nonlinear output function "sign" in (15)-(16) was removed. However, the cell remains nonlinear through $w(\mathbf{u})$. The effect is a *continuous* state space for each cell, instead of the binary state space for the seed cell described by (15)-(16). The parameters underlined with the symbol "*" in (15)-(16) were mutated and consequently underlined with the symbol "+" in the cell described by (17)-(18).

The influence of initial condition

Let us now choose two initial conditions and run the generalized cellular automata described by cell equations (17)-(18) for a sufficiently large number of iterations.

a) Initial condition with one "seed"

In this case, 12 snapshots of the GCA evolutions are presented in Fig. 9 chosen over the discrete time steps n = 1, ..., 10000. The initial condition is represented in the first snapshot $(n = 1)$ and it consists of a single cell (the "seed") in +1 state (represented by color red), located at the center of the CNN grid. The remaining cells are all in the -1 state (represented by color blue). Observe that at the "growth" phase certain circular "memebrane-like" patterns begin to emerge at about n = 50. Inside and outside of the membranes the dynamic patterns appear chaotic in nature. At $n = 200$, a pattern of organization reminiscent of a unicellular organism has "matured". For the next few snap shots, this pattern of organization exhibits the features of a self-making system, as defined by [Maturana & Varela, 1980]. Observe that at n = 200, the "membrane-like" structures are preserved at the expense of slight deformations and interaction with the more chaotic dynamics surrounding the membranes. Like in living systems, after a "maturity" phase corresponding to $n \in [200,500]$ where all "membranes" are self-making, the system enters an "aging" and "dying" phase where some of the membranes are vanishing. For example, the two concentric "green" membranes and a "nucleus" observed until $n = 300$ eventually coalesced into a unique "green" membrane at around $n = 700$. The resulting pattern dynamically maintains itself until $n = 1000$, but at around $n = 2000$ the unique "green" membrane vanishes. It appears that the smaller the self-making pattern is, the shorter is its "life-time". This observation applies to biological systems as well. Following the same scenario, at $n = 5000$ we are left with only one "self-making" membrane and a "nucleus", the latter disintegrating at around $n = 10000$. The remaining "membrane" implodes slowly so that after $n = 20000$ the self-making patterns disappear completely, and the system enters into a stationary chaotic regime characterized by short-time moving spots appearing at apparently random positions in the CNN grid. Finally, we remark that when a zero mean random initial condition is chosen, no self-making patterns similar to the one describe above for $n > 20000$ is observed after 100 time steps.

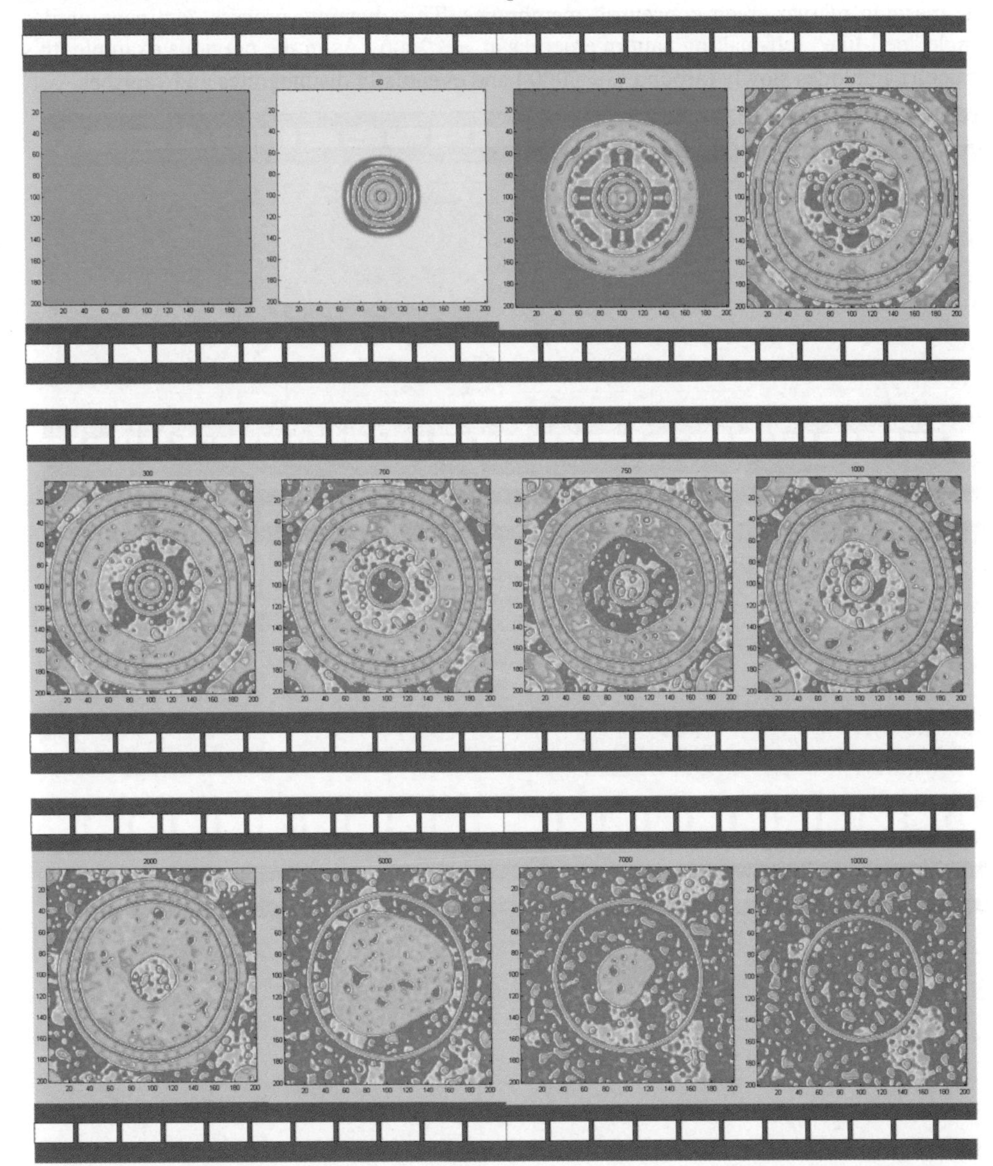

Fig. 9. Snapshots from the dynamic evolution of a *generalized cellular automaton* with the cell described by equations (17) and (18). A "one seed" initial condition determines the emergence of a self-making pattern of organization reminiscent of unicellular organisms. As in *autopoietic* pattern evolutions, these snapshots can be partitioned into approximately 3 steps, a growth stage ($1<n<200$), a "maturity" stage ($200<n<1000$) and an "aging" and "dying" stage ($1000<n<2000$).

b) Initial condition with three "seeds"

The initial condition consists now of three narrow slits of excited (i.e. in the +1 state) cells, as shown in Fig. 10 for *n=1*. As in the previous example, we can still identify a "growth" stage. This stage is followed by a "maturity" stage. Then an "aging" and a finally a "dying" stage follow. The resulting pattern is influenced dramatically by the position, number, and shape of slits in the initial condition. In our example, a face-like pattern emerges at around n=200 and 300. An "aging" phenomenon occurs at around $n = 700$ with the merging of two green concentric membranes. This dynamic process continues until a single "cell-like" self-making pattern emerges at $n = 2000$. As in the previous example, this structure implodes slowly (around $n = 5000$) and eventually disintegrates and vanishes.

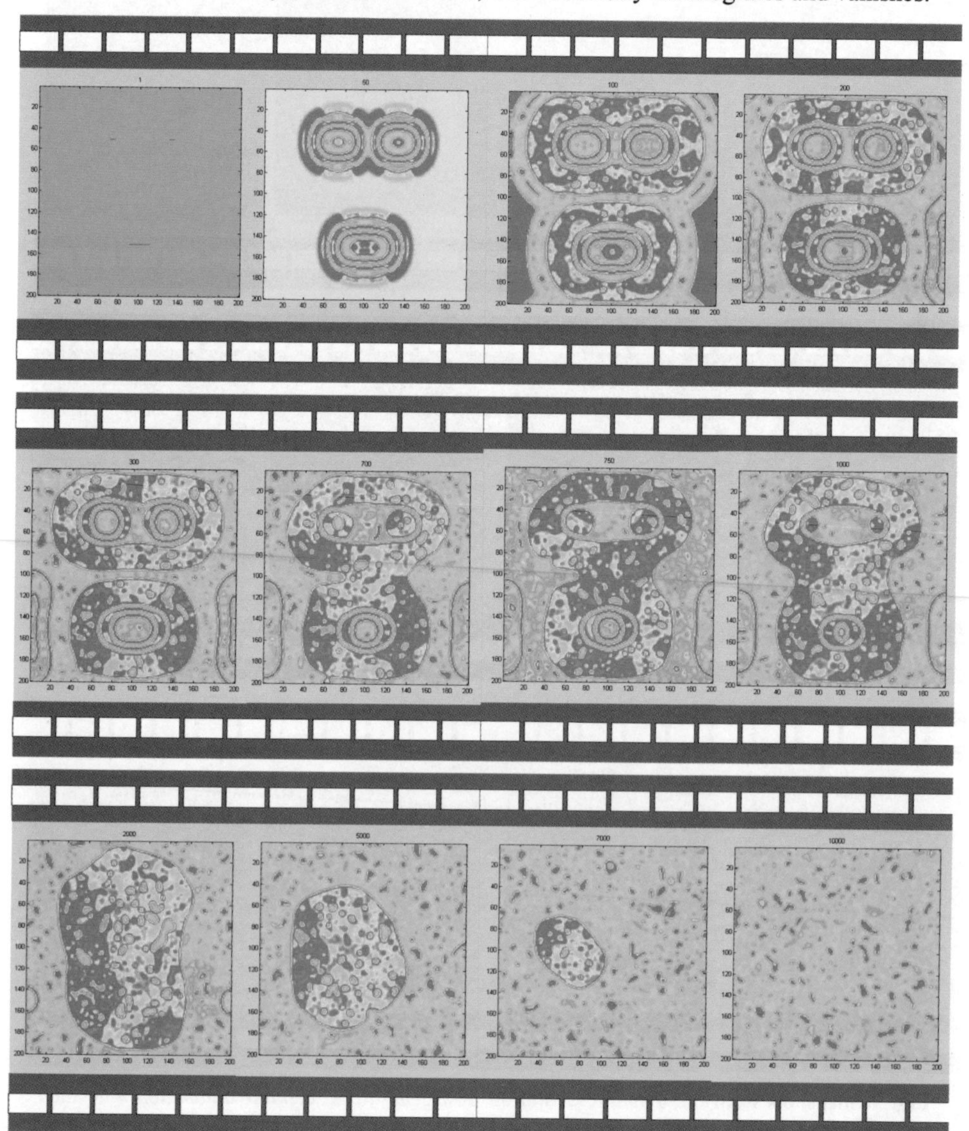

Fig. 10. Snapshots from the dynamic evolution of a generalized cellular automaton with the same cell as in Fig. 9 but with a "three-seed" initial condition. Several patterns of organization emerge from each seed during the growth stage and then coalesced into a unique self-making pattern (at around $n = 700$), which when through "maturity" and "aging" stages similar to those displayed in Fig. 9.

The influence of cells parameters

In this example, a slight change of the parameter from 0.9 to 0.91 was made in (18). The rest of the cell structure remains unchanged. For the same initial condition used in the previous example (three "seeds") the evolution of the resulting GCA dynamics is shown in Fig. 11. Except for the shape of the self-making patterns, which is altered dramatically, the same scenario of evolution is observed, and the "life-duration" is quite similar, i.e. around 10,000 time steps.

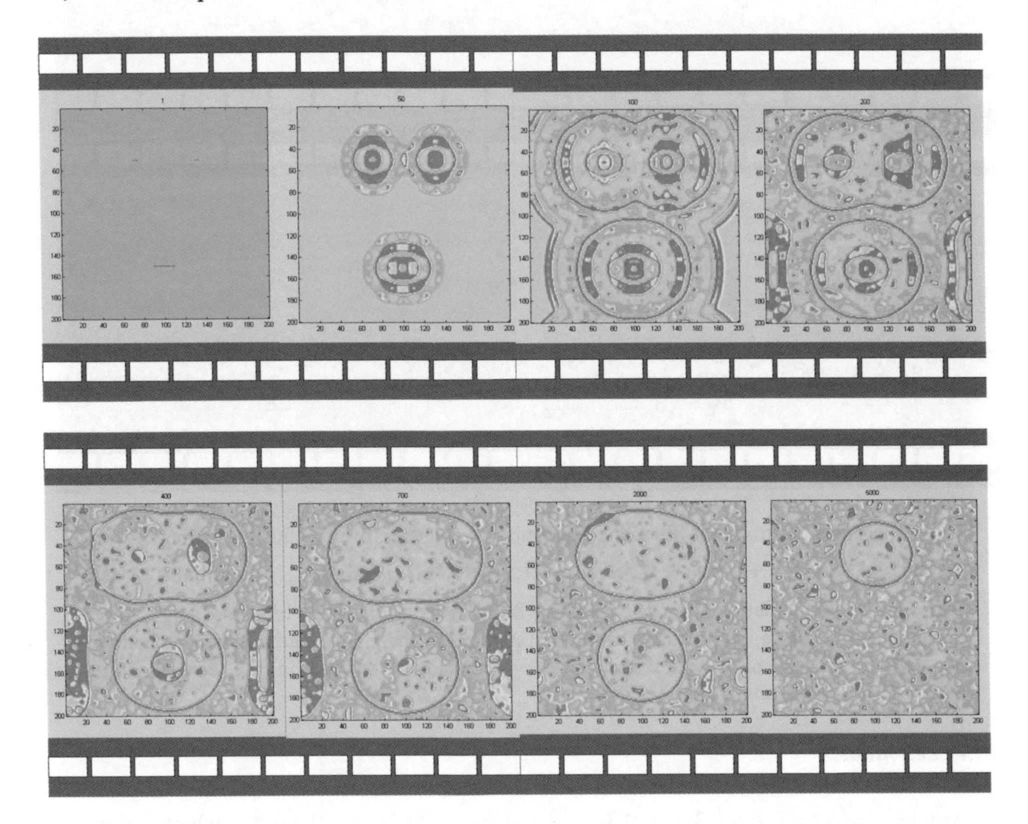

Fig. 11. Snapshots from the dynamic evolution of a generalized cellular automaton with function $w(\mathbf{u}) = 0.91 + 0.3u_5 + 0.3u_\sigma - |2 + 0.3u_5 + 0.6u_\sigma| + 0.5|u_\sigma|$ where only the first parameter was slightly mutated from 0.9 to 0.91. The rest of the parameters and the initial state are the same as in Fig. 10. Autopoietic patterns still emerge but they have a shorter "life-time" and a lower complexity.

When the same parameter undergoes a mutation from 0.9 to 0.8, the resulting dynamics becomes less complex in terms of the number of membranes, as shown in Fig. 12. However, the evolution scenario observed in previous experiments, still apply. In this case, the "maturity" phase is reached at around $n = 100$, but the "aging" process is much faster, with the self-making patterns vanishing after around $n = 800$ iterations. Further simulations using the same initial condition but with parameters chosen outside of the range $[0.75, 0.95]$ led to chaotic spatio-temporal patterns *without* any self-making property. This result indicates that there is a relatively narrow region in the parameter space where *autopoietic* patterns of the type described above can occur.

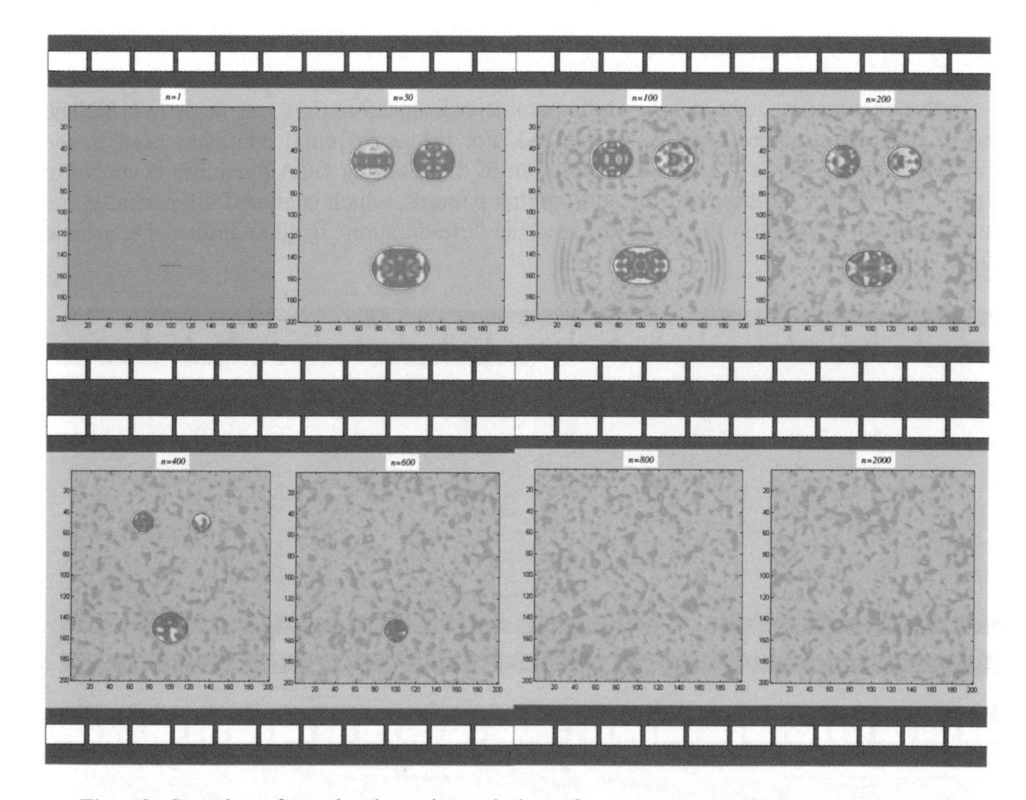

Fig. 12. Snapshots from the dynamic evolution of a generalized cellular automaton with function $w(\mathbf{u}) = 0.8 + 0.3u_5 + 0.3u_\sigma\text{-}|2 + 0.3u_5 + 0.6u_\sigma| + 0.5|u_\sigma|$. The initial states are as in Fig. 10. Autopoietic patterns still emerge but they have a shorter "life-time" and a lower complexity. Therefore, with respect to the first parameter, there is a relatively narrow band in the cell parameter space where *autopoietic* patterns can emerge in the corresponding *generalized cellular automaton.*

Other cell models

Besides $w(\mathbf{u}) = 0.8 + 0.3u_5 + 0.3u_\sigma\text{-}|2 + 0.3u_5 + 0.82u_\sigma| + 0.5|u_\sigma|$, which gives an exact realization of the Game of Life, an even simpler discriminant function $w(\mathbf{u}) = 1.5\text{-}|2.5 + 0.5u_5 + u_\sigma|$ can do the same, where u_σ is defined as in (7). Let us now consider a mutation of this cell structure leading to a continuous state cellular system:

Analog Mutation (2) of the
Generalized Cellular Automata (version 2) for "Game of Life"

$$u_\sigma = u_1 + u_2 + u_3 + u_4 + u_6 + u_7 + u_8 + u_9 \quad (19)$$

$$w(\mathbf{u}) = 0.2 \cdot (1.5\text{-}|2.5 + 0.5u_5 + u_\sigma|) \quad (20)$$

Observe that in this case none of the CNN cell parameters within the discriminant function $w(\mathbf{u})$ were altered. Only the cell output function is changed from sign(.) to 0.2(.).

The dynamics of the resulting generalized cellular automata for one "seed" initial condition is shown in Fig.13. The same scenario of evolution as in the previous examples is observed, while the shape of the self-making patterns is now more rugged. As in the previous cell model, the autopoietic patterns were found to occur in a relatively narrow domain of the cell parameters.

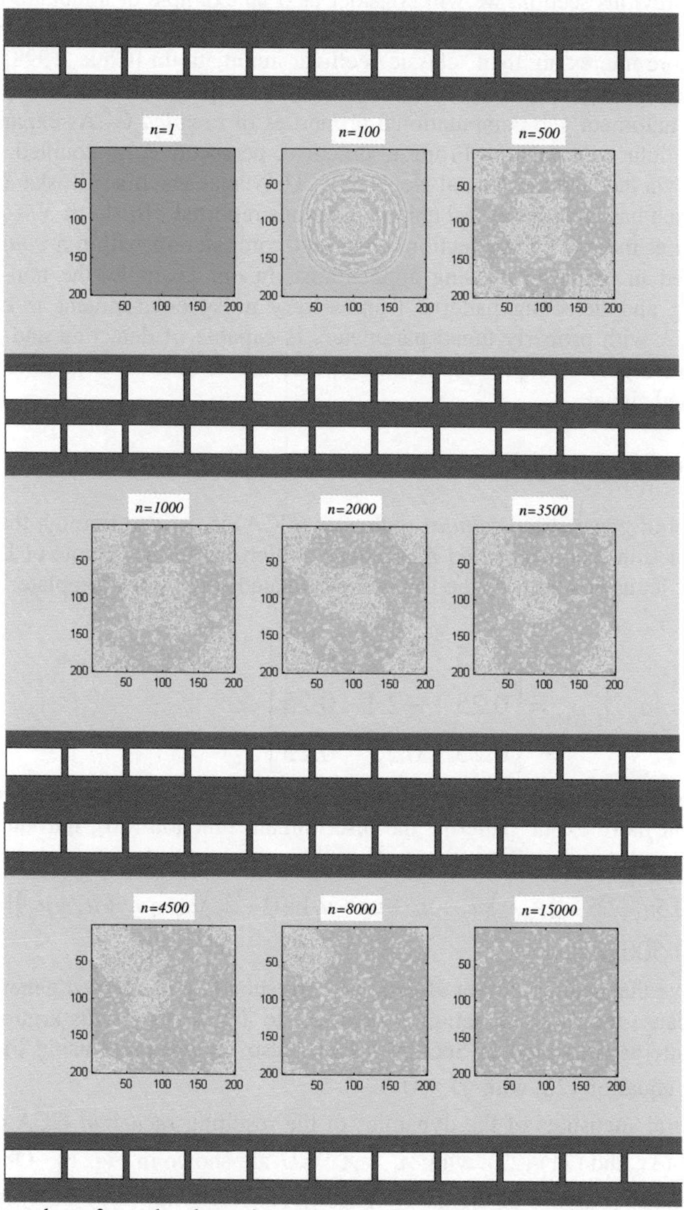

Fig. 13. Snapshots from the dynamic evolution of a generalized cellular automata which realizes the "game of life" with the simplest cell discriminant function $w(\mathbf{u}) = 0.2(1.5\text{-}|2.5 + 0.5u_5 + u_\sigma|)$. As in the previous examples, a pattern of organization is maintained over a finite "life-time" from a one-seed initial condition. Comparing with the previous examples, the pattern here has a different shape but is qualitatively equivalent to an *autopoietic* pattern of organization reminiscent of unicellular organisms.

5.3. Emergence in coupled generalized cellular automata

5.3.1 Introduction

In addition to the case of uncoupled *generalized cellular automata,* which is the main focus of the previous section, we will consider next an example of a mutation in the "Game of Life" with a *coupled* CNN cell. This class of *generalized cellular automata* (GCA) is potentially more interesting than "classic" cellular automata. In [Chua, 1998] it was proved that while uncoupled *generalized cellular automata* are equivalent with binary or continuous state cellular automata, the computational properties of *coupled* GCAs expand beyond that of classic cellular automata. From a practical perspective, a coupled GCA can be implemented via the Cellular Neural Networks – Universal Machine [Roska & Chua, 1993], a concept which has already several chip realizations reported [Roska & Vázquez, 2000a].

As shown at the end of the section, emergent computation within a *coupled* GCA cell can be applied in image processing applications. In our example, the non-trivial task of reconstructing and detecting patterns from a very noisy environment is considered. As shown, a GCA with properly tuned parameters is capable of detecting and reconstructing patterns from a highly corrupted input image where the noise level is 10 times stronger than the uncorrupted signal.

5.3.2. The cellular model

The model of *generalized cellular automata* (GCA) is defined through the equation (1)-(3). Instead of using the uncoupled **A**-template, which led to the "Game of Life" discussed in Section 5.1 let us now choose the following (mutated) coupled **A**-template:

$$\mathbf{A} = \{a_{k,l}\}_{k,l \in \mathrm{N}} = \begin{bmatrix} 0.25 & 0.25 & 0.25 \\ 0.25 & -2.1 & 0.25 \\ 0.25 & 0.25 & 0.25 \end{bmatrix} \qquad (21)$$

It leads to a *stable* dynamical behavior of the standard CNN described by equation (1) and (2). To exhibit more exotic patterns, the discriminant function (10), introduced in Section 5.1, was scaled by a factor β to obtain

$$w = \beta[1.5 + 0.5u_5 \text{-} |2.5 + (1+\lambda_1)(u_7 + u_6 + u_4 + u_1) + (1+\lambda_2)(u_9 + u_8 + u_3 + u_2)|] \qquad (22)$$

where $\beta = 0.500$.

Observe that scaling do not affects the corresponding *uncoupled* generalized cellular automata because the "sign" function is invariant to any scaling of its argument. In other words, all patterns presented in Section 5.1 can also be generated using the discriminant function w in equation (22) with $\beta = 0.5$.

Several snapshots of the dynamics of the resulting *recurrent* GCA defined by the equations (1)-(3), and (21)-(22), with $\lambda_1 = \lambda_2 = 0$ are shown in Fig. 14. Observe that as a result of the local *coupling* of the CNN cells, a completely new type of dynamics emerge, characterized by *a continuum* of states for each cell. While a CNN by itself (described by the equation (1) and (2)) can produce only the static pattern shown in Fig.8 (at time step 2) the addition of the GCA loop leads to a series of dynamic patterns, as shown in iterations 3 to 13. Such a dynamic behavior, with a *continuum of states* can be viewed as reminiscent of some simple biological formations.

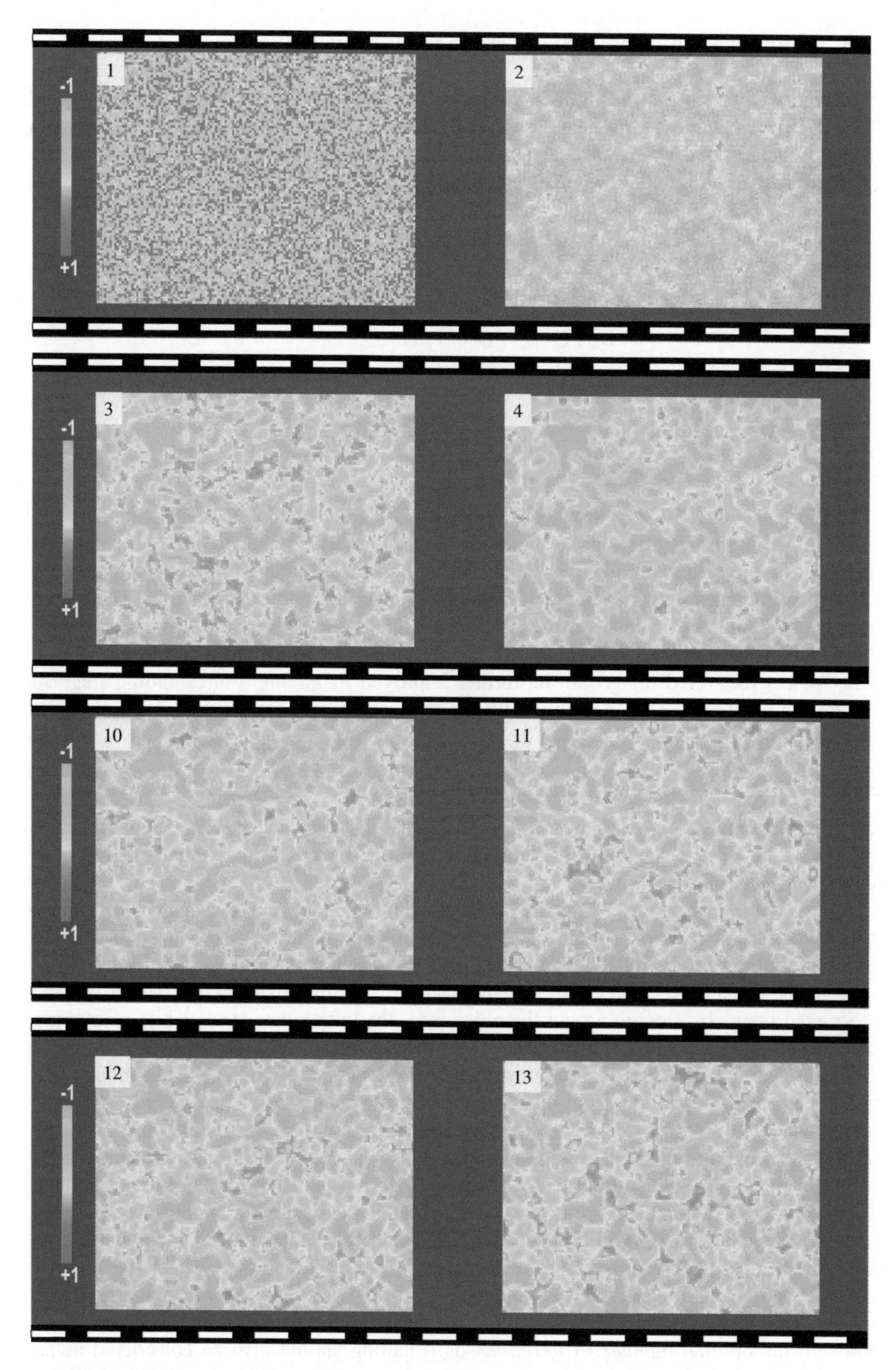

Fig. 14 Snapshots of dynamic evolution illustrating typical unstable-like Class U behaviors from a *generalized cellular automaton* with *coupled* cells. The color code used to code the cell output is shown on the left of the "film-like" snapshots.

5.3.3. An application in pattern detection from extremely noisy background

In this section we present a practical application of emergent computation in *coupled* GCA. In this application, we make a slight mutation of equations (21) and (22) to solve a difficult vision problem; namely, detection and reconstruction of patterns from highly perturbed images. Our mutated GCA is characterized by:

$$\mathbf{A} = \{a_{k,l}\}_{k,l \in \mathrm{N}} = \begin{bmatrix} 3/8 & 3/8 & 3/8 \\ 3/8 & -3 & 3/8 \\ 3/8 & 3/8 & 3/8 \end{bmatrix} \tag{23}$$

and the discriminant function:

$$w = \beta[1.5 + 0.5u_5 - |2.5 + (1+\lambda_1)(u_7 + u_6 + u_4 + u_1) + (1+\lambda_2)(u_9 + u_8 + u_3 + u_2)|] \tag{24}$$

with $\beta = 0.2$, and $\lambda_1 = \lambda_2 = -30/128$.

The corresponding GCA is therefore defined by Eqs. (1)-(3), and (23)-(24).

As shown in Fig.15, after only 8 iterations (at iteration 9) the above GCA has succeeded in uncovering the patterns hidden within the heavily corrupted image (a noise to signal ratio of 10 to 1 was chosen in this example) shown in the upper left corner in Fig. 15. The output image is coded in gray shades over the continuous range [-1/3,1/3] (color black corresponds to an output equal to -1/3, while color white corresponds to an output equal to +1/3). The quality of our recovered image can be further improved by adding a thresholding stage so that one can compare the GCA output with the reference binary image. The thresholding is mathematically defined by $y_{ij} = \mathrm{sgn}(y_{ij} - z)$, where z is the threshold value and y_{ij} is the output of cell C_{ij}. The threshold value chosen in our example is $z = 0.12$. Observe that the recovered patterns resemble quite well the original patterns in the uncorrupted image although the naked eye can barely see any pattern in the corrupted image that was fed in to the *generalized cellular automata.*

A sequence of snapshots of the dynamic simulation of the GCA is shown in Fig.16, where colors are now used to code the cell output for more clarity. The color code for the outputs is defined in the left edge of each figure. Observe that the generalized cellular automata is capable of detecting and recovering patterns that were present in the uncorrupted image by smoothing out the noise from the initial state ($t=1$) while recovering most of the original contour information. The CNN by itself (without being *nested* in the GCA loop) can only produce the output shown in Fig.16 at $t=2$. Observe that although one can recognize certain patterns from this CNN output, the quality of the recovered patterns is far from the one obtained after carrying out several additional GCA iterations. It is the additional processing via the iterated dynamics inherent in the GCA definition that provides the improvement. An inherent tendency of the *generalized cellular automata* is to produce *dynamic patterns* rather than static ones, even though the recurrent coupling defined by equation (23) defines a *stable* output of the CNN embedded in the GCA loop. Indeed, as shown in Fig. 16, the dynamic process does not stop after iteration 9, when the optimal recovered pattern has been obtained, and will in fact give rise to chaotic patterns if the iteration cycle is not interrupted. Therefore, in addition to the *gene parameters* defining the CNN cell, the optimal *number* of *iterations* until halting should also be considered as an additional parameter in defining the computational properties of the *generalized cellular automaton.* Observe that this adaptive approach provides us with more flexibility while the computational tasks are specified in a highly compressed manner reminiscent of the genetic information coding in highly complex biological systems. Just as in most biological systems undergoing a life cycle, where most of their functions are fulfilled in their youth, our GCA seems to undergo a similar process, performing the useful computations at the beginning of its "life-cycle". Since each additional time step in this example can be considered as performing an additional "dilation"-type of processing, a finite number of such additional steps may be included if the processing task calls for additional dilations. The above observations suggest that information processing problems in a GCA should be treated from an evolutionary perspective, i.e. by allowing populations of GCA to evolve until an optimal solution emerges. A heuristic search procedure (like in this case, where we started from an initial set of gene parameters defining the "Game of Life") can also be used to speed up the evolutionary process.

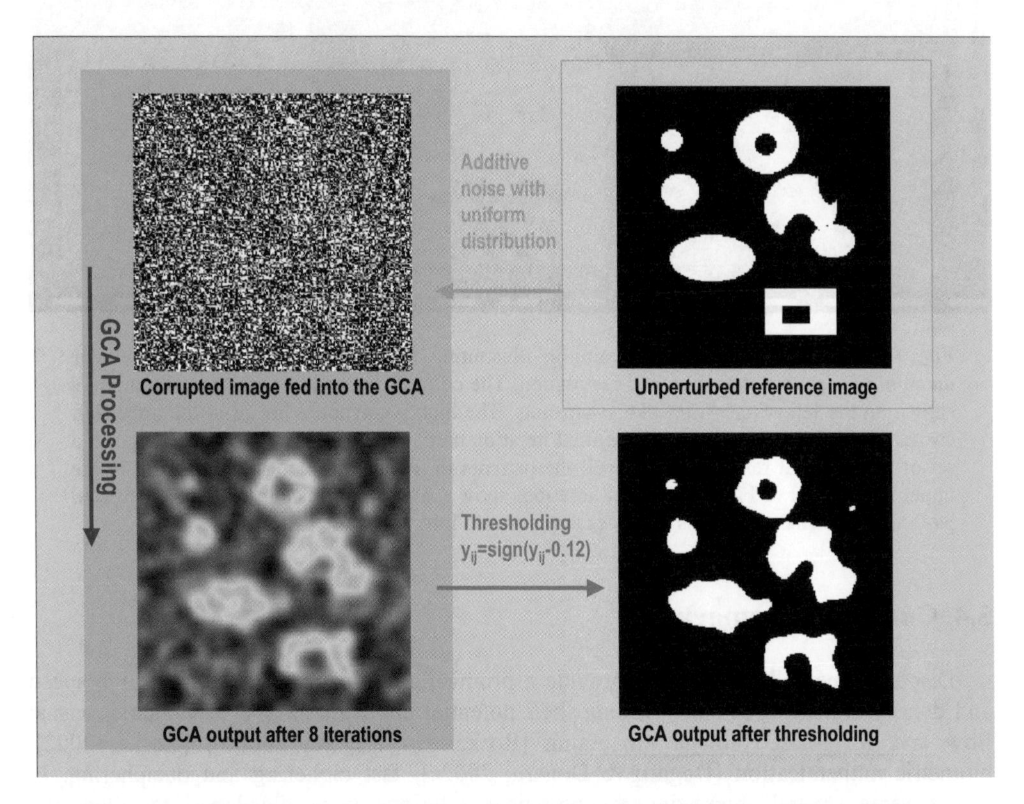

Fig. 15. Pattern detection and restoration from highly degraded image using a generalized cellular automaton with *recurrent* CNN coupling. The noise added to the original pattern in the upper right corner has the amplitude 10 times larger than the unperturbed image, and the noisy input is shown in the upper left corner. Color black corresponds to an output equal to $-1/3$, while the white color codes an output equal to $1/3$. Corresponding shades of gray are assigned to the intermediate values.

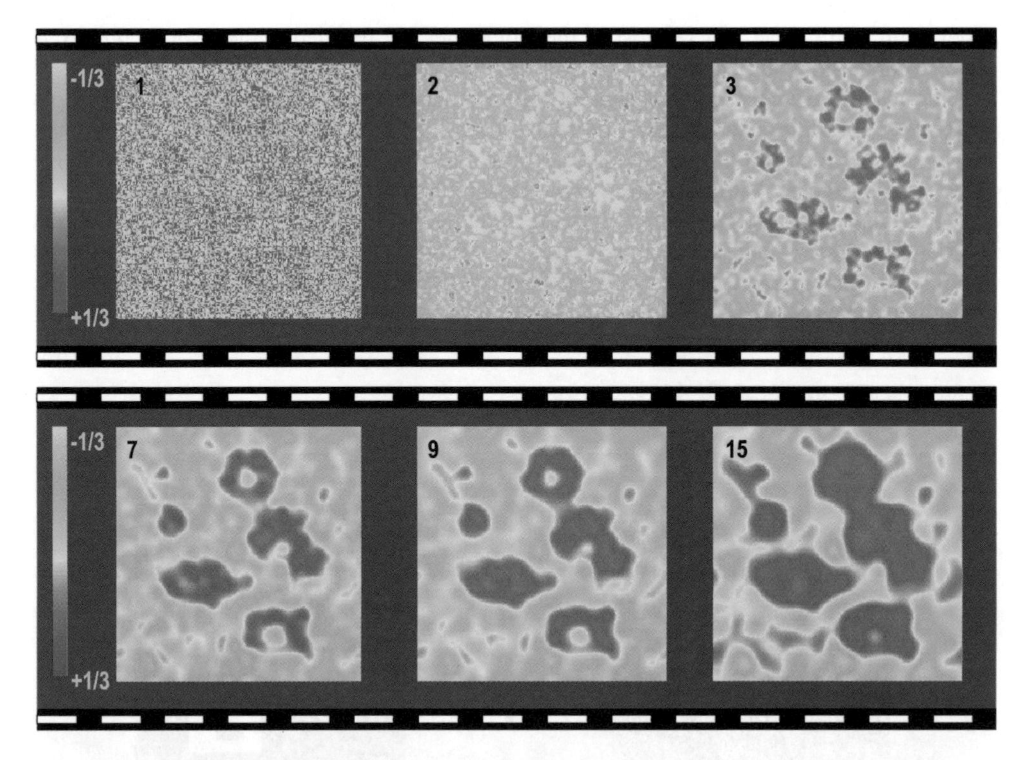

Fig. 16. Snapshots of dynamic evolution illustrating the use of the *generalized cellular automata* for pattern detection and restoration. The color code used to code the cell output is shown on the left of the "film-like" snapshots. The highly corrupted image in the upper left corner ($t=1$) evolves so that at the optimal iteration number $t=9$, one can clearly distinguish a set of patterns that resemble quite well the patterns in the unperturbed image (shown in the upper right corner of Fig.15). Further iterations show a *dilation* of this optimally reconstructed pattern. Still further iterations in this example would lead to chaotic patterns.

5.4. Concluding remarks

Discrete time cellular systems provide a promising approach for designing high speed and density signal processors. Among their potential applications one may include visual flow and wave-based image processing [Roska & Vázquez, 2000b], [Roska, 2002], biometric authentication [Dogaru & Dogaru, 2002c], fast ciphering and deciphering. In recent years several interesting technological solutions were developed to allow the implementation of generalized cellular automata discussed in this chapter [Roska & Vázquez, 2000a].

This chapter aims to answer the following fundamental design question: Under what conditions will such cellular systems support the kind of computation, which is essential for information transmission, storage and processing? Our goal was to give *precise* answers in terms the unified paradigm of *generalized cellular automata* and a set of dynamic analysis tools inspired by the theory of *local activity* [Chua, 1998] and the *edge of chaos* method introduced in Chapter 4. Unlike other cellular system models where transition tables or rules are used to define the cell, we found it very convenient to use the *universal* cell based on the piecewise-linear discriminant introduced in Chapter 3. Particularly, this method of specifying the cell allows the search for emergent computation in a cell *parameter space* where some structure can be identified. This method also allows applying evolutionary programming techniques in a straightforward manner by mutating the parameters such that emergent dynamic phenomena are more likely to occur.

In Section 5.1 it was have shown that the entire set of failure boundaries can be *precisely* determined for Boolean universal cells with piecewise-linear realization. Then some *structure* is induced into the cell parameters space by a set of dynamic criteria inspired in their definitions by their counterparts in the "edge of chaos" method in Chapter 3.

Let us remind that the "edge of chaos" method relies on the *local activity* theory, developed for the case of continuous-time cellular neural networks. It essentially says that non-homogeneous patterns can not emerge in a non-conservative homogeneous cellular system (with symmetrical boundary conditions) formed of cells coupled via passive systems unless the isolated cells are locally active. The local activity theory offers a set of mathematical tools which can be used to predict the behavior of a nonlinear cellular system modeled by a CNN having a huge state space by investigating only a much simpler dynamical system; namely, an isolated CNN cell. As shown in Chapter 4, well-defined boundaries can be defined within the cell parameter space, isolating 3 qualitatively distinct sub-domains called "locally passive", "edge of chaos" (locally active and stable) and "locally active and unstable" regimes, respectively. As long as a cell parameter point is located within the "locally passive" sub-domain, according to the local activity theory, the entire system will evolve dynamically towards a completely homogeneous pattern, and computer simulations also revealed that this dynamics tend to have very short transients. On the contrary, if the cell parameter point is located within one of the *locally active* domains, then there exists a *potential* for non-homogeneous static or dynamic patterns to occur. The boundary between the unstable and stable locally active domains and its neighborhood was found to give rise to a wide range of complex dynamic behaviors, many of them having computational potentials [Dogaru & Chua, 1998b,c].

In the absence of a corresponding *analytic theory of local activity* for the generalized cellular automata (GCA) with *discrete* time dynamics, we introduced in this chapter an *empirical scalar* function of time *m(t),* called the *cellular disorder measure,* for measuring the non-homogeneity of patterns. A completely homogeneous pattern corresponds to $m=0$. Therefore, a passive-like behavior was defined, analogously to the *local activity theory*, by a rapid decrease of *m(t)* to the steady state value 0. Consequently, any other form of time evolution of *m(t)* starting from a non-zero value will correspond to an *active* GCA behavior, where the word *"active"* is used in the same sense as in the *local activity theory.* Within this "activity" domain, we can further differentiate between an "unstable-like" behavior (corresponding to an *increase* in *m(t)* with time) and an "edge of chaos"-like behavior (corresponding to a decrease in *m(t)* with time).

Unlike other entropic measures often used to characterize the "complexity" of dynamics, our measure is inspired by the principle of local activity theory. Instead of measuring a cells' entropy as in [Langton, 1990] we take into consideration the entire cell neighborhood by calculating the *time evolution* of a *disorder measure m(t)* which characterizes the degree of non-homogeneity in the pattern as a function of time. In the examples presented herein we considered discrete-time systems with *binary* cell states. However, this measure can be easily extended to discrete-time systems having any number of distinct states per cell, and furthermore to continuous-time systems by using appropriate time sampling methods (e.g., Euler or Runge Kutta).

Using the above disorder measure and observing its tendency to increase, or decrease, towards a non-zero value, or to decrease towards zero, a specific class label and color representation is then assigned to each "paving stone" located within the cell parameter domain surrounding a *seed* cell. The CNN *seed* cell is a piecewise-linear realization of some "interesting" local Boolean function (in our examples we used the "Game of Life" function) whose unfolding yields a diverse and structured cell parameter space, as shown in Fig. 2 and Fig. 3. Similarly to results described in Chapter 4 where the *local activity theory* was applied, the most "interesting" dynamical phenomena were observed at the boundary between the *stable* and the *unstable* active domains, namely for all cell parameters characterized by either slow *increases* or slow *decreases* in the values of *m(t)*. Particularly interesting, yet rare, are those behaviors associated with a very slow increase in *m(t),* such as that observed from cell realization "114". We believe that using appropriate initial conditions, such dynamical behaviors may lead to non-trivial self-reproductions of the type described in [Langton, 1984], [Reggia, 1993], [Sipper, 1995]. Indeed, if we choose one of the above cited examples and plot the evolution of the *cellular disorder measure* in each case, we will observe an increasing tendency towards non-homogeneity generated by a "birth" and "metamorphosis" process, tempered by a "population overcrowding" phenomena reported in all artificial life experiments.

By *mutations* of cell parameters (in the sense of crossing failure boundaries), a wide range of dynamic behaviors was observed, yielding many interesting properties, such as those associated with cell realizations "45", "55", "114", and "122", described in detail in Section 5.1.3. Although the "Game of Life" function was chosen here as the *seed* function, any other piecewise-linear cell can be similarly explored. For example, one may use the same tools to determine the structure of the 5-dimensional cell parameter space (s, z_0, b_1, b_2, b_3) associated with the entire set of 256 local Boolean functions of a universal uniform multi-nested CNN cell realization defined by $w=-1.5+\left|-3+\left|z_0+b_1u_1+b_2u_2+b_3u_3\right|\right|$, for any prescribed interconnection topology.

Unlike in the application of the *local activity theory* to continuous-time reaction-diffusion systems we still need to simulate the dynamics of the generalized cellular automata considered in this paper, in order to determine its behavior class. However, we hope that *analytical* methods for determining the class of behaviors can also be developed in the future. Moreover, as long as an initial "reference" state is prescribed, the tools presented in this paper still allow one to *precisely* identify the specific class of behavior at any given cell parameter point.

We believe that our approach for studying adaptive cellular systems could lead to a more precise characterization of their dynamics. Instead of considering the four classes of behaviors proposed in [Wolfram, 1984], which is still subject to debate, we propose here a set of three well-defined classes of behaviors which can be identified by monitoring the time evolution of the *cellular disorder measure m(t).*

Moreover, adaptive phenomena like mutations can be more easily traced by precisely defining the failure boundaries in the cell parameter space. For example, one can easily observe from Fig. 3 that class E ("edge of chaos"-like) behaviors are concentrated around the $\lambda_2=-\lambda_1$ line while the perpendicular line $\lambda_2=\lambda_1$ encompasses all three classes of dynamical behaviors. The use of piecewise-linear cell realizations opens a very interesting perspective towards building non-homogeneous cellular systems where many types of behaviors can coexist and interact on the same grid. Indeed, if one introduces two additional CNN input layers associated with the two parameters (λ_1, λ_2), each cell can be controlled independently to realize one of the 148 local Boolean functions represented by the "paving stones" in Fig.1.

As shown in Section 4.5., where the *gene* parameters were chosen as a mutation of the gene producing the "Game of Life", the paradigm of *generalized cellular automata,* particularly when the recurrent coupling is considered among the CNN cells, is also very promising for engineering applications. A difficult problem in vision; namely, the recovery of meaningful patterns representing objects hidden in highly perturbed images was shown to have a simple and effective solution using a heuristically tuned generalized cellular automata.

Chapter 6
Unconventional Applications: Biometric Authentication

In previous chapters we have seen that using properly designed cells and gene parameters emergent phenomena can be easily obtained in such systems in either the form of static or dynamic patterns. What are the possible applications of such patterns? Since cellular systems process information in parallel in an homogeneous array of cells, each cell can be assigned to an image pixel and thus most of their applications are in image processing. Unlike classic computers which sequentially deal with pixels or their neighborhoods cellular systems process information in parallel. Thus the processing speed is accelerated at several orders of magnitudes when compared to sequential digital signal processors. As we already mentioned in the previous chapters for some combinations of parameters such emergent patterns occur which correspond to useful image processing tasks such as edges or corners detection, image restoration and so on.

In this chapter we will focus on a much interesting property of the emergent patterns; namely, that of stimulating human brains to have "like" or "dislike" reactions when such a pattern is presented. Such an interaction may be called a *perceptual resonance* [Dogaru & Dogaru, 2002c] and in the following it is shown how it can be used to identify a specific person in a group, introducing a novel authentication method called a "brain signature".

6.1. Introduction

Biometric identification is a novel and emerging field where the goal is to identify a person using measurements of some biological identifiers. In other words, biometrics assumes that each individual carries its own biological signature, which is much more difficult to falsify than, for example, his/her written signature. On the other hand, biometrics offers a solution for the nightmare of the nowadays widely use of password-based identification. Indeed, instead of remembering several difficult passwords one would simply offer the password of his body. Several well established biometric methods are currently in use, including fingerprint, iris and face recognition. However, all of them require intensive computation and sophisticated sensing systems [Corby, 2002].

Here a completely new biometric feature is described, which leads to a low cost reliable method of checking the identity of a person within a limited group. Such methods implemented as a dedicated microchip can be used for accessing a mobile phone or other personal computing device or to give access to a restricted area zone (e.g. within an institution). The method relies on the old Latin proverb "*degustibus non disputandum*" which essentially says that when a person is presented a sequence of perceptual stimuli (e.g. a sequence of visual patterns or a sequence of melodies) it will react to them in a personalized manner. For example he/she may be asked to classify the stimuli in one of the three fuzzy sets: {DISLIKE, NEUTRAL, LIKE} coded numerically as {-1,0,1}. The person *resonates* in his/her personalized manner with the sequence of stimuli assuming that a proper sequence of stimuli is given. Thus given a sequence of stimuli the person will let the system identify her/him through a ternary word signature as in Fig.1.

This resonance is a more sophisticated process than the simple memorizing of some conventional labeling assigned to the stimuli patterns. In fact, memorizing the responses to a sequence of 20 (typical number of stimuli in a sequence) is a very difficult task. Unlike when using a password the human subject does not need to remember what she/he likes and therefore this way of accessing a system becomes more user-friendlier. One may lose his/her memory but still be able to provide a sequence of responses, which in fact represents a codified *signature* of her/his brain to the given sequence of stimuli. Our experiments suggests that complex patterns emerging near the *edge of chaos* [Dogaru & Chua, 2000b] [Dogaru & Chua 1999b] [Dogaru & Chua 1998b] in cellular systems provide a rich repertoire of patterns where the personality of each individual in choosing "nice" patterns has a great influence.

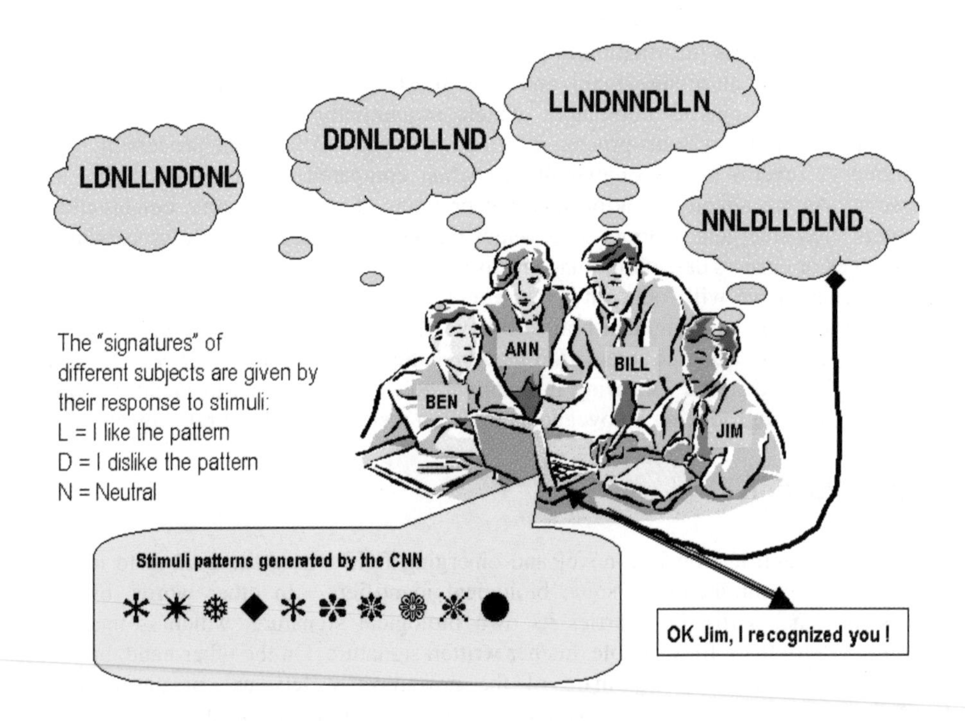

Fig. 1. A schematic example of the biometric identification system. The CNN is embedded in the system where the access has to be granted and will send a sequence of stimuli (in this example – visual) to different users. By previous presentations the system learned the preferences of Jim (also called resonances) and of the other potential users and grants the access to Jim based on his *brain signature* NNLDLLDLND. The sequence generated by Jim at other presentations of the same stimuli might be slightly different but still located within a sub-domain of the signature space which can be differentiated (separated) by the sub-domains corresponding to other people *brain signatures*, if the length of the sequence is large enough and the stimuli are propoerly chosen.

Thus, let us assume that we consider a set of N human individuals $H = \{P_1, P_2,.., P_k.., P_N\}$ and a set of M stimuli $S = \{S_1, S_2,..S_m,..S_M\}$ generated by the system, which wants to identify the persons on biometric basis.

Among different possibilities to generate the stimuli (e.g. by previously storing them in a memory) the CNN framework [Chua, 1998] provides a very convenient and cheap alternative to generate complex patterns capable to induce *human resonance* in the sense specified above. Section 2 provides an overview on the Generalized Cellular Automata and indicates why we believe that such sets of stimuli have the best properties to generate *brain signatures* based on *perceptual resonance*.

Let us now consider a set of Q presentation sessions each session being identified by an index $q \in \{1,..,Q\}$. The sequence of presentations may follow a certain time schedule

$T = \{t_{12},..,t_{q,q+1},..t_{Q-1,Q}\}$, which is characterized by the time between two consecutive presentations. At each presentation, the subject (person or a machine emulating a human intelligence) k will generate a sequence of responses defined by the matrix $R^k = \{r^k{}_{q,j}\}_{q=1,..,Q;m=1,...,M}$ where the element $r^k{}_{q,j}$ has the meaning of *the response* of subject k to the perceptual stimuli S_m in a presentation q. Although the person will have a subjective interpretation about the stimuli S_m it can be quantified numerically as shown in detail in Section 3. For example one can define a scale of grades which has to be known by all subjects and then they are asked to give their responses as particular grades within the scale. Assuming that such a *grading convention* is established, the *resonance* hypothesis says that after a reasonable number of presentations the information stored in the matrix R^k can be used to differentiate user k among all other potential users. Therefore, besides the stimuli generator, another sub-system should be designed which interprets the response matrices R^k from different users and decides which user is the one reacting when the sequence of stimuli is presented to him. Section 3 presents the details and the results of an experiment with *N=5* human users, *Q=6* presentations and up to *M=20* stimuli in a presentation sequence. The results of the experiment prove that the *resonance hypothesis* is correct and specific *brain signatures* could be identified for each user. A simple method of learning and identifying the user by using a perceptron neural network is also described in Section 3, showing that it has very good results in classifying brain signatures provided that stimuli sequences are long enough and a sufficient number of presentations were done before the system is asked to classify the responses.

6.2. Generating the stimuli

In our experiments we used 200x200 cells generalized cellular automata (GCA) to produce certain sequences of visual stimuli. A stimulus S_m is the image associated with the whole array of 200x200 cell outputs where the output of each cell is coded with black if +1 or white if -1. Color stimuli can be also generated using certain types of cells having their outputs varying within the domain [-1,1].

Let us remind that a generalized cellular automata (see Chapters 2 and 5 of this book) is composed of a CNN lattice where an additional loop was added to each cell C_{ij} (where the pair of integers (i, j) locates the CNN cell within a rectangular grid) so that the CNN cell output y_{ij} is sampled at time moments $t_n = n\Delta T$ and then fed into the cell input at the beginning of a new cycle of the dynamical CNN evolution. The duration of each cycle of CNN evolution is ΔT, where ΔT is chosen so that all transients had died out and the CNN has reached a steady state before a new cycle begins. We will assume a periodic CNN boundary condition throughout this paper. The distribution of cells around the central cell C_{ij} is denoted by

$$u_{kl}\big|_{kl\in N} = \begin{array}{|ccc|} u_{i-1,j-1} & u_{i-1,j} & u_{i-1,j+1} \\ u_{i,j-1} & u_{i,j} & u_{i,j+1} \\ u_{i+1,j-1} & u_{i+1,j} & u_{i,+1j+1} \end{array} \equiv \begin{array}{|ccc|} u_9 & u_8 & u_7 \\ u_6 & u_5 & u_4 \\ u_3 & u_2 & u_1 \end{array}$$

where the rightmost notation is a simplified abbreviation. The same notation applies also for the other cell variables (state, output).

Here we will consider only the case of *uncoupled CNN cells*, defined in [Chua, 1998] as cells where there is no recurrent connection between a cell and its neighbors. With these constrains, our generalized cellular automata is now functionally equivalent with a classic *cellular automata* except that the cell *is not defined via a transition table or set of rules but via a deterministic non-linear equation which includes certain parameters*. The hardware implementation of such a structure is far more flexible and compact than a programmable digital cell. The dynamics of this specific GCA can now be written as a discrete time equation in the discrete time variable t marking the beginning of a new cycle of duration ΔT:

For $t = 1,2,..\infty$

$$u_{ij}(t) = y_{ij}(t-1) = y_{ij}(t_n - \Delta T)$$

$$y_{ij}(t) = gene(u_{k,l}),$$

End

The sequence of stimuli patterns depends entirely by the nonlinear function *gene* and by the initial state of the array of cells. In our experiments, mostly a simple cross-like initial state was used. To generate such a pattern except a cross formed by 5 cells in the middle of the array (for which $y_{ij} = 1$), all other cells are set to $y_{ij} = -1$. By arbitrarily choosing *M* sampling moments (e.g. {345,453,567,..}) one can easily select a convenient set of stimuli for *brain signature* experiments.Fig.2 presents such a set of 10 stimuli used in our experiments. Five different *gene* functions were used as shown in Fig. 2.

Is there any reason in choosing the stimuli as above? One reason might be that both human beings (as well as other living beings) and the patterns above are generated recurrently starting from a *gene*. Although *genes* are almost similar for all entities they manifest their phenotype diversity in the same manner as many different human personalities with different brain wiring. Correspondingly, different types of patterns may match different human personalities in very different ways. Complex emergent patterns in cellular systems always impress human beings and the reason might be the one conjectured in [Kaufmann, 1995] and other recent work on complex adaptive systems. According to these authors life and intelligence itself are in fact dynamic manifestations of nonlinear systems operated near the edge of chaos. Instead of presenting CNN patterns one could consider a set of human faces or a collection of shells, all results of natural processes, but storing and presenting then may lead to complex hardware implementations. Sine we are concerned with a simple, low power consumption solution easy to embed even in mobile equipment, the patterns generated by GCAs are a better alternative. Simple devices can be used to implement the generalized cellular automata and consequently a wide range of patterns can be easily generated with an efficient use of the resources.

How should one select the proper stimuli from a large sequence of them generated by the GCA? One possible answer is that we can ask a group of people (often, the same persons who will later use the authentication system) to label as "LIKE" or "DISLIKE" the stimuli. Then one may randomly select a subset from the labeled set of stimuli having the property that the counts of "LIKE" and "DISLIKE" are sensible equal for all persons in the group. In addition one can also select a subset of stimuli producing the less inter correlation between different users. In the next experiments such sophisticated pre-adjustments of the set of stimuli were not done. In fact the method of choosing the stimuli was by subjectively picking them from a larger set by a user who was not then involved in the experiment.

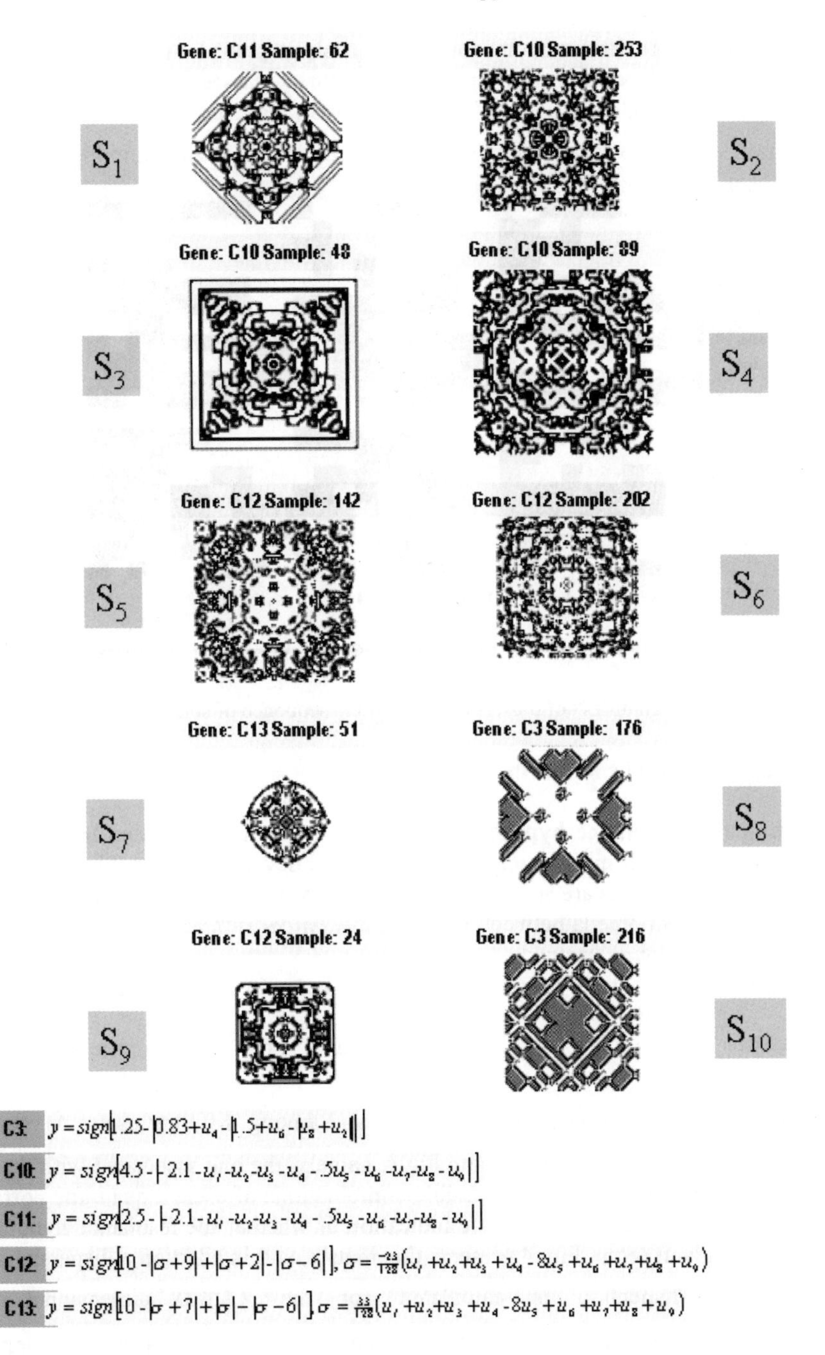

Fig. 2. A set of 10 stimuli used in our experiments. The five different gene formulae are given and for each stimuli its associated sample time and gene are specified.

6.3. Experimental results

In the next experiment the set of 10 stimuli in Fig.2 plus another set of similar but colored 10 patterns were presented to the 5 subjects in 3 sessions, each session being composed of 2 consecutive presentations. There was a week time difference between sessions such that short time memory could be entirely excluded. At each presentation the subject was asked to label a stimuli in one of the three classes {Like, Neutral, Dislike}. Their reaction was encoded in a set of 30 ternary brain signature vectors (each column in Fig.3 is a brain signature vector) composing the 5 matrices R^k as shown in Fig.3.

Fig. 3. The "brain signatures" of 5 users exposed to the sequence of stimuli S_1 *to* S_{20}.

The following observations follow the experiments:

(i) Testing the resonance hypothesis: Although a certain correlation between stimuli can be identified in Fig.3. more precise tools should be developed to establish if the "brain signatures" of the 5 subjects are some random reaction or whether they indeed are the effect of certain resonant phenomena between *artificial dynamical systems* operated near the edge of chaos (CNN) and *human brains*.

We thus define the *correlation index* between two users k_1 and k_2 $Ma(k_1,k_2)=\sum_{q1=1}^{Q\max}\sum_{q2=1}^{Q\max}\left(\mathbf{r}_{q1}^{\mathbf{k}_1}\right)^T\cdot\mathbf{r}_{q2}^{\mathbf{k}_2}$ which indicates how much correlated are the responses (or brain signatures) of different users and among different users. A vector $\mathbf{r}_{\mathbf{q}}^{\mathbf{k}}$ corresponds to a column of responses from Fig.3; namely, the one associated with the user k at presentation q. If one considers all possible combinations of users a quadratic correlation matrix *Ma* results which gives a synthetic indication of whether the resonance hypothesis is valid or not for a given setup. Fig. 4 presents this matrix, and in addition a dispersion matrix *Di* (where the dispersion is calculated for the $Q\max^2$ dimensional vector $\left\{\left(\mathbf{r}_{q1}^{\mathbf{k}_1}\right)^T\mathbf{r}_{q2}^{\mathbf{k}_2}\right\}_{q1,q2=1,\ldots,Q\max}$) for the experimental data collected in Fig. 3, after 2 sessions of 2 presentations each (i.e. $Qmax=4$) separated by one week.

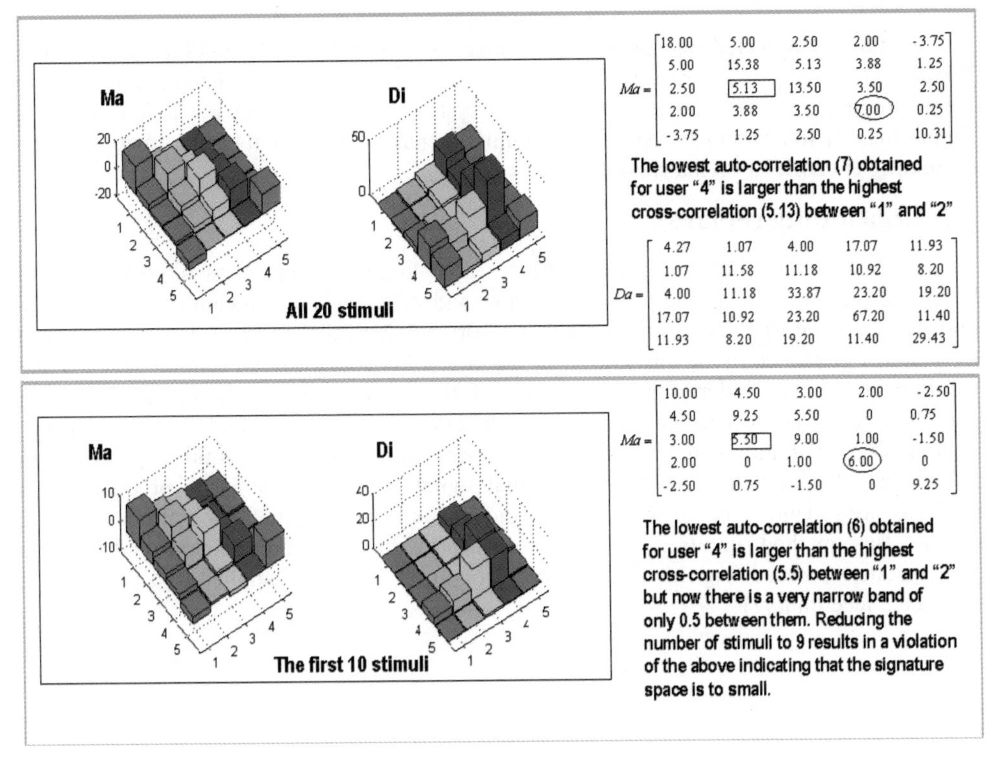

Fig. 4. The correlation and dispersion matrices after 4 presentations (in 2 sessions separated in time by one week) for 5 users and different number of stimuli. If enough stimuli are provided the *dimension of the "brain signature"* is large enough to produce resonance (the lowest value of an auto-correlation is larger than the largest value of a cross-correlation).

The diagonal elements of the *Ma* matrix are larger than the remaining elements, thus indicating that indeed there is a consistent "brain signature" of each user. The above results also reveal some psychological characteristics of the subjects, which can be also exploited in a biometric identification system. For example the subject "4" is very unstable in his reactions (the dispersion coefficient is 67) compared with subject "1" who has a very good stability of her reactions (a dispersion coefficient of only 4.27). On the other hand the results in Fig.4 indicate that for a better accuracy one may need a larger number of symbols. Indeed, the gap between the largest cross-correlation and the smallest auto-correlation become as small as 0.5 if only the first 10 stimuli presented in Fig.2 are used. The above measurements can be easily done for an arbitrary setup and the dimension of the stimuli sequence (or the stimuli) should be modified until a large enough gap in the correlation matrix is established.

(ii) Biometric identification using perceptrons: Once the size and the right sequence of stimuli is established, a simple linear perceptron neural network was trained to perform a classification (identification) of a user given his/her new "signature" vector $\mathbf{r}_{\mathbf{q}}^{\mathbf{k}}$. Thus, the first 20 columns in Fig.1 were used to train the network while the last ten column vectors, generated in a new presentation session, were used to test the network. In the case of using all 20 stimuli the classification rate was of 100% indicating a perfect identification based on "brain signatures". As expected, if the number of stimuli is reduced to 10, confusion errors between users "4" and "5" are reported by the perceptron-based classification system.

(iii) The CNN-based perceptual resonance as a new kind of intelligence test: The Turing test is well known as a possible test of intelligence. Based on the above results we might propose that one can decide whether a system is intelligent or not if it reacts in a similar manner like humans. The distinction intelligent/non-intelligent can be easily done by examining the correlation matrix *Ma*.

6.4. Concluding remarks

This last chapter introduces the concept of a "brain signature" as a novel type of biometric feature, which leads to very simple and effective authentication systems. It is postulated and verified by experiments that a sort of perceptual resonance occurs between the patterns generated in the unstable (near the edge of chaos) artificial cellular systems and the human beings. Under a proper choice of the stimuli sequences this resonance is consistent and can be detected using simple adaptive systems such as linear perceptrons. Although visual stimuli were used in the experiments described here, a similar approach can be considered with sounds or other perceptual signals which can be easily produced by a CNN as well. Unlike other biometric methods employing features of various external organs, this method involves the brain thus providing a more general and reliable authentication method (e.g. how can a face recognition system be applied to a person who suffered a face injury?)

References

[AnalogicLAB, 2002] Analogic & Neural Computing Laboratory - Hungarian Academy of Sciences, web site: http://lab.analogic.sztaki.hu/

[Arena *et al.,* 1998] Paolo Arena, Salvatore Baglio, Luigi Fortuna, and Gabriele Manganaro, "Self-Organization in a Two-Layer CNN", *in IEEE Transactions on Circuits and Systems—I: (fundamental theory and applications),* vol. 45, no. 2, february 1998, pp. 157-162

[Baba, 1989] N. Baba, "A new approach for finding the global minimum of error function of neural networks", in *Neural Networks,* Vol. 2, pp. 367-373, 1989.

[Berlekamp *et al.,* 1982] Berlekamp, E., Conway, J.H., and Guy, R.K., *Winning ways for your mathematical plays,* New York: Academic, chapter 2, vol.2, 817-850, 1982.

[Bronshtein & Semendyayev, 1985] Bronshtein, I.N., and Semendyayev, K.A., *Handbook of Mathematics* (English translation) Van Nostrand Reinhold Company, New York, (p. 119).

[Burks, 1968] Burks A.W., ed, *Essays on Cellular Automata*, Univ. of Illinois Press, Illinois, 1968.

[Calvin, 1996] Calvin, W. H., *The cerebral code,* MIT Press, 1996.

[Capra, 1996] Capra, F., *The web of life,* Anchor Books, 1996.

[Cardarilli *et al.*, 1994] G.C. Cardarilli, C. D'Alessandro, P. Marinucci, F. Bordoni, "VLSI implementation of a modular and programable neural architecture", *in Proceedings of the MicroNeuro'94 (4'th International Conference on Microelectronics for Neural Networks and Fuzzy Systems, Turin-Italy, September 1994),* pp. 218-225.

[Castleman, 1996] Castleman, K.R., *Digital Image Processing*. Second ed. 1996, Englewood Cliffs, New Jersey: Prentice-Hall.

[Chien and Kuh, 1977] M. Chien, and E. Kuh, "Solving nonlinear resistive networks using picewise-linear analysis and simplicial subdivision", *IEEE Trans. Circits Syst. I,* vol CAS-24, pp. 305-317, June 1977.

[Chua & Kang, 1977] Chua, L.O., and Kang, S.M., "Section-wise piecewise-linear functions: canonical representation, properties, and applications", *Proceedings of IEEE,* 65, 6, pp. 915-929.

[Chua, 1980] Chua, L.O., "Dynamic nonlinear networks: state of the art", *IEEE Trans. Circuits and Systems*, **27**, 11, 1059-1087.

[Chua *et al.*, 1985] Chua, L.O., Desoer, C.A., and Kuh, E.A., *Linear and Nonlinear Circuits*, McGraw Hill, New York.

[Chua et al, 1987] L.O. Chua, C.A. Desoer, E.S. Kuh, *Linear and Nonlinear Circuits*, 1987, McGraw-Hill.

[Chua & Yang, 1988] Chua, L.O. and Yang, L., "Cellular neural networks: Theory and Applications", *IEEE Trans. Circuits and Systems*, vol.35, 1257-1290, 1988.

[Chua et al., 1995] Chua, L.O.; Hasler, M.; Moschytz, G.S.; Neirynck, J., "Autonomous cellular neural networks: a unified paradigm for pattern formation and active wave propagation", *Circuits and Systems I: Fundamental Theory and Applications, IEEE Transactions on* , Volume: 42 Issue: 10 , Oct. 1995, Page(s): 559 –577, 1995.

[Chua 1998] Chua, L.O., *CNN: A paradigm for complexity*, Singapore: World Scientific, 1998.

[Chua, 1999] Chua, L.O., "Passivity and complexity" *IEEE Trans. Circuits and Systems-I*, Vol. 46, No.1, pp. 71-82, 1999.

[Chua & Roska, 2001] L. Chua and T. Roska, *Cellular neural networks and visual computing - Foundations and applications*, Cambridge University Press, 2001.

[Codd, 1968] Codd E.F., *Cellular Automata*, Academic Press, New York, 1968.

[Corby, 2002] Michael J. Corby, *Authentication: "I Know You"*, a tutorial available via the Internet http://www.mcorby.com

[Cover, 1965] Cover, T.M., "Geometrical and statistical properties of systems of linear inequalities with applications in pattern recognition", in *IEEE Trans. on Electronic Computers*, EC-14, pp. 326-334, June, 1965.

[Cronin, 1987] Cronin, J., *Mathematical aspects of Hodgkin-Huxley neural theory*, Cambridge, Cambridge University Press.

[Crounse & Chua, 1996] Crounse, K.R., and Chua, L.O., "The CNN universal machine is as universal as Turing machine," *IEEE Trans. on Circuits and Systems*, **43**, 4, 353-355.

[Crounse *et al.*, 1997] Crounse, K.R., Fung, E.L. and Chua, L.O., "Efficient implementation of neighbourhood logic for cellular automata via the cellular neural network universal machine", *IEEE Tr. on Circuits and Systems - I,* vol.44, No.4, 355-361.

[Dogaru et al, 1996a] R. Dogaru, A.T. Murgan, S. Ortmann, M. Glesner, "A modified RBF neural network for efficient current-mode VLSI implementation", in *Proceedings of the Fifth International Conference on Microelectronics for Neural Networks and Fuzzy Systems (Micro-Neuro'96),* IEEE Computer-Press, Laussane 12-14 Febr. 1996, pp. 265-270, 1996.

[Dogaru et al, 1996b] R. Dogaru, A.T. Murgan, M. Glesner and S. Ortmann, "A VLSI friendly image compression algorithm based on Fuzzy-ART neural network" in *Proceedings WCNN'96* (World Congress on Neural Networks), San Diego, September 1996, pp. 1327-1330.

[Dogaru & Chua 1998a] Dogaru R., and Chua, L.O, "Rectification neural networks: a novel adaptive architecture and its application for implementing the local logic of cellular neural networks", U.C. Berkeley, Electronics Research Laboratory Memorandum No. UCB/ERL M98/4, 15 January 1998.

[Dogaru & Chua 1998b] Dogaru R., and Chua, L.O, "Edge of chaos and local activity domain of FitzHugh-Nagumo Equation", *International Journal of Bifurcation and Chaos*, Vol. 8, No. 2, 211-257, 1998.

[Dogaru & Chua 1998c] Dogaru, R., and Chua, L.O., "Edge of Chaos and Local Activity Domain of the Brusselator CNN", *International Journal of Bifircation and Chaos,*] Vol. 8, No. 6, 1998, pp. 1107-1130.

[Dogaru & Chua 1998d] Dogaru, R., and Chua, L.O., "Edge of Chaos and Local Activity Domain of the Gierer-Meinhardt CNN", to appear in *International Journal of Bifircation and Chaos,* Vol. 8, No. 12, 1998, pp. 2321-2340.

[Dogaru *et al.,* 1998e] Dogaru, R., Chua, L.O., and Crounse, K., "Pyramidal Cells:A novel class of adaptive coupling cells and their applications for cellular neural networks", *IEEE Transactions on Circuits and Systems - I* , Vol. 45, No. 10, pp. 1077-1090, October 1998.

[Dogaru & Chua, 1998f] Dogaru, R., and Chua, L.O., "CNN genes for one-dimensional cellular automata: a multi-nested piecewise-linear approach", *Int. Journal of Bif. and Chaos,* Vol. 8, No. 10, 1998, pp. 1987-2001.

[Dogaru & Chua, 1999a] R. Dogaru and L. O. Chua, "Universal CNN cells", *Int. Journal of Bifurcation and Chaos,* vol. 9, pp. 1-48, Jan. 1999.

[Dogaru & Chua 1999b] Dogaru R., and Chua, L..O, "Emergence of Unicellular Organisms from a Simple Generalized Cellular Automata Universal CNN cells", *International Journal of Bifurcation and Chaos*,9(6), 1999, pp. 1219-1236.

[Dogaru *et al.*, 1999b] R.Dogaru, M. Alangiu, M. Rychetsky, M. Glesner, "Perceptrons revisited: the addition of a non-monotone recursion greatly enhances their representation and classification properties", in *Proceedigs IJCNN'97 (International Joint Conference on Neural Networks - Washington, DC., July 10-16, 1999)*, pp. 862-867.

[Dogaru & Chua, 1999c] R. Dogaru, L. O. Chua, "The comparative synapse: a multiplication-free approach to neuro-fuzzy classifiers", ", in *IEEE Trans. on Circuits and Systems I*, Nov. 1999, 1366-1371.

[Dogaru *et al.*, 2000a] R. Dogaru, L. O. Chua, and M. Haenggi, "A compact universal cellular neural network cell based on resonant tunnelling diodes: circuit design, model and functional capabilities", in *Proc. of Intl. Workshop on Cellular Neural Networks and their Applications,* May 2000, pp 183-188.

[Dogaru & Chua, 2000b] R. Dogaru, L. O. Chua, "Mutations of the Game of Life: a generalized cellular automata perspective of complex adaptive systems" *in International Journal of Bifurcation and Chaos*, Vol. 10, No. 8 (2000) pp.1821-1866.

[Dogaru *et al.*, 2001a] R. Dogaru, P. Julián, and L.O. Chua, "A robust and efficient universal CNN cell circuit using simplicial neuro-fuzzy inferences for fast image processing", in Proceedings of ISCAS 2001. The 2001 IEEE International Symposium on Circuits and Systems, Volume: 2, pp.493-496, 2001.

[Dogaru *et al.*, 2001b] Radu Dogaru, Pedro Julián, and Leon O. Chua, "The simplicial Adaline: a versatile VLSI-friendly nonlinear filter based on a synergy between standard cellular neural networks and simplicial inferences", *in Proceedings NSIP 2001 (International Workshop on Nonlinear Signal and Image processing)*, Baltimore USA, (CD proceedings).

[Dogaru *et al.*, 2002a] Dogaru, R.; Julian, P.; Chua, L.O.; Glesner, M., "The simplicial neural cell and its mixed-signal circuit implementation: an efficient neural-network architecture for intelligent signal processing in portable multimedia applications", IEEE Transactions on Neural Networks, , Volume: 13 Issue: 4 , July 2002, Page(s): 995 –1008.

[Dogaru *et al*, 2002b] R. Dogaru, F. Ionescu, P. Julián, M. Glesner, " Novel Methods and Results in Training Universal Multi-Nested Neurons", *in Cellular Neural Networks and Their*

Applications: Proceedings of the 7th IEEE International Workshop, edited by Ronald Tetzlaff, World Scientific 2002, pp. 601-608.

[Dogaru & Dogaru, 2002c] R. Dogaru and Ioana Dogaru, "Biometric Authentication Based on Perceptual Resonance between CNN Emergent Patterns and Humans ", *in Cellular Neural Networks and Their Applications: Proceedings of the 7th IEEE International Workshop*, edited by Ronald Tetzlaff, World Scientific 2002, pp. 267-274.

[ELENA, 1995] Technical report "Enhanced Learning for Evolutive Neural Architecture", ESPRIT Basic Research Project 6891, 1995 (Databases and Benchmarks) available from ftp://ftp.dice.ucl.ac.be/pub/neural-nets/ELENA/ (use an FTP browser to login)

[Fieleder & Seitzer, 1979] Fielder, U., and Seitzer, D., "A high-speed 8 bit A/D converter based on a Gray-code multiple folding circuit", *IEEE Journ. of Solid-State Circuits,* 14, 3, 547-551, 1979.

[FitzHugh 1969] FitzHugh R., "Mathematical models of excitation and propagation nerve", *in Biological Engineering,* H. Schwan, Ed. New York: McGraw-Hill, 1969.

[Forrest, 1990] Forrest, S., "Emergent computation: self-organizing, collective, and cooperative phenomena in natural and artificial computing networks", *Physica* D, **42,** pp. 1-11.

[Gardner, 1970] Gardner M., *The Fantastic Combinations of John Conway's New Solitaire Game of "Life",* Scientific American, 223:4, (April 1970), 120-123.

[Gardner M, 1983] Gardner M., *Wheels, Life and Other Mathematical Amusements,* W.H. Freeman, New York, 1983.

[Gierer & Meinhardt, 1972] Gierer, A., and Meinhardt, H., "A theory of biological pattern formation", *Kybernetik,* **12**: 30—39, 1972.

[Goldberg, 1989] Goldberg, D., *Genetic algorithms*, Edit. Adison Wesley, 1989.

[Haken & Olbrich, 1978] Haken, H., and H. Olbrich, "Analytical treatment of pattern formation in the Gierer-Meinhardt model of morphogenesis", *J. of Math. Biology*, **6**: 317--331, 1978.

[Haken, 1994] Haken, H., "Synergetics: From pattern formation to pattern analysis and pattern recognition", *International Journal of Bifurcation and Chaos*,**4**(5):1069--1083, 1994.

[Harth & Pandya, 1988] Harth, E., and Pandya, A.S., "Dynamics of ALOPEX process: application to optimization problems", *Biomathematics and Related Computational Problems, L.M. Ricciardi (Ed.),* Kluwer Academic Publishers, 459-471, 1988.

[Hartman, 1982] Hartman, P., *Ordinary Differential Equations,* Boston, Birkhäuser.

[Hassoun, 1995] Hassoun, M.H., *Fundamentals of Artificial Neural Networks,* MIT Press, 1995.

[Haykin, 1994] Haykin, S., *Neural networks: A Comprehensive Foundation*, 1994, New York, Macmillan.

[Haykin, 1999] Haykin, S., *Neural networks: A Comprehensive Foundation –second edition*, 1999, Prentice Hall.

[Hänggi, 1998] A Java-based CNN simulator, web page: http://www.isi.ee.ethz.ch/~haenggi/CNNsim.html

[Hänggi et al., 1999] M. Hänggi, S. Moser, E. Pfaffhauser and G. S. Moschytz , "Simulation and Visualization of CNN Dynamics", in *International Journal of Bifurcation and Chaos*, Vol. 9, No. 7, July 1999, pp. 1237-1262.

[Hänggi & Moschytz, 2000] Martin Hänggi, George S. Moschytz, *Cellular Neural Networks: Analysis, Design, and Optimization*, Kluwer Academic Publishers Pub, 2000.

[Ionescu & Dogaru, 2002a] Felicia Ionescu, Radu Dogaru "High performance distributed architecture for multi-nested cellular neural network cells design", *in Proceedings of Twentieth IASTED International ConferenceApplied Informatics (AI 2002), Innsbruck, Austria*, February 18-21, 2002, volume "Applied Informatics", pp. 41-46.

[Ionescu *et al.*, 2002b] Ionescu, Felicia, Dogaru, R., Tosca, A."Hybrid Genetic Algorithm for Multi-nested Universal Neuron Design", to appear *in Proceedings SCI 2002 (The 6-th World Multiconference on Systemics,Cybernetics and Informatics),* July 2002.

[Jahnke & Winfree, 1991] Jahnke,W & Winfree, A.T., "A survey of spiral-wave behaviors in the oregonator model", *Int. J. Bifurcation and Chaos*, **2**, 445-466.

[Julián *et al.*, 1999] P. Julián, A. Desages, O. Agamennoni, "High level canonical piecewise linear representation using a simplicial partition", *IEEE Transactions on Circuits and Systems - I* , Vol. 46, April 1999, pp. 463-480.

[Julián *et al.*, 2000] P. Julián, A. Desages, B. D'Amico, "Orthonormal high level canonical piecewise linear functions with applications to model reduction", *IEEE Transactions on Circuits and Systems - I*, Vol. 47, May 2000, pp. 702-712.

[Julián et al., 2001] P. Julián, R. Dogaru, and L.O. Chua, "A piecewise-linear simplicial coupling cell for CNN gray-level image processing", *in Proceeding ISCAS 2001 (IEEE Symposium on Circuits and Systems), Sydney, May 2001*, Page(s): 109 - 112 vol. 2.

[Julián et al, 2002a] Julian, P.; Dogaru, R.; Chua, L.O., "A piecewise-linear simplicial coupling cell for CNN gray-level image processing", *in Circuits and Systems I: Fundamental Theory and Applications, IEEE Transactions on*, Volume: 49 Issue: 7, July 2002, Page(s): 904 –913

[Julián *et al*, 2002b] P. Julián, R. Dogaru, M. Haenggi, L. O. Chua, "A Search Algorithm for the design of Multi-nested Cellular Neural Networks", in Proceedings of ISCAS-2002 (IEEE Symposium on Circuits and Systems), vol.1, pp. 617-620, 2002.

[Julián *et al*, 2002c] P. Julián, R. Dogaru, M. Itoh, M. Haenggi, L. O. Chua "On the RTD implementation of simplicial cellular neural networks", *in Cellular Neural Networks and Their Applications: Proceedings of the 7th IEEE International Workshop*, edited by Ronald Tetzlaff, World Scientific 2002, pp. 140-147.

[Kahlert & Chua, 1992] Kahlert, C., and Chua, L.O., "The complete canonical piecewise-linear representation-Part I: The geometry of the domain space", *IEEE Trans. Circ. and Systems-I,* 39, 3, 222-236.

[Kauffman, 1993] Kauffman, S.A., *The origins of order: Self-organization and selection in evolution,* New York, Oxford University Press.

[Kauffman, 1995] Kauffman, S.A., *At home in the universe: the search for laws of self-organization and complexity*, Oxford University Press.

[Kennedy, 1995] Kennedy, M.P., "A nonlinear dynamics interpretation of algorithmic A/D conversion", *Int. Journal of Bif. and Chaos,* 5, 3, 891-893.

[Kirkpatrick *et al.,* 1983] S. Kirkpatrick, C.D. Gelatt jr., and M.P. Vecchi, "Optimization by simulated annealing", *Science* 220, pp. 671-680, 1983.

[Koza, 1994] Koza, J. R., *Genetic programming II,* MIT Press, Cambridge MA, 1994.

[Kuhnn, 1968] H.W. Kuhnn, "Simplicial approximation of fixed points", in *Proc. of the National Academy of Sciences U.S.A.,* vol. 61, 1968, pp. 1238-1242.

[Langton, 1984] Langton C.G., "Self-Reproduction in Cellular Automata", *Physica D*, **10**, 135-144, 1984.

[Langton, 1986] Langton, C. G., "Studying artificial life with cellular automata", *Physica D,* **22**, 120-149.

[Langton, 1990] Langton, C.G., "Computation at the edge of chaos: phase transitions and emergent computation", *Physica D*, 42, pp. 12-37, 1990.

[Lin & Lee, 1996] C-T. Lin, and C.S. George Lee, *Neural Fuzzy Systems: A Neuro-Fuzzy Synergism to Intelligent Systems,* Prentice Hall, 1996.

[Liu & Liu, 2001] Ming-Huang Liu and Shen-Iuan Liu, "An 8-bit 10 MS/s Folding and Interpolating ADC Using the Continuous-Time Auto-Zero Technique", *in IEEE journal of solid-state circuits*, VOL. 36, NO. 1, pp. 122-128, January 2001.

[Lohn, 1996] Lohn, J.,"Automatic discovery of self-replicating structures in cellular space automata models", *Dept. of Comp. Sci. Tech. Rep. CS-TR-3677, Univ. of Maryland at College Park,* August 1996.

[Maheshwari *et al.*, 1997] R. Maheshwari, S.P. Rao and E.G. Poonach, " FPGA Implementation of Median Filter", in Proceedings of the Tenth International Conference on VLSI Design: VLSI in Multimedia Applications, 1997, pp. 523-524.

[Mar & St. Denis, 1996] Mar, G., and St. Denis, P., "Real life", *Int. Journal of Bif. and Chaos,* **6**(11), 1996, pp. 2077-2086.

[Mange & Tomassini,'98] Mange D., and Tomassini, M. (Eds.), *Bio-inspired Computing Machines,* Press Polytechniques et Universitaires Romandes, Lausanne, 1998.

[Maturana & Varela, 1980] Maturana, H, and Varela, F., *Autopoiesis and Cognition,* D. Reidel, Dodrecht, Holland.

[Matyas, 1965] J. Matyas, "Random optimization", *Automation and Remote Control,* vol. 26, pp. 246-253, 1965.

[McMullin & Varela, 1997] McMullin, B., and Varela, F., "Rediscovering computational autopoiesis", *Santa Fe Institue working paper 97-02-012*.

[Meinhardt & Gierer, 1974] Meinhardt, H., and Gierer. A., "Applications of a theory of biological pattern formation based on lateral inhibition", *J. Cell. Sci.,* **15**: 321-346.

[Meinhardt, 1977] Meinhardt, H., "The spatial control of cell differentiation by autocatalysis and lateral inhibition", *Synergetics,* (H. Haken, ed.), Berlin: Springer, 1997, 215-223.

[Min, L., *et.al.*, 2000a] Min, L., Crounse, K. R., and Chua, L. O. "Analytical criteria for local activity and applications to the Oregonator CNN," *Int. J. Bifurcation and Chaos*, 10(1), 25-71.

[Min, L., *et.al.*, 2000b] Min, L., Crounse, K. R., and Chua, L. O. "Analytical criteria for local activity of reaction-diffusion CNN with four state variables and applications to the Hodgkin-Huxley equation," *Int. J. Bifurcation and Chaos*,10(6), 1295-1343.

[Mitchell, *et al,* 1993] Mitchell, M., Crutchfield, J.P., and Hraber, P.T., "Dynamics, computation, and the "edge of chaos": a re-examination", Santa Fe Institute Working Paper 93-06-040, 1993.

[Mitchell *et al.* 1996] Mitchell, M., Crutchfield, J.P., and Das, R., "Evolving cellular automata to perform computations: A review of recent work", *in Proceedings of The First International Conference on Evolutionary Computations and its Applications (EvCA'96),* Moscow; Russia: Russian Academy of Science (reprint available from http://www.santafe.edu/projects/evca).

[Mizuta and Tanoue, 1995] H. Mizuta and T. Tanoue*, The physics and Applications of Resonant Tunneling Diodes*, Cambridge University Press, 1995.

[Murray, 1989] Murray, J.D., *Mathematical Biology*, Berlin: Springer-Verlag.

[Muñuzuri,A.P. *et al*, 1995] Muñuzuri,A.P., Pérez-Muñuzuri, V., Gómez-Gesteira, M., Chua, L.O. & Pérez-Villar, V., "Spatiotemporal structures in discretely-coupled arrays of nonlinear circuits:a review", *Int. J. of Bif. and Chaos* **5**(1), 17-50.

[Nicolis. & Prigogine, 1989] Nicolis, G. and Prigogine, I., *Exploring Complexity*, W.H. Freeman, New York.

[Neumann von, John, 1966] Neumann von, John, *Theory of Self-Reproducing Automata*, University of Illinois Press, Champaign, IL, 1966.

[Nemes et al, 1998] L. Nemes, L.O. Chua, and T. Roska, “Implementation of Arbitrary Boolean Functions on a Cnn Universal Machine”, in *Int'l Journal of Circuit Theory and Applications* 1998, Vol 26, Issue 6, pp 593-610.

[Ota & Wilamowski, 1999] Y. Ota and B. M. Wilamowski, “Analog Implementation of Pulse-Coupled Neural Networks”, in IEEE Transactions on Neural Networks, vol. 10, no. 3, pp. 539-543, May 1999.

[Ott, 1993] Ott,E, *Chaos in Dynamical Systems*, Cambridge University Press, Cambridge, UK.

[Packard, 1988] Packard, N., “Adaptation toward the edge of chaos”, Center for Complex Systems Research Technical Report, University of Illinois,
CCSR-88-5, 1988.

[Prigogine, 1980] Prigogine, I., *From being to becoming : time and complexity in the physical sciences*, W. H. Freeman, San Francisco.

[Reggia *et al.*,1993] Reggia, J. A., Armentrout S. L., Chou, H-H., and Peng, Y., "Simple systems that exhibit self-directed replication", *Science,* **259**, 1282-1287, 26 February 1993.

[Rennard, 2000] Jean-Philippe Rennard, *Introduction to Cellular Automata*, available from http://www.rennard.org/

[Ronald *et al.,* 1999] Ronald EMA, Sipper M, Capcarrere MS “Design, observation, surprise! A test of emergence”, *Artificial Life,* Vol. 5, no. 3, pp. 225-239, 1999.

[Roska & Chua, 1993] Roska, T.; Chua, L.O. “The CNN universal machine: an analogic array computer”, *Circuits and Systems II: Analog and Digital Signal Processing, IEEE Transactions on*, Volume: 40 Issue: 3, March 1993, Page(s): 163 –173.

[Roska, 2000] Roska, T. “Analogic computing: system aspects of analogic CNN sensor computers”, *Cellular Neural Networks and Their Applications, 2000. (CNNA 2000). Proceedings of the 2000 6th IEEE International Workshop on* , 2000, Page(s): 73 –78

[Roska & Vázquez, 2000a] Roska, T.; Rodríguez-Vázquez, A., “Review of CMOS implementations of the CNN universal machine-type visual microprocessors”, *Circuits and Systems, 2000. Proceedings. ISCAS 2000 Geneva. The 2000 IEEE International Symposium on*, Volume: 2 , 2000, Page(s): 120 -123 vol.2

[Roska & Vázquez, 2000b] T. Roska and A. Rodríguez-Vázquez (editor), *Towards the Visual Microprocessor: VLSI Design and the Use of Cellular Network Universal Machines*, John Wiley & Sons, 2000.

[Roska, 2001] Roska, T, "AnaLogic Wave Computers-wave-type algorithms: canonical description, computer classes, and computational complexity ", *in Proceedings of Circuits and Systems, 2001. ISCAS 2001. The 2001 IEEE International Symposium on,* Volume: 2 , 2001. Page(s): 41-44 vol. 2

[Roska, 2002], T. Roska, "Computational and Computer Complexity of Analogic Cellular Wave Computers", *in Cellular Neural Networks and Their Applications Proceedings of the 7th IEEE International Workshop,* edited by Ronald Tetzlaff, World Scientific 2002.

[Sayama, 2000] Hiroki Sayama, Structurally Dissolvable Self-Reproducing Loop & Evoloop: Evolving SDSR Loop, available via http://necsi.org/postdocs/sayama/sdsr

[Schatten A., 1999] Alexander Schatten, Cellular Automata – Digital Worlds", http://www.ifs.tuwien.ac.at/~aschatt/info/ca/ca.html

[Schrödinger, 1967] Schrödinger, E., *What is life? the physical aspect of the living cell & Mind and matter* , University press, Cambridge.

[SCNN Simulator, 2000] CNN Simulator from the Frankfurt University – the Institute of Applied Physics, web page: http://www.uni-frankfurt.de/fb13/iap/e_ag_rt/SCNN/

[Serrano and Vázquez, 1999] T. Serrano-Gotarredona, Rodríguez-Vázquez, "On the Design of Second Order Dynamics Reaction-Diffusion CNNs", *Journal of VLSI Signal Processing*, Vol. 23, pp. 351–371.

[Sigmon, 1993] Kermit Sigmon, *MATLAB Primer, Third Edition,* available online: http://ise.stanford.edu/Matlab/matlab-primer.pdf

[Sigmon and Davis, 2001] Kermit Sigmon, and Timothy A Davis, *MATLAB Primer, Sixth Edition*, CRC Press, 2001.

[Sipper, 1995] Sipper, M., "Studying artificial life using a simple, general cellular model",*Artificial Life Journal,* **2**(1), 1-35, MIT Press.

[Standish, 2001] R. K. Standish, "On Complexity and Emergence", *in Complexity International,* Vol 09, 2001. Available from http://www.csu.edu.au/ci/vol09/standi09/

[Sbitnev et al., 2001] Sbitnev VI, Yang T, Chua LO, "The local activity criteria for "difference-equation" CNN", *Int. J. Bifurcation and Chaos* 11 (2): 311-419, February 2001.

[Toffoli, 1987] T. Toffoli, N. Margolus, *Cellular Automata Machines*, Cambridge, MA: MIT Press, 1987.

[Toffoli, 1998] T. Toffoli, "Non-conventional computers", *in Encyclopedia of Electrical and Electronics Engineering* (John Webster ed.), Wiley and Sons, 1998.

[de la Torre & Mártin, 1997] A.C. de la Torre, and H.O. Mártin, "A survey of cellular automata like the "game of life", *Physica A,* **240**, 560-570, 1997.

[Turing, 1952] AM Turing, "The Chemical Basis of Morphogenesis", *Philosophical Transactions of the Royal Society B* (London), 237, 37--72, 1952.

[Vapnik, 1998] V.N.Vapnik, *Statistical Learning Theory,* New York: Wiley, 1998.

[Varela *et al.*, 1974] Varela,F., H.Maturana and R.Uribe, "Autopoiesis: The organization of living systems, its characterization and a model", *Biosystems* 5:pp.187-196, 1974.

[Weimar, 1996] Jörg R. Weimar, *Simulation with Cellular Automata - Lecture Notes, available from:* http://www.tu-bs.de/institute/WiR/weimar/ZAscript/ZAscript.html

[Widrow & Hoff, 1960] Widrow, B., and Hoff, M. E., Jr., "Adaptive switching circuits", in *IRE Western Electric Show and Convention Record*, part 4, pp. 96-104, 1960.

[Widrow & Stearns, 1985] B. Widrow, and S.D. Stearns, *Adaptive signal processing*, Englewood Cliffs, Prentice-Hall, 1985.

[Wolfram, 1984] Wolfram, S., "Universality and complexity in cellular automata", *Physica D,* 10, 1-35. , 1984.

[Zeleny, 1977] Zeleny, M., "Self-organization of living systems: a formal model of autopoiesis", *Int. J. General Systems,* **4**, 13-28, 1977.

Index

R

S

T

U

W

About the Author

Professor Radu Dogaru is attached to the Applied Electronics and Information Engineering Department of "Politehnica" University of Bucharest, where he teaches courses in Neural Networks, Computational Intelligence, Cellular Systems, Numerical Methods for Bioengineering.

He is the recipient of the Fulbright award (1996) and a co-recipient of the Romanian Academy Award – Tudor Tanasescu for research in computational intelligence for signal processing (received in 1997).

Dr. Dogaru earned a Ph.D. in Electronics (1996) at "Politehnica" University of Bucharest with a thesis focusing on neural network architectures for VLSI implementation. He is the recipient of a Graduation Certificate (1994) of the Tempus Postgraduate School in Computer Aided Electrical Engineering (VLSI design) and is the recipent of the M.Sc. degree (Diploma de Inginer) in Electronics and Communications (1987) from the same university.

He is also the recipient of a grant starting in March 2002 for a two-year joint research program funded by the "Volkswagen Stiftung" (Germany) on compact neural architectures for intelligent signal processing in mobile applications.

Between September 1996 and August 1998 he was a Postdoctoral Research Engineer with the Nonlinear Electronics Laboratory at the University of California at Berkeley, sponsored in the first year by a Fulbright award. He joined the same laboratory again between August 1999 and April 2000, having been invited as a Visiting Research Scholar.

He was the recipient of several other research scholarships including a six-month Tempus grant — at I.N.P. Grenoble, France (1994) working at T.I.M.A. on a project dealing with neural architectures optimized for mixed signal implementation, Technical University of Darmstadt, Germany — the Institute of Microelectronic Systems, (October 1995, May 1996, December 1998, December 2000, September and December 2002) where he did research on various topics in the area of intelligent signal processing.

Dr. Dogaru has published more than 60 journal and conference papers, books and research reports in the areas of neural and neuro-fuzzy architectures, cellular neural networks, complex adaptive systems and artificial life. More than 30 papers were published in highly recognized international journals and conference proceedings.

His scientific interests are in the areas of intelligent systems, bio-inspired computing architectures, nonlinear signal processing, complex adaptive systems, cellular systems, emergent computation, and neural architectures for micro and nano-technologies.

He can be contacted at radu_d@ieee.org; http://atm.neuro.pub.ro/~radu_d

About the Author